Praise for *The Nature of Drugs, Volume Two*

"This anticipated second volume of lectures captures the playful and informative way that Sasha Shulgin brings the chemistry of plants to life. Richly illustrated and interspersed with students' questions, this book makes a timeless contribution to our unquenched curiosity about drugs in nature."

— Erika Dyck, PhD, Professor and Tier 1 Canada Research Chair, University of Saskatchewan

"The best class you will ever audit! Professor Shulgin is in all his glory, whether bringing in bottles of cocaine for 'show-and-tell,' explaining how to get a DEA Schedule I license (persistence and personality), or dissecting the legal intricacies of our drug policy. He is playful with his students' questions, and Ann, his wife, is somehow miraculously present to add incisive commentary. All this, along with colorful drawings of the plants and fungi which punctuate his lessons make for an amazing book. Highly recommended!"

—Julie Holland, MD, Author of *Good Chemistry: The Science of Connection from Soul to Psychedelics*

"Although the lectures presented here began as a pharmacology class, these talks also reveal Sasha's deeper insights: his discerning take on the social future of psychedelics, his remarkable fund of miscellaneous information, and the linkages he recognized among physiology, pharmacology, psychology, and philosophy. Ann's voice is also heard here, in a questioning dialog that prompts and probes for deeper and more detailed information. She represents the audience as a whole, encouraging Sasha to keep tethered to their level of understanding. It's a singular gift to be able to hear their voices so clearly today."

— Mariavittoria Mangini, PhD, FNP, Women's Visionary Council

"In gratitude and laughter for another intimate collection of teachings from the mad chemist himself, full of enthusiasm and reverence—like being in the front row of a lecture, or psychedelic sermon."

— Danielle M. Herrera, LMFT, Psychedelic Psychotherapist

"For those that never got to meet Sasha or his amazing wife, these lectures are not just records of conversations with students; these are cultural documents. Through science, humor, insight, and some incredible storytelling, Shulgin manages to deliver a snapshot of the life of a psychonaut before Decriminalization, before major clinical studies, and after a generation of psychedelic wild children had their way and lost it. From caffeine to LSD, Shulgin takes the time to lay out the biochemical and cultural uses of these compounds that have shaped our lives. In doing so, he highlights not only psychedelics but our deeper pharmacological relationships with substances as well. . . . *The Nature of Drugs* should be required reading for all those interested in psychedelics and human interactions with all substances."

— Ayize Jama-Everett, MDiv, MA, MFA, Associate Professor at Starr King School for the Ministry and Author of *The Liminal* series and *The Last Count of Monte Cristo*

"*The Nature of Drugs* would serve well for any motivated autodidact to gain a firm grasp of chemistry and pharmacology. But they are so much more than that. To read them is like reading a fascinating science-centric mystery story. To be able to view these subjects through the eyes and brain, humor, and heart of one the most brilliant scientists of the 20th and 21st centuries is an incredible experience, and one not to be missed."

— Dennis McKenna, PhD, Author of *The Brotherhood of the Screaming Abyss: My Life with Terence McKenna*

"In *The Nature of Drugs*, Sasha expertly articulates the whole gamut of chemistry, pharmacology, drug experiences and the ancient and contemporary culture that surround them. A true gem. The only downside is the intense envy felt—I wish I'd actually been in those lectures! But this wonderful book brings Sasha and his motivational presence right back for everyone to experience as if you were there. Very highly recommended."

— Ben Sessa MBBS (MD) BSc MRCPsych
  Psychiatrist, Psychedelic Researcher & Author of *The Psychedelic Renaissance*

"This is the much-anticipated second volume of Sasha Shulgin's *The Nature of Drugs*. While Volume One serves as general background information, Volume Two presents 15 chapters focused on specific classes of drugs such as stimulants, depressants, psychedelics, cannabis, and more. Sasha was an amazing teacher and these two volumes will be accessible to everyone, including those who have little or no science background. His style of teaching in both volumes is humorous and sprinkled with anecdotes. For those who never knew Sasha personally, these two volumes may be the best introduction to his unique personality and teaching style."

— David E. Nichols, PhD
    Distinguished Professor Emeritus, Purdue University School of Pharmacy

"There is magic in him. Beloved by generations of aspiring medicinal chemists and therapists, Sasha enthralls, teases, and inspires his classes through anecdotes, word play, irreverent asides, and information unobtainable from any other source. Sasha was, and remains, the chemist's chemist. . . . With great hearts, Sasha and Ann invited into their shining world many who recognized joyful science and inquiry: students, researchers, those on the fringe who had the vision but no refuge. In these collected lectures, we join this most precious family of seekers. At last, we are home."

— William Leonard Pickard, Author of *The Rose of Paracelsus: On Secrets and Sacraments*

"Sasha Shulgin was a dear friend, a true pioneer, and one of the leading lights of psychedelic science—carrying the torch throughout the more prohibitive era of the late 20th century. In this volume, his wide breadth of knowledge about the consciousness-altering effects of a range of substances is on clear display, including on the psychedelics for which he is perhaps best known. What a privilege to be able to read his fascinating lectures now, so full of the energy, wit, and charm for which he was known and loved, and set in the context of the Reagan Administration of the time and its blinkered 'Just Say No' policies. It feels as though one is right there in the room with him and his students, learning from one of the greats."

— Amanda Feilding, Executive Director of the Beckley Foundation

— The Nature of Drugs —

The Nature of Drugs
History, Pharmacology, and Social Impact

VOLUME TWO

ALEXANDER SHULGIN

Copyright © 2023 by Alexander Shulgin
All rights reserved.

No part of this publication may be reproduced, stored in any retrieval system, or transmitted, in any form or by any means, electronic, mechanical, photocopying, recording, or otherwise without the prior permission of the publisher, except for the quotation of brief passages in reviews.

Co-published by
Transform Press | P.O. Box 11552, Berkeley, California 94712
Synergetic Press | 1 Bluebird Court, Santa Fe, New Mexico 87508
& 24 Old Gloucester Street, London, WCIN 3AL, England

Library of Congress Control Number: 2023932638

ISBN 9780999547250 (hardcover)
ISBN 9780999547298 (ebook)

Cover design by Amanda Müller
Book design by Brad Greene
Cover and interior illustrations by Donna Torres
Transcribed by Stacy Simone
Transcription corrections: Keeper Trout
Design and Production Manager: Amanda Müller
Managing Editor: Noelle Armstrong
Production Editor: Allison Felus

Printed in the United States of America

TABLE OF CONTENTS

LECTURE 9: Stimulants I 1
The psychotropic states and their five classifications
Stimulation in the nervous system
Caffeine in widely consumed substances
Methamphetamine, amphetamine, and other synthetic stimulants
Possession, conspiracy, intent, misdemeanors, and felonies

LECTURE 10: Stimulants II 35
Tobacco, nicotine, and blood levels of carbon monoxide
The history of smoking as a practice
Cancer
The origins of tobacco and its spread around the world
Cigarette advertising

LECTURE 11: Stimulants III 67
Cocaine, crack, and coca leaves
Bureau of Narcotics drug licensing and scheduling
The production of cocaine in South America
The accuracy of autopsy analysis
The medical use for cocaine

LECTURE 12: Depressants I 99
The opium poppy: origins, botany, and effects
Treatments for heroin addiction
Sertürner, Davy, and self-experimentation

Morphine and its ability to cause indifference to pain
Pathways of opioid metabolism

LECTURE 13: Depressants II 129
"Narcotics are not particularly damaging to the body."
MPTP and its connection to Parkinson's disease
Designer drugs and analogs to fentanyl
Drug abuse versus drug laws

LECTURE 14: Depressants III 159
An order of magnitude
Observable signs of intoxication
Chromatography
Alcohol analyses via blood or urine
The five major barbiturates

LECTURE 15: Intoxicants.................................. 189
Why people do drugs
The origins and uses of nitrous oxide
Amyl nitrite and ephedrine; chloroform, ether, and phosgene
Nutmeg, parsley, dill, apiole, and cloves
The active alkaloids in betel nut

LECTURE 16: Deliriants 223
The chemistry of neurotransmitters
Love apples, thorn apples, Angel's trumpet, datura, and belladonna
PCP and ketamine
The notion of personal choice as it pertains to drug use and addiction
The government regulation of controlled substances

LECTURE 17: Peyote 253
The biology of peyote
Indigenous uses of peyote
The Native American Church and peyote ceremonies
Peyote alkaloids
The difference between peyote and mescaline

LECTURE 18: Psychedelics I............................ 283
Phenethylamine and the misuse of terminology
The question of drug potency and metabolic disposition
Effectively doubling dosages
The 10 Ladies of unique hydrogens
Clinical use of MDMA

LECTURE 19: Psychedelics II........................... 317
Drug scheduling and the Emergency Scheduling Act of 1984
Short-term and long-term effects of MDMA
Mescaline analogs
Indoles and hallucinogenics
Drugs originating from South America and their traditional use

LECTURE 20: Psychedelics III.......................... 351
Continued discussion of indoles
Set and setting
Ergot and St. Anthony's Fire
Sansert, LSD, "Narcotic Farm," and Harris Isbell
Talking a person through a "bad trip"

LECTURE 21: Prescriptionals 383
Ergot alkaloids, lysergic acid, Albert Hofmann, and Sandoz

Wholesale costs of prescription drugs; stress clinics and Quaaludes
The two-story narcotics laboratory in Stockholm, Sweden
DNA, proteins, enzymes, hormones, amino acids, polypeptides
The science of the morning-after pill

LECTURE 22: Cannabis 411
Experiment to determine genotypic knowledge in marijuana plants
The Bureau of Narcotics and Dangerous Drugs, the Marijuana Act of 1937
A compliance visit from the DEA
Marijuana's origins in China
The government's attempt to synthesize THC

LECTURE 23: Cannabis and LSD 445
Field test kits for evidence of drug use
Crime labs and drug exemplars
Voir dire and being called as an expert in a criminal case
LSD flashbacks
Refractory periods with LSD and cocaine

Afterword ~ by David E. Presti 475
Index 481

LECTURE 9
February 26, 1987

Stimulants I

SASHA: Today, we are going to go into more detail on *Excitantia*, one of the five classifications[1] that I gave three or four lectures ago. These are the stimulating drugs referred to by the up-pointing arrow in Ups, Downs, and Stars.[2]

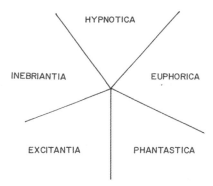

The psychotropic states, Lewin, 1931.

As with all the different classifications, we've learned a lot from nature, from the plant world, what the natural world has given us. I'll go through these in time, as I cover different plants. These are basically the bulk of the plants that are used in one form or another as stimulants in the world.

I would like to go back to where we were about two lectures ago in the nervous system, because stimulation is expressed in the body, in essence, as increased nerve conduction. With those nerves that are stimulatory nerves, it's kind of a begging issue.

The nerves in the body don't all do something. Sometimes nerves *don't* do something. In fact, by activating a nerve you sometimes keep something

---
[1] The five categories of Louis Lewin's *Phantastica* were defined beginning on page 223 in *The Nature of Drugs*. Vol. 1.
[2] See page 93 in *The Nature of Drugs*, Vol. 1.

from happening. One has the impression that you'll push a nerve and ring a bell. Sometimes the bell is ringing, you push the nerve, and you stop the bell from ringing.

The whole process of sleep, for example, is thought to be serotonergic—dealing with serotonin, one of the neurotransmitters—a serotonin system in which sleep is an active stimulation of the nervous system; the nervous system produces sleep. Sleep is not a non-state. Sleep is just another state and it can be produced as an active thing, just as wakefulness can be produced as an active thing.

In the same way, there are many interconnections in the brain in which neurons feed upon other neurons in the area known as the synapse, where things come back and create a feedback to the presynaptic area, or the presynaptic feeds on itself, creating feedback loops, inhibitory loops, and reinforcing loops. The entire thing is a resonating connection of neurons that are all kind of touching one another. These neurotransmitters are saying, "Conduct; don't conduct," and some signal goes through there, they change their status, and a new concept or a new thought, or an action or a state of mind or state of body takes place.

It is not a simple wire, synapse, wire, synapse, wire. It is an interactive, reiterative structure of neurological connection. The one I have drawn on the board is just the very simplest concept of the neuron. The direction of the signal is this way, through the synapse. Not that it wouldn't go both ways. It happens that the receptors are on this portion of the neuron, on that side of the synapse. There's no receptor on this side. [Indicating the presynaptic side.] So let's take and sprinkle a neurotransmitter down into the synapse; in this case we're dealing with an adrenergic or excitatory synapse, so the neurotransmitter will be norepinephrine.

I have mentioned that norepinephrine and noradrenaline are the same compound. Since this is a synapse, it is often called a norepinephrine synapse or a noradrenaline synapse, or simply an adrenergic synapse, leading to the term "adrenaline." So, if this were a neuron, one side is presynaptic, the other is postsynaptic. This is before the synapse and after the synapse. Let's take a neurotransmitter and just let it wash down through this space that keeps these two nerves from touching. The nerves don't touch one another in the body. They are very close, but they do not touch. You need a

LECTURE 9 ~ *Stimulants I*

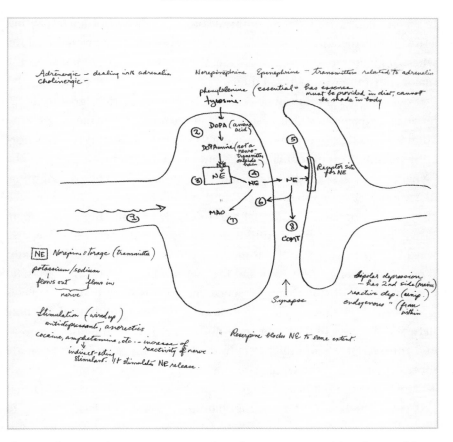

chemical to wander across to complete the connection. The chemical here is norepinephrine. If you put the chemical down in here, norepinephrine would serve as a direct acting transmitter, it would activate these receptors, and this nerve would fire and go. There are no receptors this way. And so, by putting a neurotransmitter down in here, this nerve would not know the difference. [Referring to the presynaptic side.]

But this is it, in essence, the stimulant and the adrenergic system which are the nerve connections that transmit signals from here to there through the body. There are a bunch of amino acids in the body that give rise to these neurotransmitters. Tyrosine is in the body, and part of this system. It is one of the non-essential amino acids of our metabolism. Tyrosine gives rise to a compound called DOPA (dihydroxyphenylalanine). Tyrosine is mono-oxy; DOPA is dioxy, which goes to dopamine, which we have talked about in the brain, but it also occurs as an intermediate; not as a neurotransmitter, but as an intermediate in the actual adrenergic receptor, the neuronal synapse, where it gives rise to norepinephrine, then norepinephrine becomes the major transmitter to be found there.

So if this neuron functions properly, along comes the signal to depolarize and you get to a little vesicle that stores norepinephrine. Something about that vesicle gets shaken up, it gets disturbed, then it releases norepinephrine. Norepinephrine leaks out of the neuron into this gap, the "synaptic cleft" it's called, and it wanders around. It's kind of a casual thing, like wandering across the street except that you've got a lot of speed and the street is very narrow. It's only .04 microns wide or something, very small. It may be destroyed by an inhibitor, by a deaminase or an oxidase, as a way of getting rid of it. But if it makes it across to the other side, literally, then it triggers this other side and this neuron fires and down it goes. So the whole area of stimulus, in essence, makes this work better, or makes it work more often, or makes it work more easily. If you have this whole network of nerves in the brain and some of them are stimulatory neurons, and you activate them preferentially, the brain is more active and you are more active. This is the whole concept of a stimulant.

How can stimulants act? They can act by knocking out the thing that gets rid of the transmitter, the monoamine oxidase, and indeed many of the antihypertensives contain a part of the hypertensive agents. The materials that are antidepressant are indeed antidepressant; they tend to up the mood by increasing the stimulation of the body. When you get stimulated, there are a number of ways of being stimulated. You can be stimulated by not being unstimulated, namely, being below normal and coming up toward where you would think normal would be. This is the function of an antidepressant.

You can be stimulated by wiring up the system. I mean, people after that eighteenth to twentieth cup of coffee are saying, "My golly," you know ... "Look ... fingernails ... wall ..." You're all wired up, the hair is standing up on end. You have a feeling of a pressor; you have a pressor effect, your heart's pounding. You can hear your heart beating in your ears if you listen. Your blood pressure is high. You try to go to sleep, you're looking at the ceiling in the dark all night long. You cannot sleep. You want to eat, you have no

LECTURE 9 ~ *Stimulants I*

appetite. All of these are symptoms of this type of stimulation. I mentioned the antidepressants often work in this area. You can do it by going directly into the cleft and acting as if you're on norepinephrine. These are called "direct acting stimulants," there are some, they're usually catechols. I'll write this out in a handout. But things that are catechol-like, I suspect that the material we were talking about three or four lectures ago on MDMA/Adam/Ecstasy acts by being a direct-acting stimulant because it becomes a catechol in the body.

You have things that act by inhibiting the reuptake. How does norepinephrine get removed? Once norepinephrine gets in this cleft and begins wandering across, you don't want it to stay there forever. One of the really serious poisons that lives in the body is a neurotransmitter that does its job and then you can't get rid of it. There is acetylcholine, for example. Acetylcholine is the neurotransmitter of a choline, a cholinergic synapse, and once it's done its job, you want to get rid of it. You have to destroy it. Norepinephrine, once it's done its job, has to be swept out. This is a function of sweeping out by monoamine oxidase. There is a methylation catechol, methyltransferase—don't worry about the word—that destroys it by chemically modifying it. Or there is a presynaptic thing that can take it back up again, can snag it out of there and stick it back in the vesicles to reuse it. That can be interfered with by what is known as a reuptake inhibitor, something that keeps it from being taken back up; as an inhibitor it will act as a stimulant because these things are around more. Cocaine probably acts in this route of keeping things from being taken back up.

There are agents that are direct-releasing agents. They shake this little capsule up and cause the capsule to release more material. These are called "direct-acting agents" or "releasing agents." Amphetamine, which we'll be talking about today, probably acts by that route. There are other drugs that act in this general area to influence stimulation, often by removing it. One of the more dramatically effective and widely studied one is reserpine, which is a depressant.

Reserpine drops blood pressure. Reserpine drops the emotion, drops the affect. How many people know the word "affect"? One, two, lots... "Affect," the way you look, the way you feel, your attitude toward the world. People tend to confuse the word "affect" and "effect." It's a tricky little combination.

THE NATURE OF DRUGS

"Effect" is something like: I twisted the chalk and the effect was to break the chalk. I effected a change in the state of the chalk. I affected the status of the chalk, but my affect has no bearing on cause and effect. My affect is my attitude, my spirit, my approach to the world, what you see in my face by looking in my face; what my personal point of view is and what my spirit is. Reserpine will depress that affect. A person will become very quieted, low blood pressure, low spirit, depressed. What reserpine does is interfere with the uptake of norepinephrine that is made in the cell, into the actual capsule that stores it. And all these things can work in concert. All these things, in essence, work to make a neuron more or less effective as a transmitter. Caffeine, which we'll talk about today, works in here somewhere; no one knows the actual mechanism of caffeine action. It's probably in and around this area, but how it actually works is not known. Let me give that in a handout that we'll use later. These are the general routes by which transmitters, neurotransmitters, work and how they work as far as we know.

What are the stimulants? The most broadly used is caffeine. We mentioned this in the first lecture. I want to go back to it more. Coffee is the primary vehicle. Tea contains it. Maté contains it. Other materials, other stimulants might give rise to it by body conversion, but to a large measure, caffeine is the major stimulant used in the world. I have estimated 3 billion people have used it or use it. That's over half the world's population. Coffee first grew wild probably in southern Ethiopia, Abyssinia. It was a scrubby brush, a short tree that grew up in the rocky areas. How was it discovered? The fruit contains two seeds and the seeds themselves are the source of the coffee. The seed itself is not particularly tasteful, not particularly flavorful, it's sort of a yellowish, yellowish-green seed. I don't think there's anyone alive who has not seen coffee ground up or as an unground thing, these are roasted seeds. The actual chemical in this, caffeine, is a white solid. Here's a half a kilo of it. [Pointing to bottle.] It's buyable by the kilo.

I know people who are very much against using coffee, but they have no objection to using caffeine. It's an interesting balance of accepting the drug and not the vehicle that contains the drug. This brings up an interesting point. Are the plants that contain drugs and the drugs that are in the plants having the same effect on the body? Not at all. It's easy to say, well, caffeine is the active component of coffee. I'm going to take milk and add caffeine to

LECTURE 9 ~ *Stimulants I*

it and I'll get the same results as if I had a cup of coffee. No, coffee contains lots of other things. All of the natural plants contain other things. And these other things themselves have action. In coffee you have theobromine. You have theophylline. You have methylxanthine, you have different methylxanthines. You have aspects of uric acid. You have tannic acid, you have fats, you have flavors, you have essential oils, you have fixed oils, all of these. And God knows how many other things are in coffee. To remove from coffee that one chemical and take that one chemical and say, "This is the active component of coffee" is not correct.

Andy Weil, the author of the book that I trust you all read the chapter on stimulants for today of. I love ending sentences in strange prepositions. I once had worked out six, but five is usually the maximum you can get at the end of a sentence. I'll write that up for you some time, if you don't know it. To end a sentence with a preposition.

Andy Weil espouses the theory that in some ways, in fact, in many ways, plants provide a protection against the misuse of drugs, and the misuse comes from its removal from the plant. How many people have eaten really raw sugar? Sugar cane? Good, almost everyone. Try eating a pound and a half of sugar cane. I mean, there's just no way that's going to stay down. It's got a sweetness to it, but it's got a lot of other stuff too, and really raw sugar, the stuff that's expressed out of the cane, is an interesting taste, but you can't handle that much of it. You process it, you get rid of the fiber, you get rid of the bitter stuff, you get rid of all the colors, the dead ants, everything else that's in there and get down to our sugar: refined sugar. Well, I don't think there are many people who can't go through much of that.

I was reading an essay about the amount of sugar that's in breakfast food. They're changing the names from sugar loops, sugar snacks, sugar this, sugar that to "Strong Snacks" and "Fruity Loops" and god-knows-what-all. They're taking the name *sugar* out of the names, but the sugar is still in there. There are breakfast snacks in which more than half of their entire weight is sugar, and they are appealing. They are used, they are advertised strongly, and they have that marvelous sweet-tooth requirement that apparently is built into our needs. Sugar can be grossly abused as a pure material, but as the raw material, not so. You see brown sugar, you think that's a raw sugar. Not at all. It's pure sugar that's been made impure again by the addition of some of the materials that had been taken out. It is, in essence, molasses and other factors added back to sugar. Brown sugar is not an intermediate in the purification. It used to be. It is not a by-product now.

Take the example of the raw material opium. You can certainly abuse heroin. I mean, half of our whole urge toward getting drug abuse under control is the fact that heroin is widely used and is widely abused. Opium

LECTURE 9 ~ *Stimulants I*

is very hard to overuse. Very, very, hard. This should be in a later lecture, but it's here now. If you take a pipe of opium, you get into this very strange little dreamlike place and you're waiting for whatever the drama is, and opium doesn't carry drama. Opium if anything is anti-drama. It puts you away from the drama, it puts you into your mind and into your little retreat and lets your mind wander away. Take a second pipe and you'll find that you can't quite stay awake and you're tending to dream but you're not quite asleep. Greed will get you a third pipe and you're going to be vomiting for 24 hours. I mean, there is intense poison. There are other things in there. There are perhaps a dozen alkaloids that are removed, perhaps as many as 20, many of them biologically active, some of them actually promoting nausea as their pharmacological property, out of which morphine is taken, from which is made refined heroin, with which you don't have these defensive guards.

Take mescaline, which is a white crystalline solid that comes out of the peyote plant. In the peyote there are over 50 other alkaloids, some of them are quite toxic, some of them are quite pharmacologically active, some of them interact with one another. The plant very often has a safeguard that is lost in the refinement or in the synthesis of what is called the "active" component of the plant.

So, coffee contains . . . I think I'll start writing up . . . I'll take this off for the moment. [Referring to a "dirty picture" on chalkboard.] Coffee contains as its major component, caffeine, or "caffein" as it's called in the old chemical, old pharmacological literature. How many people have taken a course in organic chemistry? One, two . . . nah, it doesn't work putting structures on the board. Anyone who's interested in structures, I'll write them out again and hand them out.

Caffeine is a great big purine. It's called a "purine." It's got a big ring, and a not-so-big ring, and has methyl groups sticking out. And it's got nitrogens everywhere. It is a water-soluble, organic-soluble, easily-extracted alkaloid, a weak base. The coffee, the source of it that I mentioned, started in Ethiopia, it moved up into Persia, now Iran, about 600 or 800 AD. From there it moved into Arabia, maybe 300 or 400 years later moved into Turkey, and came into England, came into Europe, probably in the 1400s. I believe the first coffeehouse in England was in the middle of the 1500s. As with all

THE NATURE OF DRUGS

drugs, when they are first introduced in society, there's a rebellion against them. There are edicts and proscriptions about them, and coffee was very strongly condemned as being whatever drugs are condemned for being after they're first introduced. And then it was accepted, and now it's used so broadly that it's not even thought of as being a drug. It is a drug. It is a stimulant. Too much coffee, too long usage of coffee, many people have withdrawal problems from it. Headache is a very common withdrawal symptom. And irritability. Coffee itself has an active property of being an irritant. It's a stomach irritant, and a bladder irritant. It is a diuretic; diuretic meaning to *di*: across; *uresis*: flow. The urine; flowing across. It tends to increase the flow of urine. And caffeine is the principal active component.

Thé (I'm getting into the wrong language). Tea. Thea. *Thea*, the source of *Thea* is "Goddess," as *Theo* is the source of "God." And you'll find Thea and Theo in these things, in these words. Theo and Thea, what do you have; theology—the study of God. *Theobroma* is the food of the gods, and *Theobroma* comes from cocoa, which was considered to be a royal and divine food in the New World. In fact, it's worth putting up these words, they are confusing. Let me put up these four words that are sometimes interchanged. "Cocoa," which we'll get down to in a moment, is the material that is made by grinding up cacao seed and is a brownish powder. We know it as a chocolate-flavored drink. It's the powdered seed containing the fat but not the husk of the seed. "Cacao" is the botanic name of the plant. *Theobroma cacao*.

"Cola" comes from the name of an African seed, which we'll get to in a minute. The word "Kola," this name and its accuracy we'll also get to.

"Coca" is name of the plant that gives rise to cocaine, another stimulant, which we will not touch at all today. We're going to have a separate lecture that is entirely on cocaine.

Very, very close words to one another, but each has a distinct meaning.

Coffee is largely an equatorial tropic type of plant. Except for some areas down on the southern part of Argentina, Uruguay, and some areas in the northern part of Mexico, where coffee can be raised. Almost all the coffee is raised within a few degrees of the equatorial belt around the world.

Tea, on the other hand, is grown largely away from the tropics. Most of the tea is raised in semitropic areas. China is still a major source of tea, the southern part of Brazil, Argentina, northern Argentina, Uruguay, Paraguay,

LECTURE 9 ~ *Stimulants I*

but largely in Asia. Tea is probably even more ubiquitous, more broadly used than coffee. I know the name in several languages. *Thé* in French but *chai* in Russian. I worked with some people who are Indian and I humorously used the term "chai" and they served me a cup of tea. *Chai* is the Indian name, Hindustani name. In northern India it's called "cha." In China it's called "cha." Chai, tea, thé, cha. This one-syllable thing is substantially available around the world.

Tea is a shrub. It's the entire leaf that's used. Again, the actual leaf itself has very little taste, and it does not make a satisfactory drink. It has to be roasted. It has to be heated and changed by a process of preparation. There's green tea, there's black tea. I think everyone . . . Who here has never had a cup of tea? I think it's most unusual, as there are no hands. No one here; tea is around the world. Great. Yes?

STUDENT: All these teas you just named, black, green, whatever . . . Are they all the same plant?

SASHA: That becomes a tricky question. Probably originally, yes. But now, what constitutes a species? Things have been made into cultivars, have been modified by relocation to new areas that are not their native areas. They tend to grow in new ways. Probably they're all from the original *Thea*. Yes?

STUDENT: The Celestial Seasoning tea they make with the orange blossom and all with the almonds, they're all the basic tea?

SASHA: On the contrary. None of them have any tea. They're herbs, other sources. In fact, one of their sales points is that they do not contain caffeine.

STUDENT: They have something like Black Thunder, though.

STUDENT: Morning Thunder. It's got plenty of caffeine and it's got tea in it.

SASHA: It has maté. In fact, this use came up with a word, a four-letter word that is twice the strength of Coca Cola . . . zoom, boom, bang, bam . . .

STUDENT: Does it have tea, or does it have maté?

Student: Jolt!

Sasha: Jolt! [Laughter.] I think some of these actually have maté as the source. Tea is not as strong. In fact, when you make a cup of tea, it does not have as much caffeine as a cup of coffee. But actually you use much, much less tea in weight to make the cup of tea. I mean, you use a little bit of dried-up leaf as opposed to coffee, you grind up a bunch of beans. And so, the tea itself is much richer in caffeine as a plant, as an actual weight. But you use much less of it and the net result is probably a little bit weaker of a drink. How much caffeine is in a cup of tea? Oh, depends how strong it is.

LECTURE 9 ~ *Stimulants I*

Fifteen to 100 milligrams. How much is in a cup of coffee? Depends upon the strength. One hundred, maybe 100 or 150 milligrams. How much is a bottle of Coke? Depends on the size, probably around 100 or 150 milligrams. You're dealing, in essence, with something like a tenth of a gram of caffeine. How much is a tenth of a gram of caffeine? Probably the amount you can put on the tip of a teaspoon. The amount that you could balance on your thumbnail. That's the kind of quantity of drugs we're looking at. Here is a half kilo. I brought it out of the lab at home. I have four of these bottles, so I have two kilos of caffeine. It is a cheap, easily available material, because if nothing else, it's obtained in quantity by the decaffeination of coffee.

How is coffee decaffeinated? What is meant by taking caffeine out of coffee? You're actually leaching out the caffeine, either by an organic solvent, there is a dichlor or chloroform method, or there is a water method that's used in Europe. It's patented there. It's not used in this country. I don't know what the water method is, how it's used. There is a method that uses carbon dioxide and supercritical carbon dioxide, which actually infuses and removes the caffeine by solubilization.

Yes?

STUDENT: Is there any residue from that process?

SASHA: Oh yes. You take out a lot of the oils, the fats, the flavors that are processed back in. So in essence, you have a . . .

STUDENT: Like toxic residue from the chemicals . . .

SASHA: In principle, from the water it would be negligible. From the carbon dioxide it would be negligible because it's a natural thing. From the dichlor yes, there is a residue and the amount that's left in has been limited by law to be, say, one or seven or some number of parts per million, and in truth, dichlor is sufficiently volatile that almost by definition it's gone by the time you finish the process and you've got the material. I don't believe there is a risk on the residual solvent in decaffeinated coffee. How much caffeine is there in coffee that does not have caffeine? Quite a few percent. Some of it can be 1 or 2 percent; some could be 5 percent or more. Decaffeinated coffee is not free of caffeine. If you're sensitive to caffeine, your decaffeinated coffee will not relieve you of that exposure.

Where does caffeine come from in the environment? In foods. It's contained in a lot of foods in small amounts. There are a lot of ways of preparing foods that introduce caffeine without its being known. Soft drinks, almost all soft drinks that are darkly colored are caffeine containing, and classically the light drinks, the 7-Up types, Sprite types, do not contain caffeine. Okay!

Oh! Someone mentioned green and black tea. Same plant, just different method of processing. Tea, when it's gathered from the actual plant, is roasted virtually immediately, within an hour or two, before it has a chance to darken like an apple or a pear will darken. When exposed to the air, tea picked from the plant will darken. It's roasted and heated up until it gets soft, then it's rolled, then it's reroasted, then it's rerolled, it's reroasted, back and forth until there is a general fixing of the color, and the texture has become quite brittle, quite fragile. It can be broken into the fine tea powder, not powder but less than a leaf and more than a powder, whatever that small stuff is called, tea! This will retain its color. It's called "green tea." The black tea has been allowed to stand and be exposed to the air and be air oxidized. It is then roasted and rolled, and produces black tea. The caffeine content is comparable. The tastes are quite different. Many different tastes—and this applies to coffee almost as much as it does to tea. A lot of the taste is dependent on how the coffee is prepared, not where the coffee is grown. The plant does not have as much of an effect on the quality, on the acceptability of the final product, as does the preparation of the plant.

STUDENT: What is the taste difference? Is one stronger?

SASHA: Oh, the flavor. Different flavors, different ways of roasting will retain or throw away different degrees of the flavorful aspects. The oils that are in there add quite a bit to the flavor and what are also called "volatile" or "essential" oils. Gosh, will we ever get into essential oils? Possibly not in class, so this is a good time to do it. The oils that are in plants, these are the materials that are not water soluble. Usually they are liquid, fluid, and float on water. The concept of oil you're familiar with. There are two general types of oils in plants. Very crudely, those that you put someplace or spill and they'll disappear because they're volatile, and those that you can pour or spill and they'll stay there because they're fixed. Volatile oils usually carry a smell. And for that reason are often called "essential" oils, because

LECTURE 9 ~ *Stimulants I*

they have an essence. Not as you would have essential amino acids that are needed, not in the sense of "essential," being needed, but the essence of "essential" being they have a smell. And we'll get into this when we get into intoxicants because there's a whole host of materials known as spices that are known primarily for their smell, and not for their food or drug value, but they almost all have (either in their history or in their manipulations) some structures of compounds that are capable of altering the state of where you are. The whole spice trade becomes an interesting chapter of the economics of people knocking their tails off to bring spices across 10,000 miles of wild savagery just to hide the taste of spoiled meat in England, and people couldn't afford it. Ridiculous.

Yes?

STUDENT: What happens when you smoke tea?

SASHA: I have no idea. Caffeine is volatile in the sense that it sublimes. I imagine you get the caffeine in you. I have never tried it. I have no idea. Anybody here ever smoke tea? Good. What comes out of it?

STUDENT: I don't remember the effects, but a very small amount. It felt exactly like marijuana.

SASHA: Anybody else contribute? Any kick out of it at all?

STUDENT: I had a buzz but I think it was just because of the smoke. [Laughter.]

SASHA: If you take a good deep breath of something that does not contain oxygen, you go to a strange place, like flat on your rear.

Yes?

STUDENT: Yeah, I know some people who use it if they want to get sober really quick from alcohol. They smoke tea and apparently they snap right out of it, right out of their drunken state.

SASHA: Well, you're certainly going to get caffeine, caffeine is volatile. In fact one of the original isolations of caffeine was putting tea on a piece of glass with a paper cone on the top of it, and putting the piece of glass down into where it can get very hot with a flame and heating up tea leaves. You can do this as an experiment. Heat up tea leaves, and when it gets hot enough you can take that paper cone and tap on something like the hard table that's clean, and you'll see these fine white crystals drop out. You've actually sublimed. The concept of sublime. Not in the emotional liking-of-something sense, but in the physical sense. *Subliming* is going from a solid to a vapor without going through the liquid phase. An ice cube melts, you have water. Water boils, you have steam. If you go from ice to steam without having a liquid phase, it's called "sublimation." And caffeine is a compound that will go on heating from a solid to a solid by going through a vapor phase with no melting in between. And you can actually get caffeine, white crystalline caffeine, out of tea by just doing that—putting it in a hot area with a paper cone above it. It's a very early demonstration of the isolation of an alkaloid.

STUDENT: What will pure caffeine do to you?

SASHA: Pure caffeine? A slightly bitter taste. It's a stimulant. As I said it's an irritant. You definitely will get a slight diuretic effect from it. And probably sleep problems. Easily handled, obviously easily tolerated in quantity by half the world.

STUDENT: Like natural opiates?

SASHA: No. The natural opiates are especially strong bases that are present in the plant as salts. You have to get it out of the plant. Caffeine yeah, opium no. Morphine, codeine, no. They don't volatilize well.

STUDENT: You can't heat it up?

SASHA: Oh, you can heat it up and volatilize it, and you get an aerosol, as you do with tobacco. The aerosol will carry the alkaloids. But they are not necessarily in the vapor phase, they're in the aerosol phase. It's like smoke. When you inhale from a cigarette, you take this big bluish cloud of not totally transparent material. It's not gas. It's particulate, and in there are the tars that the alkaloids are usually absorbed in. So nicotine yes, codeine, morphine, thebaine, no. They probably would be transmitted by aerosols. Yes?

STUDENT: MSG is another transmitter. Would that be considered . . . ?

SASHA: What is?

LECTURE 9 ~ *Stimulants I*

STUDENT: MSG.

SASHA: Monosodium glutamate ... ah, okay, is it a neurotransmitter? There is what I call the Neurotransmitter of the Month Club. Every time you dig in closer into the brain, you come out with a new chemical, by golly, that plays some role. Glycine, monosodium glutamate, glutamic acid, amino acid. Monosodium glutamate is the material that is being used as a taste enhancer in much of Chinese and Japanese cooking. Some people are extremely sensitive to it. It has given rise to what has been called the *Chinese Restaurant Syndrome*, the flushed face and general rashiness that some people experience when eating Chinese food. I know that the percentage of people who are sensitive to it is small, but real.

Is glutamic acid a transmitter? It might well be. I think the current tally, there are about 20 to 25 materials that are suspected of being transmitters. And it may well turn out that there are neurons that can use many, many things. Polypeptides, amino acids themselves, glycine has been found to be a transmitter involved with convulsion in some way. Strychnine interferes with glycine's integrity. Monosodium glutamate, you mentioned. Gamma aminobutyric acid fits in a role that may play a neurotransmitter role. Epinephrine may. Norepinephrine we've already mentioned. Acetylcholine. There are polypeptides, Substance P may be a neurotransmitter in the conduction of pain. I don't know what their roles are, but they are argued as playing a role in neural transmission. There is a collection of requirements that a compound must meet to be called a "neurotransmitter" and this changes continuously too. I am not aware of what the current fad is for the definition of neurotransmitters.

STUDENT: Wouldn't that be considered a stimulant, though?

SASHA: Not necessarily. Neurotransmitters need not be stimulants. Neurotransmitters can be depressants. Neurotransmitters can put you to sleep. Neurotransmitters make nerves conduct, but the conduction of nerves need not be a stimulatory thing. It could be a depressant thing. It could be a feedback that inhibits. So not all transmitters are stimulants by any means. Okay, other questions in the area? Yes.

STUDENT: MSG is a stimulant?

SASHA: No, it enhances taste, enhances the sense of taste. It does not have appreciable taste itself. And it causes an irritation, a neural and a physical irritation in those people who are sensitive to it. Its function in the body is glutamic acid as an amino acid. It's one of the twentysome amino acids in the body's function. So it is a biochemical component of protein. If it plays a neurotransmitter role, it would be a very obscure one. Certainly not a well-known one.

Yes?

STUDENT: Are there any times when a person would be more sensitive to a stimulant than at another time? I've seen effects where the chest gets really red, and it looks like it's a rash and on the insides of the arm the face gets . . .

SASHA: From monosodium glutamate?

STUDENT: No. Just from stimulants. Like a really bad headache, like a . . .

SASHA: We're going to get toward that when we get into the ephedrine and norephedrine area. By the way, just give me a call on time. Where are we . . . What time is it roughly?

ANN: Quarter to two.

SASHA: Quarter to two. Okay a half hour to go. A lot of the stimulants that are sold—the *High Times* type of stimulants that are sold as uppers, as legal speed and what have you—contain a material that is known as phenylpropanolamine, or pseudoephedrine, or norpseudoephedrine. I'll get into these terms in a few minutes. They are stimulants, but they don't get into the brain very easily. I think the best example is ephedrine itself. I brought in a bottle of it, you can see what it looks like. Almost all organic compounds are white solids. This, by the way, is a material that is often used as a chemical precursor to amphetamine and methamphetamine, but is in itself a stimulant.

How many people use Sudafed to clear up their whoosies? Yeah, most people have. Try a half a dozen of those some time, you'll find yourself in a very wired-up kind of place. It's not a particularly nice one. And I think

LECTURE 9 ~ *Stimulants I*

that is what your question is getting into. To take enough to get into the brain, you're overwhelming that which is outside the brain. I mean, your nostrils are not up there, they're out here. And you're taking this to clear, to decongest, to open up, and a little bit of it gets into the brain. So you are maybe getting a toxic response from a relatively ineffective stimulant. So it's a matter of using, or overusing, a relatively ineffective stimulant. Other questions on the approaching the plant stimulants.

Question, yes?

STUDENT: What is a black beauty? What is it? They are capsules.

SASHA: Yes. Are they Dexedrine or propanolamine?

STUDENT: I think I wrote Dexedrine sulfate.

SASHA: Okay, Dexedrine is an actual stimulant. I don't think ... are they obtained by a prescription or by *High Times*, so to speak?

STUDENT: The black beauties?

SASHA: Yep. Okay. I suspect they're the lookalikes. They're called "pea shooters" in the trade, which are things that look like things that are obtainable by prescription and they're obtainable by the hundreds of thousands or tens of thousands by writing a postcard to somewhere in Kansas advertised in *High Times*. Yep. And you're probably getting phenylpropanolamine, which is a mediocre, relatively ineffective, stimulant.

STUDENT: Black beauties are replicates, so a lot of these black beauties are not really ...

SASHA: They're not what I call black beauties.

STUDENT: They're like 500 milligrams of caffeine with a little bit of phenylpropanolamine in there. But they're not real amphetamines. Real amphetamines are hard to get.

SASHA: Real amphetamine is hard to get. Real methamphetamine is extremely hard to get.

STUDENT: So a lot of stuff on the street isn't real.

SASHA: Yeah. There is a term I should put up here. Caffeine. Let me write up the other two terms that are in these plants. [Sound of chalk on chalkboard.] Theobromine and theophylline. And caffeine. Caffeine, theobromine, theophylline are the active components, caffeine being the most active, of coffee, of tea, of maté. Maté is often called "Paraguay tea." How many people have had maté? I have too. It's really a potent coffeelike drink.

Maté, *Ilex paraguayensis*, botanically speaking, is a holly plant. The leaves are gathered. They are dried and roasted, but not charred. And as a roasted leaf, they are powdered to a very fine powder. The powder is gathered and stored in wet skins. You take the hide of an animal that is wet and fill it with this powder and it's stored that way and kept until the whole thing is dry. And when it's dry, you can drop it and it will go down through three concrete floors. It is concrete. It's solid. Then this in turn is ground up and used in an infusion to make the tea. I believe it's available in health food stores. It's broadly used in South America. Originally Paraguay, it's gone to Argentina, then it's gone to Brazil and swept through all of South America. They've made drinks out of it, maté drinks, I think I mentioned, somebody mentioned this, this boom. Why am I blocking on what's that four letter . . . ? Once again, what's the . . .

STUDENT: Jolt!

SASHA: Jolt! Good. That, I believe, is a maté drink.

What is up next? Cocoa. Now, here again you have a material, cocoa, that is the seed of a plant *Theobroma cacao*, but the seed is filled with fat. It's a brittle seed that is very fatty and it's ground up and infused with water. It is usually not roasted. It is a ground up seed, it's again New World, the drink was called "xocolatl" by the Indians, the name of chocolate, which is of course the candy that we know that flavor by.

In actually making chocolate, you take the seed and I believe the seed is fermented. And the husk is removed and then what is inside is crushed up and smashed. It's roasted and it then gives the powder that is cocoa, and when it's mixed with vanillin, or vanilla, and sugar it gives the material that is known as chocolate. This contains some caffeine, but largely theobromine, the food of the gods.

Take caffeine; I mentioned caffeine is a big structure with the methyl

LECTURE 9 ~ *Stimulants I*

groups sticking out like rabbit ears. If you take *this* methyl group off, you have theobromine. If you take *this* methyl group off, you have theophylline. If both methyl groups are on, you have caffeine. So the three compounds are virtually identical to one another except they contain one more or one less methyl group.

The kola nut, I mentioned briefly. Cola is a very common name from the Coca-Cola approach. But the kola nut is a nut from Cola trees[3] that are raised in the northern part of Africa along the Mediterranean coast. Its use moved down the west coast of Africa bordering the Atlantic. This is a nut that is often eaten raw. It's chewed, although it can be roasted and the roasted nut can be infused into a drink. Again, the components in it are caffeine and theobromine.

There is a question somewhere. Yes?

STUDENT: What about yohimbe . . . yohimbe bark . . .

SASHA: Yohimbine comes from yohimbe bark. This is an African material. I don't know if it's ever been made into a stimulant drink. It's certainly been used in Africa where it has a reputation as an aphrodisiac. Everything has at one time had a reputation of being an aphrodisiac and promoting fertility, or for starting the menses. I think you can go through the entire world of plants and find something that gives each of these categories. It's a vasodilator. It will cause a drop in blood pressure. It passed in and out of the drug-oriented scene for quite a while, went out for quite a long while, then came back in along with interest in the scientific area for its effect on circulation. It causes vasodilation and as such it may have some effect on the erectile tissue and maybe that would warrant its arguments as an aphrodisiac. But I don't know of its native usage outside of Africa where it's used. Anyone here worked at all with yohimbine? I've explored it only to see that the effect on blood pressure is very mild.

ANN: Is it legal?

SASHA: Yohimbine is . . . the term "legal" is worth a comment. I'm often asked, "Is a drug legal?" I'd rather say a drug is not illegal. Because if you take a material that is not illegal, but use it in a way that has an illegal

---

[3] *Cola acuminata* and *Cola nitida*.

character to it.... Let's say you take something like sodium chloride, which is salt, which is not illegal, and sell it as if it were methamphetamine, then the selling of that is a violation of public health laws and you're guilty of a felony. Is salt legal? Not if it's done, handled in an illegal way. Yohimbine is not illegal. It is not specified in the legal structure. Okay, cool.

Intermediate plants—plants that do not contain caffeine or theobromine or theophylline, but still are stimulants. I've put up two, in essence, khat, or chat, as it is pronounced, you spell it with a K, with a C . . .

STUDENT: With a Q.

SASHA: Or the Q, right. Was it Q-H? What's the second letter? Q-A. Q-A.

STUDENT: Q-A-T.

SASHA: Q-A-T. It's one of the two or three words that has a Q without a U following it.

An H with K or C. Either one. Both interchangeable spellings, as *khat* has a variety of spellings. Khat is used in Yemen, used in the area around the Red Sea. It is used in Abyssinia. Here is a case where the active ingredient is cathinone. These terms are really awkward, let me start out with an amphetamine.

Methamphetamine and amphetamine are synthetic stimulants. They are man-made stimulants. The hydroxy counterpart of amphetamine is known as norephedrine. The hydroxy counterpart of methamphetamine is known as ephedrine. The active components in khat are cathinone and ephedrine. Ephedrine is the active components in ephedra. In the *Ephedra* species, by the way, whatever the species may be. It doesn't really matter if you don't know, or it may be one of several. The *Ephedra* species contain ephedrine and norephedrine. These are the mediocre stimulants that we're talking about in the area of the over-the-counter pills. Norephedrine, if it is not a specific material from a plant and a specific isomer, can also be called phenylpropanolamine.

These are not the same, but they are very closely allied. Ephedrine is one of a variety of isomers. Phenylpropanolamine is the racemic mixture of those isomers. But in this case, methamphetamine has an N-methyl,

LECTURE 9 ~ *Stimulants I*

hence the term "meth," and amphetamine has an NH_2. This kind of category allows these classifications to be made. *Ephedra* species includes Ma Huang of China, an ancient, ancient, material used in China. It is a stimulant, used as an agent to induce sweating, used an agent to treat many different types of illnesses. It is, in essence, a plant that contains ephedrine as its major component.

Yes?

STUDENT: What's the word right there? The last one?

SASHA: Cathinone, from "*Catha*," the genus name of khat. Ephedrine and norephedrine. The term "nor" originally came from the German "N ohne radikal," "N without a radical." Nitrogen was something removed from it, and indeed "nor" in this case indicates the nitrogen has lost this methyl radical. And so, ephedrine and norephedrine are related by that one carbon relationship. I mentioned that ephedrine itself is one of two isomers. The other isomer is called "false ephedrine" or "pseudo ephedrine," and hence you have the term ephedrine and pseudoephedrine. Sudafed is the trade name for pseudoephedrine. It's the other isomer of ephedrine and pseudo-ephedrine, norephedrine or norpseudoephedrine, that would be the other isomer of this compound. Mixed isomers would be phenylpropanolamine.

Now, the *Ephedra* species, I brought in the example that I collected in Utah when I was back there a few years ago. This is what ephedra looks like. It's a plant that looks like a bunch of straws hooked together with little blossoms at the junction of the straws. It was used throughout the Rockies, under the name of "Mormon tea" or "Brigham tea." The Mormons are not allowed to use stimulants, so they drank teas from the native plants. The native plants contain ephedrine, and they got their stimulant from drinking Mormon tea with the innocence of not knowing that it was a stimulant. But these ephedra plants are found throughout their area. Once you see the plant in the in the wild, it'll be recognizable immediately, very characteristic type of stemmed, long, leggy, straw-like stem structure. You can tell by looking at that. Okay now, the function of . . .

Yes?

STUDENT: When you classify ephedrine and norephedrine with this hydroxy group as mediocre stimulants, what about cathinone?

SASHA: Quite a bit more effective. It does not have the hydroxy group to block its entry. Hence it has the carbonyl group and is quite a bit more of a stimulant. But you cannot cook it. You cannot make it. It contains an amine group and a carbonyl group in the same molecule, and anyone in chemistry knows that's a no-no. When you heat it up or let it stand, they tend to polymerize and it loses its activity. It can be made into an infusion. Khat tea has been made. It's usually eaten fresh. When it's transported and sold, it is actually in something like plant leaves, inside of banana skins. It's the leaf, the fine leaf, and the growing buds that are eaten, as a raw plant. It does not keep; it does not dry satisfactorily. The counterpart N-methyl is not recorded in nature. Well, the role of this hydroxy compound is a very important one in the drug scene, because although ephedrine and norephedrine or pseudoephedrine and norpseudoephedrine are stimulants found in plants, this hydroxy group can very easily be reduced chemically to form the non-hydroxy group.

In short, ephedrine has a very large market not only as pills and poppers and such in the nonprescription trade, but as a chemical precursor of methamphetamine. In fact, one of the major routes to methamphetamine is the reduction of ephedrine. And for that reason, ephedrine has quite a high finder's fee. Again, it is a very large item of commerce. This is an example of its appearance.

STUDENT: Is methamphetamine produced industrially from them?

SASHA: Yeah! Not only do you get that, you get the D isomer. You get the correct optical isomer. Yeah! [Laughter.] Ephedrine is the legal isomer, the pseudoephedrine, which is the Sudafed, the "plus" isomer. But both have the amino group in the S-orientation, and both give rise to D or "plus" methamphetamine chemically. How is reduction done? The hydroxy group can be reduced by platinum and charcoal with hydrogen. No solvent at all. It can be done with hydrogen iodide. It can be done in a number of ways catalytically. It can be done with a thionyl chloride reduction with sodium borohydride. The scientific literature is filled with methods of doing it; you can also buy one for $5 from some post office box in New York. These are

LECTURE 9 ~ *Stimulants I*

the standard bearers of the industry. One of the major uses of ephedrine or pseudoephedrine is the manufacture of methamphetamine. Yes?

STUDENT: So how did you get this jar? [Pointing to a jar of ephedrine.]

SASHA: Aldrich Chemical Company.

STUDENT: It's not illegal?

SASHA: Nope.

STUDENT: The corresponding amino acid you can find is tyrosine.

SASHA: Not tyrosine. Phenylalanine.

STUDENT: Phenylalanine and reduce the carboxyl?

SASHA: Right. You're getting quite a bit of chemistry. There is a far better route. In fact, there's another tie between all these things. Time, about 10 minutes to go? Where am I? Perfect. Another tie of all these things is another way of getting to methamphetamine. Might as well get this term on the board because this is probably the only place it will come in. The material called "P2P" is phenyl-2-propanone. It's commercially available, also called "phenylacetone." This is probably the major chemical precursor to the illicit synthesis of methamphetamine. This material can be purchased from any number of supply houses, but it has, not too recently, become restricted. Position point: materials to be placed in a scheduled list must have a high abuse potential. I've given you a handout of requirements that are stated there. But also, if a material is an immediate precursor of a scheduled material, it can be made a scheduled material in its own right, if indeed that conversion can be made easily.

Yes?

STUDENT: So how come ephedrine is and not these other ones?

SASHA: Not federal, but interestingly state. The state passed a law just at the end of last year that made phenylpropanolamine, specifically, a material that is not illegal, but you have to keep record of its sale and inform the attorney general's office that you have purchased it. Methylamine is in this category, phenylacetic acid is this category. There are about 20 chemicals

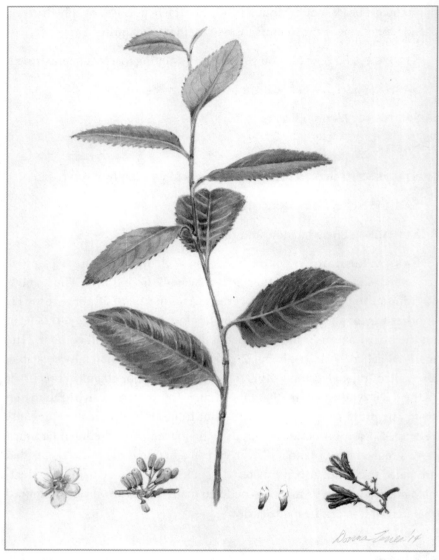

in that category. Yes, they are not illegal, but if you buy them, you must be able to assure the person from whom you buy it that you're not going to do anything naughty with it. And if you get it from out of state, it's your responsibility to inform the Attorney General that you bought it. It's one of the "gotcha laws." If you do it, they know where to look for the manufacturer of methamphetamine. If you don't do it, and they come across you, they get

LECTURE 9 ~ *Stimulants I*

you for not having done it. So it's kind of a hassle law, but phenylpropanolamine was put on roughly November of last year.

STUDENT: Ephedrine itself?

SASHA: Ephedrine is not so listed. It falls in the category that indeed it should be. They haven't looked at it yet or haven't noticed it yet. They are aware of the fact that ephedrine is a precursor, but they don't have a legal handle on it. But possession of ephedrine, charcoal, palladium, and a tank of hydrogen, and I assure you, 10 years in prison, because that is possession of the paraphernalia, or the componentry that would imply the intent to manufacture.

In fact, the whole argument of conspiracy, intent, and act is a very interesting point of law that really has to be in mind at some stage along the process. There are acts that are felonies—acts that are sufficiently ugly against society, against your fellow man, against the state, against the people, that the commission of them is considered a felony, which means you can be put away for a year or more in a prison. But there are acts that are called *misdemeanors* that are kind of, nah, a slap on the wrist, nothing serious. "You shouldn't have done it. Don't do it again." You are punished lightly. But there are interesting transitions between them that are not commonly appreciated.

One of my favorite discoveries is to commit a petty theft, a petty crime. Let's say you and I are walking along the sidewalk and we're going past a store where they're selling jelly beans, and we scoot in, grab a jelly bean, scoot out, and run down the sidewalk eating a jelly bean. We've committed a misdemeanor. We've stolen something with a value of two cents or five cents or something. There is no great ending of the wheels of industry or commerce because we've stolen the jelly bean. Naughty, but not criminal in that sense. Let's say we are walking down the street and we glance into that candy store and we say, "Hey, why don't we scoot in and steal a jelly bean?" The act of talking to another person to consider the commission of a misdemeanor is a felony. So there is a conspiracy, that's a conspiracy, to intend to work with another person to commit a crime. It doesn't matter what the crime is. It can be stealing a jelly bean. That talking with another person becomes a felony. So the act may be a misdemeanor. The intent to commit the act, or the conspiracy to commit the act may be a felony. So you don't know exactly where you stand on these if you get into the law. What constitutes conspiracy? It takes two people or more. One person cannot be involved in a conspiracy. If you want to commit a crime, act alone and you're not involved in a conspiracy.

STUDENT: If you ever want to commit a crime just say, "We don't talk. We don't like each other."

SASHA: That's a good way to stay away from conspiracy. You avoid that one charge, however, they'll make 18 more charges, so they can drop 16 and get you on the remaining two. Look at the charges that come down the pike, the act. The intent to commit the act. The conspiracy to commit the act. The . . . I don't know what it is. They hoist these things together, they make a variety of acts. You have a number of chemicals that you can make different drugs out of. Each drug becomes an intended act. So they can drop these three and take the fourth that's left. Does the intent actually require setting the thing into motion? Never been legally clear. If you have nine components of a ten-component omelet, can you have the intent of making the omelet missing one necessary component? Not totally clear. Can you be manufacturing if you simply have all ten components of the omelet? Can you be manufacturing the omelet if the first egg has not been broken? Not totally clear. These are areas that are judged in each case, in each court, in each hearing, on their own rights. They are not written in law, and hence the interpretation of the law comes from judicial framing. It's a tricky area quite apart from where I was going, where I'm going. I'm almost to where I was going to go.

Phenyl-2-propanol, phenylacetone, are immediate precursors of methamphetamine or of amphetamine, and they have been made specifically illegal. Now, you may possess them, there are a lot of practical uses for them. But to have them you have to either assure the buyer that you're not going to do something illegal, and the buyer can send information to the Attorney General in California at least, or you must inform the seller that you have a scheduled drug license to possess. But they in turn have precursors. They have a precursor known as phenylacetic acid from which you can make phenyl-2-propanol, from which you can make methamphetamine.

LECTURE 9 ~ *Stimulants I*

Phenylacetic acid has a precursor, but that also has a precursor, which also has a precursor. So how far back along the line do you go? Do you get to the carbon dioxide and suddenly into the water, how far do you go to get precursors that go on to be illegal chemicals? You go as far as is practical by law, and by arresting based on intent.

Phenylacetic acid. How many people have actually smelled it? I should have brought phenylacetic acid. It's got an incredible smell. Very, very light, very subtle, almost nothing at all. And once you get it in your nose, it's in your nose for the rest of the day. It's one of these very not-quite-apparent smells that you can't shake. Phenylacetic acid is a precursor, it is not regulated. I came across a fantastic ad, two ads in fact, they were sent to me by one of my finders out in the field who loves finding things and sending things to me, in a hunting and trapping magazine, which is where you buy lures, you buy essence of skunk, you buy essence of this, essence of that, from which you make things that are called *lures* to go out and trap animals. Why you want to trap animals, I guess you like fur coats, I don't know. But there are magazines dedicated to hunting and trapping. Two agencies, one in Iowa and I forget where the other one is. Pop's Lure Store I think was one of them. Doc's, Doc's Lure Store, and the other was something like that, were selling amongst the lures everything kosher except at one point they had phenylacetic acid. An ounce from one place was $2.50 and the other was $2.95 an ounce. You buy it from Aldrich Chemical Company, it costs about $2.50 an ounce. So you're getting stuff from a lure company that is identical in price to a commercial thing. It is a precursor to a precursor to methamphetamine, and I got especially amused by the very lack of subtlety, at the bottom of the ad they had a little fine print that said, "Available in 55-gallon drums." [Laughter.]

Fantastic. I mean maybe they were really in big business of lures and hunting. I don't know. But the idea of using essence of skunk and twenty 55-gallon drums of phenylacetic acid, it's a mystery. But this is the continual flow of material in this market. These materials are available in tonnage amounts, they're being sold continuously, and they are, I'm sure these are, strategies. Phenylacetic acid. The reason I tie it together is that the ester of phenylacetic acid, which can go through amyl ester, a phenyl acetate, is a chemical that happens to be the artificial smell, artificial oil, of chocolate, and the chocolate you buy that is not really right out of the plant, but comes out of a place in Hayward that makes chocolate inexpensively, uses phenylacetic acid to make the amyl ester to make the chocolate flavor that flavors chocolate. So in a very funny way, chocolate as a stimulant, with its caffeine and theobromine, and amphetamine, as a purely synthetic stimulant, are tied together.

Am I close to the end of the hour?

ANN: Ten after.

SASHA: One more material. One thing I should mention about stimulants in general, in fact, this is an outgrowth of the discussion of using phenylpropanolamine as a mediocre stimulant in excess to get it into the head. Amphetamine itself and methamphetamine itself are very intensely abusable. I should give some sort of picture, I don't know if I did this earlier. I'll do it again because it's worth the emphasis.

What happens in chronic amphetamine and methamphetamine usage? I have seen this. I saw this during the Haight-Ashbury days in the late 1960s in San Francisco. The use of amphetamine, not so much as methamphetamine. Methamphetamine is much preferred in street use as being a little bit more smooth, a little bit more mellow, a little bit more clean, not as jangly, and a bit more intellectual. What all these terms mean I don't know, but that's the terminology you get when it's used.

As intravenous injection, a typical naive person using amphetamine or methamphetamine may be able to handle 10 or 15 milligrams. If they're really willing to go through some jangles they may handle 20 or 30. But a typical dose is around 5 milligrams. You can take it orally as Benzedrine, or inhale it from the old Benzedrine inhaler. Probably most of it is used by intravenous injection. The usage of it will grow with time; you like where it goes, where it puts you. You're into a sort of a laid-back stimulant, almost orgasmic-like state; it is very close to cocaine. Methamphetamine and cocaine have very similar types of responses when they're taken intravenously. Cocaine is a very dangerous one to take intravenously. It's usually taken by inhalation, up the nose, or smoking. Methamphetamine is often taken by injection.

The usage will cause you to search out and repeat that experience, and you become refractory, which means you don't respond as much. You

LECTURE 9 ~ *Stimulants I*

become tolerant. You tend to use more. I have seen usage of it in San Francisco in quantities where a syringe contains 1 to 2 milliliters of a solution that is so thick that it will barely go through the needle. It has the consistency of honey. And probably the quantity of going in is in the order of a gram being injected in a single shot, at a single time, several times a day.

A naive person taking a gram of methamphetamine I guarantee will be dead. The heart will not take it. The cardiovascular system will pound, the heart will pound. If you have a small, loose vein somewhere, it is ruptured, and you are dead. You get used to it. You get tolerant of it. You can use more. People can use a gram, a gram and a half, and then sleep between injections. You get totally tolerant to it. But you get wiggy. I mean you are absolutely nutso. You are psychotic. You do not respond normally, diet goes to pieces, no eating. Hepatitis is among the things that usually comes along with it, just from dirty needles and forgetting to eat. But somewhere around 100, 200 milligrams of amphetamine, you develop what's called "amphetamine psychosis." And it is not easily distinguished from a real psychosis, to the extent that people coming into the emergency ward as active florid psychotics almost always will have urine taken to look for amphetamine, and quite often you find amphetamine there, so you know you've got amphetamine psychosis. You treat it as a psychosis; the person is really wacko. But if you know, you can taper the person off of the amphetamine and since the problem is organic, it will clear. It is not endogenous and coming from somewhere in the inner soul. It comes from an external chemical that's used in excess.

Yes?

STUDENT: You said earlier that cocaine is very dangerous to inject. Why is it so dangerous?

SASHA: The difference between the amount that is effective as a stimulant and the amount that is anesthetic and will put you out is very, very narrow. You tend to pass out.

STUDENT: It puts you out as in a coma? Or just out . . .

SASHA: Unconscious. As in a general anesthetic. If you put it in slowly or you're used to it, you can handle large amounts. But if you put it in

VOCABULARY

Route of Administration

Oral: by mouth (per os)
Parenteral: by bypassing the intestines
Intravenous: within a vein (I.V.)
Subcutaneous: beneath the skin (Sub.Q.)
Intramuscular: within a muscle (I.M.)

Metabolism (detoxification)

Oxidative: phenobarbital → hydroxyphenobarbital

Reductive: CCl_3CHO → CCl_3CH_2OH
chloral → trichloroethanol

Alkylation: norepinephrine → normetanephrine

Dealkylation: methadone → normethadone

Conjugation: morphine → morphine glucuronide

Hydrolysis: cocaine → ecgonine

LECTURE 9 ~ *Stimulants I*

quickly or if you're not particularly used to it, you can knock yourself out and you can become unconscious. The unconsciousness, the anesthetic state is not an easy one to handle. There are respiratory problems that come with it. It has that danger. There are much safer things to put in the veins.

Another question somewhere. I saw a hand waving. Yes?

STUDENT: Which of the two is stronger, methamphetamine or amphetamine?

SASHA: Toss up. For actual potency they're about the same. Methamphetamine is smoother and easier to use and contains less of a push on the heart and on the nervous system. So methamphetamine is probably a little safer, but they're comparable in strength.

Questions? I think it's the time. Oh, yeah. One . . .

STUDENT: What about diet pills like Dexatrim and stuff like that? What are . . .

SASHA: Oh, there are literally hundreds of amphetamine derivatives. In fact, I have mentioned the term "amphetamines," which I don't like. These are materials that are derivatives of amphetamine, they're used as anorexics. There is a desire to find something that stops the appetite, keeps you from eating, but it doesn't keep you from sleeping. It's not been a very satisfactory search. Most things that will knock the appetite off will wire you as well.

Good, let's take a break. It's the hour and anyone who wants to ask questions is welcome.

LECTURE 10

March 3, 1987

Stimulants II

[Segueing from preclass conversation and questions.]

SASHA: I have no idea where it's going to be at the end of the semester, and I do need to look at it as I'm seven lectures behind, or at 30 percent of where I should be. If I think seven lectures behind, then at end of the semester it's kind of nice because I'll only be seven lectures behind. If I work on a percentage basis, it's going to be grim. But I'm going to try my best to get a lot of stuff out to you since now you're not taking notes, and this will give you the information you would have had if you had been taking notes. Plus it's a chance to review what I said and correct some mistakes. I'm going to try to get one out every week if I can, each written up from a previous lecture.

SASHA: Okay! Who is here? Two, 4, 6, 8, 10, 12, 14, 16, 18 ... And how many minutes behind are we?

ANN: Two.

SASHA: Two minutes behind? Okay, so we're off into the lecture. What I would like to do is spend this entire lecture today on a single topic, tobacco. Someone had talked about having given up and then having gone back recently, and were wondering about side effects. I uncovered quite a few. There were more than I thought. There is nausea, there is a lightheadedness, there is an irritability. And some of them—pallor, sweating, nausea, and actual vomiting are the most common—are expressed by over 50 percent of the people, and women more than men.

STUDENT: Yeah, that's the reason I went back on.

SASHA: To alleviate the withdrawal symptoms. How many people actively smoke today? I want a tally. There is one, two, three, four, five, six. How many people have smoked and have given it up? One . . .

STUDENT: How many times?

SASHA: Who do not smoke now, but smoked at one time. One, two, three... How many people have never smoked? About half. That is just about the tally in the United States. About one-third of the people smoke. About half of the people have never smoked. And about one-sixth of people have smoked and have given it up. So they run about a 30 to 35 percent smoking population. I mean, that's newborn infants and 128-year-olds. Everyone right across the board, counting heads. It is probably—if you go back and do exactly what I said I didn't want to do in the second lecture or first lecture, to use the term "addicting"—it's probably the most addicting drug that I'm aware of. It is a thoroughly compelling drug. About four out of five of the men who smoke, and about three out of five of the women who smoke, smoke over 15 cigarettes a day.

I've heard it said, and I see no reason take issue with this, that tobacco use is probably the most damaging preventable health problem that exists in the world. In the United States each year, the number of deaths is huge. This is always a tricky thing, because it's a neat one to use for showing how bad things are, to call statistics out and to use them. You can't really tell how many people die, let's say, of alcohol-related use or heroin-related use or tobacco-related use, because you'll find a lot of people spin them all together, and the person who wants a particular statistic will pick it out to support their argument. But the estimate is around 300 thousand deaths in the United States are related somehow to tobacco use annually. Three hundred thousand a year. That's approximately one tenth of 1 percent of the population. Out of about 300 million people in the country, about one tenth of 1 percent die annually from something related to tobacco use. This exceeds alcohol, people having car accidents, and everything in there.

The culprit in tobacco is unquestionably nicotine. For those who love dirty pictures, I've drawn nicotine on the board. Tobacco is *Nicotiana tabacum*. There is a *Nicotiana rustica* that was probably the early one used in the United States and the Caribbean and in South America, another

LECTURE 10 ~ *Stimulants II*

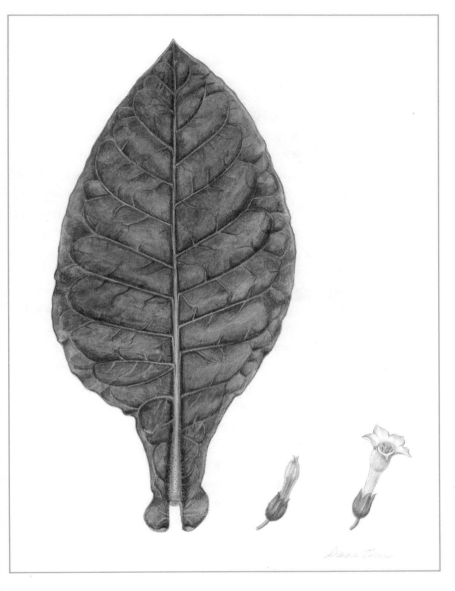

species. But the one that has caught commercial use now is the *tabacum*.

It runs about 1 or 2 percent nicotine. *Nicotiana rustica*, the earlier one that went to Europe with the explorers in the 1400s or 1500s, ran about 4 or 5 percent nicotine. There is a second nicotine plant, *Duboisia hopwoodii*. The material is known natively as *Pituri*. I might as well put this in here, because it is not tobacco. Pituri and tobacco. Pituri is *Duboisia*; used by the Aborigines in Australia, they've used it from time immemorial. They still use it, and it has never really gotten much outside of the Outback. Again, it's used much like tobacco is used throughout the world—through the mouth. Its leaves are ground up into what's called a *bolus* or a *plug* and put somewhere in the mouth and allowed to seep in. Describing nicotine; here is the chemical, show-and-tell. [Pulling out a jar.] This is the chemical nicotine, if you want to take a look at it, smell it; be careful of it. In tightening down the cap to make sure that it would not leak, I managed to break the cap, which means that you handle it incorrectly, you'll get it on you.

Nicotine, by the way, goes in through the skin. It is absorbed transdermally quite readily. We had a really a wild case at the General Hospital about three, four months ago in which a woman came in, in very bad shape. She was nauseous, she was vomiting, she had a pallor. She had a very almost ashen face. And what she had done, she had had mites, and mites were on her. So she was using Black Leaf 40, which is about 40 percent nicotine insecticide. People have heard of nicotine, I'm sure, used as an insecticide. It's an effective one. She used it and rubbed it on her skin to get rid of the mites that were on her, and it absorbed through the skin. She was running blood levels that I had never seen before. Had she not been a smoker, she would have been a much worse case, because she had a fair amount of exposure to nicotine and could handle quantities of it. But even with the smoking ability and smoking background, she was in a very difficult spot. They released her about eight hours later, relatively symptom-free. She was still running a monstrous level of nicotine and of cotinine. I'll put that name on the board. It is one that you should know just to be familiar with the territory. Nicotine is this chemical; cotinine is a major metabolite. It's an amide, not an amine. Cotinine. I'll spell it. C-o-t-i-n-i-n-e. So yes?

STUDENT: What's a metabolite? Have we gone over that?

SASHA: Oh yes, we have. Okay, go back and review. These are the kind of questions I want. A metabolite is something the body makes from another thing. It is a biotransformational product. I had mentioned that the original term had been "detoxification." In this case it *is* detoxification.

Cotinine. You can take 200, I have personally taken 250, milligrams of

LECTURE 10 ~ *Stimulants II*

cotinine just to see what its metabolites are. No problem. No action, no activity whatsoever. You take three or four milligrams of nicotine and you're probably going to be in the corner feeling very ill and wishing you hadn't done it. Nicotine is very potent. Cotinine is totally impotent. It is a metabolite.

Detoxification, a common term that had become "metabolism," which is the change of one thing to another in the body. The general term now is "biotransformation," just to indicate that you are changing something in the body rather than in the test tube or in the frying pan.

The body metabolizes, changes, biotransforms most of the things it gets. I'm going to write this on the board. In fact, you'll get it in the handout on Thursday. It does all of this primarily for one purpose—to get rid of the compound. The body does not know that it's going to be taking in nicotine, or taking in a phenylalanine tablet, or taking in a fried egg. The body has this whole machinery, largely in the liver, to some extent in the lung, to some extent in the blood, to some extent in the kidneys, that has an ability to take everything that comes to it, be it a food, be it a drug, be it a metabolite of a food or a drug and, if necessary, change it to make it more soluble. The way you get rid of things from the body is usually in the water system. It goes through the blood to the kidney, it gets filtered, it goes into the urine, the urine gets dumped. You get it out. So if you take in something like paraffin, which is totally water insoluble, or oil, or what have you, this sort of thing can't get excreted through the kidneys, and the body says, "Here is something I can't get into the water system. Do something to it." So the body oxidizes it, reduces it, tears it apart, adds something to it, modifies it in some way to make it more water soluble, regardless of what it is.

The mechanism of oxidizing nicotine to cotinine existed in the body before nicotine ever was known to the body. The mechanism of oxidation is totally nonspecific. It will handle everything that comes its way, including foods. Foods, if they are used, are fine. What we do not use or turn into protein is metabolized and excreted. So it's not a specific thing toward drugs. It's any and all chemicals in the body. So this is a major metabolite, this cotinine, and it is not an active compound.

The experiment that tied nicotine to the action of tobacco was done around 1940 in England. A doctor, his name was Lennox Johnson, in 1942 actually, took a whole bunch of volunteers. Some of them smoked, some of them did not. He gave intravenous injections using a low indwelling catheter. He left the catheter in and gave intravenous injections of nicotine, one or two milligrams. The people who smoked liked it. They found a good feeling from it. The people who didn't smoke developed a syndrome that initially is the new smoking syndrome: pallor and nausea, "I feel light-headed, I don't know if I'm going to keep my lunch down" type of feeling. And he continued this over several days. At this point in time it would be considered a totally irresponsible type of research, but it was done then, and you know, it had a good question to ask, namely, is nicotine responsible for the action of tobacco? The answer is yes. After about 80 injections, those who smoked preferred the injection to the smoking. It was less harsh, less dry, less irritating, and less messy. Those who did not smoke, but had this course of nicotine, on ending the course wanted it to continue. In essence, the experiment carried the whole impact of tobacco itself. And hence supported the feeling that nicotine explains the action.

People have argued that nicotine is not addictive, because it does not have withdrawal, it does not have tolerance, and it does not lead to antisocial behavior. All of these are nonsense. It certainly has withdrawal. We were just talking about it—the nervousness, headache, nausea—one or more of these is seen by over 50 percent of the people who abruptly withdraw. How many people have used and are not now using? I am amongst them. We had one. Others? What was your withdrawal? Do you remember any?

Student: I used to smoke about three-fourths of a pack a day and I stopped for about three months. I'm smoking again, lightly. But I start getting nervous and like . . .

Sasha: Irritability? Difficulty sleeping?

Student: Yeah, and having a hard time breathing.

Sasha: Oh, having a hard time breathing. That is now in the statistics. In a collection of about 20 withdrawal symptoms, the difficulty breathing is way down. Again, more common by women than by men. Interesting thing, in cutting down on smoking, you'd like to think there's what's called in pharmacology a "dose-response curve," fewer cigarettes, less nicotine, less tar, less risk, and what have you. A nice slopey line. It doesn't work.

LECTURE 10 ~ *Stimulants II*

A study was just recently completed, in fact, I had my oar in that particular stream, in which people would smoke 20 or 10 or 5 or no cigarettes a day. They were people who normally were smokers so you didn't have any loading from the experiment itself and making the new blood systems and measuring nicotine and cotinine in the blood. And, by the way, other measurements for smoking that are very common in the laboratory are for carbon monoxide and cyanide. You get both very high levels of carbon monoxide and reasonably high levels of cyanide from cigarettes.

One of the big problems you have in any research dealing with smoking is to find out what the people really did when they went into the bathroom and locked the door and pretended they were sitting down doing something else. And actually, you get this brown cloud that comes around from the toilet. Cheating in a smoking experiment is something that almost always has to be reckoned with. It is very common. It's so hard to titrate because you are trying to look at blood levels to find out what the effects are of smoking or not smoking, and meanwhile the person is not following the regimen. That person is cheating and smoking. You find the blood levels have gone wonky; it's because there's something you don't understand or because there's a little hanky-panky going on. And so you use carbon monoxide . . . Heck, I'm not even going to be able to talk about carbon monoxide. I'll talk now about carbon monoxide because it's not really a drug that people use. Very few people carry tanks of carbon monoxide to come to parties. I mean, it's not one of those abuse drugs in that sense, and yet probably the third major cause of death in this country is carbon monoxide poisoning. Odorless. You do not smell it. You do not taste it. You do not see it. It comes from incomplete combustion.

This is really a nice aside. It has nothing to do with smoking, except people who smoke will run blood levels of carbon monoxide of 10 or 15 percent. Which means the term in the clinic is "carboxyhemoglobin." Hemoglobin ties in with oxygen. For those who are completely new to the area of biology, hemoglobin is the compound in the blood that ties oxygen to it, moves oxygen to where it's needed, oxygen comes on and off quite ready. So you have hemoglobin and you have oxyhemoglobin. Oxyhemoglobin is the transporting form. Hemoglobin is the one that is free of oxygen. It goes to the lungs, picks up oxygen, takes oxygen to the brain, dumps the oxygen, and comes back without the oxygen, gets more oxygen. It is the transporter of oxygen. Carbon monoxide, C-O, binds to the hemoglobin about a couple hundred times more tightly than oxygen. And when it gets on there, it binds, it goes on and is hard to get off, so oxygen can't get on.

So all the carbon monoxide in the body, and here the term is "carboxyhemoglobin," will block that hemoglobin from being able to carry oxygen. How do you compensate for it? You get by with less oxygen. Or if you don't have enough oxygen, you don't stay conscious. And if you are not conscious long enough for the lack of oxygen, you are dead. Carbon monoxide is a lethal gas. It is hard to get off. It's easy to go on, it goes on very readily, but it's hard to get off. It takes, say, about 200 times longer to get rid of carbon monoxide compared to oxygen. It is run off through the lung.

Usually you have carbon monoxide in your blood chronically as a person who smokes or a person who commutes. You drive across the Bay Bridge coming over here from Oakland or from Marin County to work here. You're driving down a little tunnel and this tunnel is loaded with carbon monoxide, with asbestos, with all kinds of marvelous polycyclic aromatic hydrocarbons that cause cancer. We'll talk about that later in this lecture. And you're driving through this tunnel, you roll up your window, where has the air come from? From that little tunnel. You're gathering an immense amount of carbon monoxide. If a person who commutes about 20 to 30 minutes to work or to school would go in and give a blood sample, they would run about 4 percent carbon monoxide in the blood just from the commuting. At the end of the day, if they are not exposed to this on 19th Street, which probably is a tunnel in its own right, they will drop to about 1 percent or 2 percent.

A person is never zero. You make your own carbon monoxide. It is a natural metabolite—back to this term again—a metabolite of hemoglobin, interestingly. In the decomposing and getting-rid-of, you're all the time making new hemoglobin and getting rid of the old. In getting rid of the old, it releases carbon monoxide, which is carried in the blood, about half a percent perhaps, maybe a percent.

People who smoke may run 10 or 15 percent carbon monoxide. They are not particularly robbed of oxygen, though, because the body generates more hemoglobin, generates more red cells. It builds up what's called the "hematocrit," builds up the quantity of red cells to compensate for the loss of

LECTURE 10 ~ *Stimulants II*

a certain amount of function. The net result is probably comparable oxygen availability. Again, with carbon monoxide, toxicity is not due to how much you have, but how much you've gotten in excess of what you are normally accustomed to having. So a person who smokes, although they have a fairly good inventory of carbon monoxide, are not unduly susceptible to carbon monoxide poisoning. So it is not negative in that sense. Carbon monoxide is a way of titrating whether a person has smoked or not recently, because you'll get a sudden surge of carbon monoxide with a single cigarette. Cyanide is a little bit harder to assay, and cyanide is present from a lot of food sources also, so it's not as good at what's called a "marker" for smoking.

Anyway, what I was going to start on was an experiment we did in which you had people smoking 20 or 10 or 5 or no cigarettes. Oh, pardon, we also had a 40 because we had a couple of heavy smokers. We had them on 40, but we did not put the 20s on 40, because that is a pattern that is just not ethical. So 20 was the main large category. And we measured the amount of nicotine and tar. Tar is a little tricky to measure. You could measure tar in these experiments that the Federal Trade Commission—F-T-C, I think those are the initials. Am I wrong? What's the thing that measures tar and nicotine in cigarettes? Okay, anyway, it's a trade commission of some kind. It's the one when you pick up a pack of cigarettes, the whatever those three letters are that says this cigarette contains 1.2 milligrams of nicotine and 0.2 milligrams of tar.

STUDENT: FTC?

SASHA: FTC. What they do . . . let me get back to the story. The thought is if you smoked so many cigarettes, and you then smoke half the cigarettes, you get half the amount of nicotine, half the amount of tar. Not so. You get maybe a little less. You smoke 5 cigarettes, you think you'd get a fourth the amount of nicotine and tar than if you smoked 20. You don't. You get somewhat less. What these people did is they smoked entirely differently. If you're a 20-cigarette-a-day smoker and you're down to 5 cigarettes, boy, watch how you smoke those 5 cigarettes. You're down to the filter, or into the filter. You hold things longer. You take deeper breaths. You don't throw so much away. You smoke more often. You milk those 5 cigarettes very, very differently than if you smoke 20.

THE NATURE OF DRUGS

You might smoke lights. People go to the extremely low. Carlton, I think that's one that had the name of "ultra low." What they do in the smoking evaluation is they put the cigarette filter into a dingus that does its business periodically every three or four seconds, it draws out around 40 milliliters of air, I think, through the cigarette, through the filter. You measure the amount of nicotine chemically. You measure the weight of the filter once it's been dried to get the amount of tar. You weigh those and you ascribe it back to the cigarette.

What Carlton did was cut holes down the side of the cigarette, down the paper. Just little fluted cuts, razor blading. And so when the machine that smokes cigarettes pulls air through the cigarette, air comes in the side and you get a diluted amount of smoke, and of course you don't get as much nicotine and you don't get as much tar because two-thirds of what you take in is air and air doesn't contain nicotine and tar. So if a person smokes these, to get nicotine and tar they glom onto those little air vents and cover them with a hand or mouth, not knowing what they are doing. They know that's a more satisfactory way of smoking. Watch a person smoking Carlton sometime. "I'm really helping my lungs—I'm not smoking so much. Have a Carl!" Watch them smoking. They get their hands up on those little flutes that allow the air in, so they get a full inhale. They're getting a lot more than the company would say they are getting because the company is measuring it through a device that lets air come in through these vents, so this is a major maneuver.

You milk to about five cigarettes. When you get below five cigarettes or at about five you're getting at the break point. You suddenly have no more way of milking more out of what you're getting, and the amount of tar and amount of nicotine will be dropped. An interesting comment, about half the people who went to the five said they'd almost rather not smoke at all than try to make it through on five cigarettes, because apparently it was just about as agonizing, as far as the strong urge, the need of smoking. Interesting experiment. So it's not a linear relationship. It's a milking out and then a precipitous drop-off. Yes?

STUDENT: I think it's harder just to smoke at one time.

SASHA: You mean one on occasion?

LECTURE 10 ~ *Stimulants II*

STUDENT: ... at a time.

SASHA: How do you smoke? How many do you normally smoke at a time?

STUDENT: At a time?

SASHA: Yeah.

STUDENT: Probably like seven or something.

SASHA: I couldn't even begin to smoke seven. You mean, I hope, in sequence not at the same time. [Laughter.]

I saw a picture of a person who smokes cigars that had cigars coming out in all directions. He held Guinness Book of Records for getting 31 cigars going at once or something. But the idea of smoking occasionally, just a pacing, I couldn't do it. I love the idea. In fact, Brezhnev, I believe, the leader of Russia, had this little cage, this little cigarette case, that would be locked until once an hour it would go, "Bing!" and he would get one cigarette out, as a method of monitoring the amount he would smoke. There was sort of a titration against his will. I think he ended up smashing the thing. [Laughter.]

STUDENT: I guess some people don't chain smoke and they can just have a few during the day. I'm not one of them.

SASHA: I'm with you entirely. There are a lot of people who can indeed smoke occasionally, and smoke modestly. There are people who can drink occasionally and drink modestly. There are people who can use heroin occasionally and modestly. There are people who use cocaine occasionally, modestly. There are people who cannot. That's exactly the whole argument of a relationship with the drug. Yes?

STUDENT: How come when we drink we are more inclined to want to smoke?

SASHA: Fantastic. There is a connection between drinking and smoking. What you're observing is correct. A person will tend to smoke more when drinking and tend to drink more when they smoke. They go together. The reason is totally unknown.

Student: How about nervousness?

Sasha: An unknown. The connection is real.

Why does a person, when they stop smoking, tend to put on weight? It is fact. A person who is a chronic smoker stops smoking, average weight gain: 10 pounds. Some not at all, some 20, average 10 pounds. Why?

The truth is not known. Studies have been made. I might have mentioned this, I'll mention it again if I had . . . We actually ran an experiment in which we kept the people at a constant level of nicotine. These are smokers. You put a venous catheter in the arm. You put a drip up here. You let nicotine go in at such a rate that they have a constant level and then you eat. Is the food changing the nicotine level? Does nicotine influence the food metabolism? No, there's no connection. At least there's nothing observed that correlates. Is it possible that nicotine interferes with hormonal equilibrium and balance? And when you stop smoking, your hormones are wonky for a while, and the adding of weight is hormonal. Possible, not known.

Yes?

Student: People that are on the pill aren't supposed to smoke.

Sasha: Are what? They're smokers?

Student: They smoke, on the birth control pill.

Sasha: I'm not familiar with the connection and the argument. Who can help?

Student: What it does is, smoking cigarettes makes your veins more sensitive to nicotine. Does something to your veins, and so does estrogen. So the two of them together are synergistic.

Sasha: Uh-huh.

[Unintelligible audio.]

Sasha: And tend to cast off bits of debris that would be embolus. It could be. It's reasonable. I'm not aware of it.

Student: . . . the estrogen and the nicotine can cause, I guess, an embolism, like a brain clot or a blood clot. Just because they make the veins . . .

LECTURE 10 ~ *Stimulants II*

SECOND STUDENT: It makes the veins very sensitive to clotting.

SASHA: Very, very believable. I was not aware of it. I did not know that the connection had been established between the birth control pills and smoking.

There is definitely a connection between smoking and cardiovascular disease. There is a large connection between smoking and lung cancer. The tendency is to connect tobacco and lung cancer and that's not so. But let me get to the health issue in just a moment. What I did uncover is a little story of some of the efforts to keep smoking, or to keep tobacco, from being used. We need to keep things separate. We're talking about tobacco. Smoking, in the case of the use of cigarettes as a vehicle for getting nicotine inside of you, is less than 100 years old.

Smoking was known long ago, and the concept of smoking was known mostly in the form of cigars. Now smoking cigarettes is much more common. However, up until this century, the predominant usage of tobacco had been oral. Snuff up the nose or tobacco in the mouth, but not the smoking concept. That was an outgrowth about 1900, of the development of what was called "flue-cured tobacco" or "bright tobacco." Instead of taking the plant, letting it grow to its fullest, harvesting the plant, letting it dry, grinding up the leaves, and using that, or humidifying the leaves and using that, but air drying the plant. The flue-curing takes leaves as they develop (just as they become fully developed from the bottom of the plant up) and puts them in a dark, heated environment. The leaves are cured in a warm, dark flue, and cured for a period of time. It changes the entire acidity. The tobacco becomes quite acidic instead of basic. It becomes more smokable. It is less irritating and this was the source that went into cigarettes. With that, cigarettes became the mainstay about 1900. I'm a little bit ahead of myself, but I'll mention it anyway.

Oh, and the autopsy of a person who smokes is quite a sight, the inside of the lung.

Yes?

STUDENT: Do you know if it was menthol cigarettes or regular cigarettes?

SASHA: Menthol is one of the additives I'll get to, at least if we get time. Menthol apparently has no negative medical relationship that has been reported. Additives to cigarettes, by the way, are put in by the tobacco companies with no regulation at all. No one has a voice saying, "Study menthol and see if it's good or bad." It is a taste that has been put in there. A lot of materials are put in for dyeing, for flavors. There's a host of additives, it's worth a point.

STUDENT: I heard saltpeter was added.

SASHA: Saltpeter has the classic reputation of quieting the libido in the Navy, in the military. I do not know if it's being put in cigarettes.

I've heard it's ineffective, but if they say, "We're going to put saltpeter in your breakfast and it's going to keep you from getting a hard-on," you don't get one. I mean, that's effective. The mind has a superb influence on the libido.

Things that are added do not have to meet any requirement. [Pointing to stuff on a table.] This is some tobacco debris to look at, but I didn't bring any cigarettes. Most people know what they look like. But there are two types that have recently come on the market. One is the clove cigarettes that have gotten very popular in Berkeley. And they're popular over here; people talked about them? Okay, so-so.

Another is a type of false cigarette called Favor. How many people have seen that in the boozeries? Yeah. A tube, no smoking. You don't put a match to it, you merely suck through it. Favor is a superb example of showing where the administration sort of lets things go between the cracks.

Is nicotine a drug? Well, if you look in the pharmacology texts, of course. It does this, the heart rate goes up, it does that, it modifies a host of things. It definitely has drug effects, but it is not administered as a drug. It's administered as tobacco, and hence it doesn't fall under the edicts of the FDA, the Food and Drug Administration. There is a group that's known as the tobacco, firearms, and alcohol: A-T-F. ATF—alcohol, tobacco, and firearms—that regulates alcohol, tobacco presumably, and firearms, presumably. It has its own agents and it does its own licensing and what have you.

There were some people in Texas, they approached the hospital for us to do some work for them, we kind of let it go. But this group in Texas wanted

LECTURE 10 ~ *Stimulants II*

to put out nicotine in the form of a tube that you could suck through and get your nicotine, hence not have the battle of tar. Nicotine has its problems, but you would avoid the problems of tar and the irritants that come with smoking, and they wanted to see what was necessary for regulations. So they went to the FDA and said, "We want to put nicotine into a white tube with a little air agitator at the filter." They put some other things in there too. "And see if we can use it as a way of helping people stop smoking or as a substitute for smoking." The FDA said, "Nicotine is not a drug. We have no authority in this area whatsoever. Talk to the alcohol and tobacco and firearms crowd, ATF." They went over to ATF and said, "We're interested in this as a tobacco substitute for using nicotine," and described it all. "Not our problem! Check with the FDA. It's really a drug."

Both groups shuffled it onto the other group. They said, "Okay, we're starting manufacturing." They're selling it. They don't have permission. No one wants to give them permission. No one denied them permission. This is on the market for drug purposes, and completely without any regulation, without anyone's authority, without anyone saying, "I wonder if it is harmful?" They're putting in piperidine, an interesting chemical because it has a slightly mousy, amine-like flavor or smell to it. They're putting piperidine in as an additive, but no one regulated any product, no one regulates the additives.

Tobacco. The tobacco manufacturers will add one thing after another. Menthol was an example. They'll also add things that are used for humidifying. They are, what I call, "humectants." A humidifier is something that adds moisture to something. A humectant is something that keeps moisture from escaping from something. They'll add glycols, propylene glycol, or glycerol, or things as moisture retaining agents.

I was talking recently to a person who's working for some pharmaceutical house in Sweden. Leif Erikson, an unexpected name, but he is a pharmacologist, a chemist there, who is working on bronchodilators. They had developed a new bronchodilator that can be administered by spray, by inhalation. And they had an appeal from one of the large tobacco companies for rights on the patent because they wanted to add it to cigarettes. Because by adding it to cigarettes, you would dilate the bronchi and get more nicotine in from a small amount of nicotine consumed. Each cigarette becomes more efficient, and you get more from your low nicotine thing. An additive to make the drug more effectively available, in essence, to sell a less drug-laden product for the same effectiveness, it's got a competitive edge.

There's no regulation on putting drugs into cigarettes. There's no regulation on the additives in cigarettes. A company does not have to share its research. All these tobacco companies have immense research groups. They don't have to publish. They don't have to share. They do not make any of the information available. They are doing that perhaps for their own protection, perhaps for their own salability of what they sell. There is no regulation on the tobacco industry whatsoever. It's a very, very interesting thing that slips through the cracks of authority. Very much as, for example, driving under the influence of alcohol (I'll mention this in an alcohol lecture later) slips between the cracks of felony, misdemeanor. Driving under the influence is neither a felony or misdemeanor. It is something in between, in a world of its own. Tobacco is something in between, as a world of its own. It is not under either drug nor any other federal regulation.

I mentioned clove cigarettes. Why clove cigarettes? Clove is a spice. It contains a material called "eugenol." Everyone has heard at one time or another that cloves can be used for a toothache. You put a little clove oil on a tooth, it tends to numb the toothache. The oil of clove is a rather effective local anesthetic. Local, meaning where it touches goes numb. Anesthetic means it robs the sense of feeling and hence the sense of pain a little bit. You smoke a clove cigarette, you lay this eugenol down the bronchial tracts and into the lung. You are numbed. It is an anesthetic. You don't get the irritation. You don't get the body's rejection, "enough," from smoking because you're partly anesthetized and hence you smoke more. It's strictly a matter of getting more tobacco, more nicotine into you. Yes?

STUDENT: They burn so slowly, you know, they're really dense.

SASHA: In the class over in Berkeley, one girl asked me—she smoked occasionally—"I really got carried away last night at a bar." Same idea, alcohol, tobacco. Why do alcohol and tobacco link? I don't know what they do. "I went out drinking. Had quite a few beers, and someone had these clove cigarettes," and she said, "I went through a pack or more of them. Didn't feel it hit me." She was miserable the next day! Vicious, vicious hangover. It had

LECTURE 10 ~ *Stimulants II*

nothing to do with beer. There was not that much beer. It was just strictly way too much nicotine. She got poisoned on nicotine because they're pretty strong but you tend to lose that natural, "and let's take it easy," because the anesthetization, the quieting of the pain response from the—

STUDENT: Are they any worse poison . . . Are they any worse for you than regular cigarettes?

SASHA: Probably not. The bad things that come in from cigarettes are basically the nicotine and the tar. Tar comes in general from burning the plant. It is still speculative whether carcinogenic compounds found in cured tobacco contribute to cancers known to come from smoking.

During flue-curing, nitrites react with nor-nicotine and a nitroso group is put on. This is a product that is generated in both the cigarette, in the tobacco, and in the body from nicotine. In the parts of flue-curing, the nitrates that are present in nicotine generate nitrites. Nicotine is quite rich in nitrate ions and this chemistry occurs in the process of curing the tobacco. Let's not cure it and get rid of this cancer-causing agent? People won't smoke because it doesn't have a good taste and is too harsh to smoke. Yes?

STUDENT: Is it true that uncured tobacco as used by the American Indians is a powerful stimulant?

SASHA: It is a powerful stimulant, and it definitely will cause psychotropic effects.

This product can also be created in the body because we have nitrates and nitrites in us. There is a very small amount there, but it's a very potent carcinogenic material. It may be responsible for some of the cancer arguments in cigarettes. And if so, it's probably more akin to the cancer of the bladder. The relationship of cancer in the lung to cigarettes is exceptionally good, the correlation is very hard to argue with. That with bladder cancer is much weaker, but apparently is real. That with cardiovascular effects, cardiovascular damage, and heart attack is not as good, but it looks pretty good. The very strongest connection between health problems and cigarettes is lung cancer. The cigarette companies will say it has not been proven. They're quite right. It's not been proven, but it will never be proven. You cannot prove something statistically or epidemiologically. But there is no argument that can be raised about that connection being real.

Before 1900 lung cancer was virtually an unknown form of cancer. In 1910, one out of every thousand autopsies showed lung cancer. At that time there were four billion cigarettes that were consumed in the United States. Before 1900 there were almost no cigarettes. Cigarettes arose as the tobacco could not be inhaled very easily. The tobacco was not the flue-cured tobacco. It was a dried tobacco. It was harsh, it was basic, it was astringent. People absorbed it through the mouth or up the nose. It was called smoking through the nose, the use of snuff. In 1920, nine out of 1,000 autopsies were lung cancer and 30 billion cigarettes were used. In 1950, 32 out of 1,000 autopsies—that's a factor of almost two orders of magnitude more—were lung cancer and 350 billion cigarettes were used. We are now at around 600 billion cigarettes usage, and lung cancer has gone from a virtually unknown form of cancer to a major one. Fact: it is predicted this year, on the way the statistics have been going, that lung cancer will exceed breast cancer for women, which shows that it's gone from an unknown cancer to probably the major form of cancer. It's preventable.

I mentioned these nitroso compounds. They're probably associated with bladder cancer. The ones that are associated with lung cancer are probably closer to what are called PAH, polyaromatic hydrocarbons, or polycyclic aromatic hydrocarbons. Those who have played at all with chemistry know the benzene rings. Fuse the benzene rings and you can get a whole bunch of this, what I call bathroom-tile-type chemistry, where all these hexagons are jammed together. These things are generated in the course of pyrolysis of organic material, especially if there is not totally a full amount of air present, as you would have in a cigarette. A cigarette is a slightly less-than-efficient pyrolyzing-of-organic-material machine.

STUDENT: Does pyrolyzing mean—

SASHA: Pyrolyzing means "burning." "Lysis" means "to take apart." "Pyro" means "fire." To take apart by means of fire. The engine of the Muni bus is a superb generator of polycyclic aromatic hydrocarbons. You walk behind one and that big brown cloud, blue cloud that comes behind it has a bit of a bad smell, and it contains an awful lot of particulate matter that contains these materials. Our environment is loaded with them. Our industrial

LECTURE 10 ~ *Stimulants II*

environment is loaded with them. These are materials that are extremely stable. That's one reason they're generated. You take long chain hydrocarbons, you take complex molecules, and you go through this pyrolytic step, and you generate these rather extraordinarily simple clean molecules, because they're very stable. They have high chemical stability, but biologically, they're attacked by specific enzymes in the body because they are things that are not water soluble. The body says, "Here's something that's not water soluble—I want to get rid of it." And they attack it to oxidize it, and in the course of oxidizing, they happen to make materials that are extremely carcinogenic. The body, instead of detoxified, is toxified. The material, the hydrocarbon itself, is actually without any risk and it doesn't touch anything. It's totally insoluble. But as it is attacked by the body, it is oxidized to an epoxide, to a dihydroxy epoxide.

Weird chemistry, but this is something that latches on to cells and apparently has been correlated with cancer's development within those cells. It comes from pyrolysis, it comes from heating organic materials very strongly with not a sufficient amount of air. The argument of grilled steaks—you put a steak on a barbecue and you get all this black char, a nice good flavorful goop loaded with these materials, and they're more carcinogenic than the steak would be raw. And in the classic development of man, he ate his sabertoothed tiger as he caught it, without going through the grilling routine. The grilling routine definitely has added a hazard to the eating because it does generate these materials that are carcinogenic. Not much, but then again cancer is a very unlikely thing to develop from any carcinogen. It's just that if you're exposed to it for an extremely long period of time, or you are one who is susceptible to it—it's a statistical thing—you do have increased incidence of it when you're exposed to these forms of body insults. Including passive smoking.

Oh, I was going way back on the attempt to keep smoking and to keep tobacco from being used. You'll hear the same thing in the world today with other drugs that come in, "We're going to keep them from being used." Tobacco came over from the New World, from the Caribbean, where it was really first found, the West Indies, the Antilles, as they could be called, over to Portugal and Spain about the middle of the 1500s. From there it moved into England. The first of the smoking houses went right along with the coffee houses. From England to Holland, to the continent, to France, from Holland to Java, which got it started in the Asian area. From the European area down into Turkey into Arabia. From Asia into China, from Europe into Russia. In essence, in the course of about 50 years, tobacco and the use of tobacco swept around the entire world. Every culture it came into said, "No, it is savage. Only savages smoke. Only savages use tobacco. It is a savage thing. It is not written about in the Bible and hence it's against the Christian or any religious ethic."

I found in 1590, Pope Urban VII actually threatened excommunication for people who acknowledged smoking. The people in both the Turkish and the Arabian armies executed people who were caught smoking. There were less simple, less vigorous things such as cutting noses off in Russia, because they had sort of a strange way of approaching drugs for quite a while. And I did find the entire story of its development in Japan. In 1540 the first introduction to Japan came with a Chinese pirate ship that washed ashore containing some Portuguese seamen who were users of tobacco. It spread in its use.

In 1603 came the first edict against smoking in Japan; it was totally ignored. In 1607 and 1609 cultivation became a penal offense. People grew their own, so the seeds became a real item of commerce. Tobacco won't keep forever, it is an annual plant. In 1612, the property of a tobacco seller was forfeited to the state, to the government. In 1615, the attending officers of the Shogun, of the various local administrators, all smoked. It became a very embarrassing thing, so in 1616 they added fines and imprisonment for any smoking. But now all the attendants in the imperial palaces smoked. By 1625, the princes who were responsible for the edicts, they smoked, and everything was kind of ignored. They said, "Okay, but you can't raise it in rice fields and in vegetable fields." And by 1639, which is just almost exactly 100 years after its first introduction, tobacco was an accompaniment to the ceremonial cup of tea. It took 100 years from its introduction and everyone rebelling it with the most frightful laws against it, including that of seizure of property, and that of fine and imprisonment, until it became a socially accepted thing. Tobacco has never entered a culture, to my knowledge, and been accepted by that culture. And never entered the culture without being accepted by the culture itself. But it has never been accepted by the culture

LECTURE 10 ~ *Stimulants II*

and then gone out again. It is truly a commanding thing once it gets into the structure.

There are many forms of use. We mentioned cigarettes, but those are only very recent. The inhalation was a very minor, minor route of abuse. The original Indians did make cigars out of tobacco. They used corn, maize leaves as the wrappings of cigars—the Spanish were the first to introduce paper. But the original tobacco was extremely potent, extremely harsh, and when used by pipe, there was only the absorption, to a large measure, from the mouth. There was very little inhalation. Inhalation was virtually unknown. People who smoke pipes let the mouth be the absorbing tissue. People who chew, for example, chewing tobacco, or now they call, "snuff dipping." I've heard the term "snuff dipping." I've been trying to get the origin of the term, I cannot. People stick it as a bolus, as a wad, into the mouth.

But they call it "snuff dipping." And I'm curious where the "dipping" came from. "Snuff," I found came from "sniff," because they used to have these pouncet containers, little pouncet jars. They were used back in the 1500s, 1600s. In fact, they were used way before that. Little boxes, as if a falcon had landed on it with its claws—they had little holes in them. They're called "pouncet boxes." And they were taking them and [sniffing sound], did this out of them, and the particulate, very finely ground up vegetable matter was in there—largely spices, even more so herbs, they're in there to ward off infection. They were known to ward off infection. And it was a very short move to put very finely ground up tobacco in there, which then went in this way, and eventually from sniff to snuff.

Snuff is a very finely ground material. Here's an example of snuff, if you want to see it [pointing to jar], this is ground extremely fine. It's taken up into the nose, "smoking through the nose," sticking it up into the nose where it is absorbed. The nicotine is absorbed from the nares, from the inner nares. In the chewing of tobacco, we have found to our amazement that when people chew, they will chew for a period of time, perhaps half an hour, and the nicotine level in the blood really rolls up. Then they stop chewing, spit out the quid, and even rinse out the mouth so there's no particulate bits of tobacco leaf in the mouth. The nicotine level keeps going up for another half an hour. It continues rising as though you're still getting it directly in there. Well, maybe some of it is trickling down to the stomach, being absorbed through that marvelous portal system we talked about a few lectures ago. No, because you can tell material that's gone through the stomach and through the liver because you see a lot more of the cotinine, the metabolite. The metabolite is generated in the liver, but it's not generated when you go directly from the mouth into the blood, into the brain. The blood level keeps going up, but the cotinine level doesn't go up as rapidly. So apparently what happens is the nicotine gets into the tissue of the gum and in the mouth, and it continues to absorb from this tissue over the course of another half hour or an hour. So it's the same idea as this woman who came in who had washed herself with the Black Leaf 40. She washed all nicotine off, but her blood level was high and stayed high. It kept rising and stayed high for hours because it's in the tissue. And from the tissue it absorbs. The actual nicotine, you can smell when it's fresh, it's a colorless liquid. It's a viscous liquid. Do not get it on you because it will absorb through the skin and cause poisoning.

I mentioned advertising; there's a point I wanted to talk about.

Is it about the hour now?

ANN: Almost.

SASHA: Okay, so we have about 15 minutes to go. Does advertising effectively reduce the amount of smoking? I don't know the answer. Any ideas? Yes?

STUDENT: I don't think so, because for whatever advertising they do against smoking, there's plenty of advertising that is more seductive that makes it look glamorous and fun.

SASHA: Mmm-hmm. Yeah.

STUDENT: They never use some ugly bag ladies smoking. They always show—

SASHA: Yes, jazzy—legs crossed, drink in hand, you know, Caribbean cruise kind of thing. You know, they put the whole appeal on the smoking, in advertising.

Interesting thing. I was told this and I began observing it. Look for cigarette and smoking advertising and see how often you find the blue smoke

LECTURE 10 ~ *Stimulants II*

around the cigarette. It's almost as if it has been tailored out, it has been sort of brushed off. They don't want to let you know that this aspect of the cigarette is part of the trip. It's like people who advertise beer drinking on TV never show it going down the mouth. It's always out here—click, click, click, hooray kind of thing—but never drinking. And same thing with cigarettes—you never see the smoke coming out, as if the cigarette has never really been lit.

Is it effective? Is anticigarette advertising effective in discouraging smoking? Is cigarette advertising effective in promoting smoking? I don't know. I have a feeling that there's not much virtue either way. There is the example in 1964, I believe, when the Surgeon General came out with his antismoking report; smoking is correlated with cancer. The grim report of the first real connection of smoking and cancer. There was a dip in the amount of smoking that lasted about a month. This was sort of a national antismoking move, and there was a true dip in the amount of smoking that occurred, followed the next month by what's known as a national smoking rebound. It went right back to the same amount that was there before; it's as if that had not occurred.

There has been a metering of the amount of smoking, a stabilization of the amount of smoking in this country in the last three or four years. There has been a fantastic propaganda campaign against it—I don't know where it's all coming from—where people are saying, "No smoking areas. It's really bad, passive smoking." It's a very strong propaganda move against it. I don't know if it's effective or not. There's still an increased amount of smoking at the teenage level. But at the adult level, there may be a dropping off. The overall amount may be holding static for the first time. So there may be some effectiveness. Is it from an awareness? Is it from education? Is it from propaganda? Is it from negative advertising? Is it in spite of positive advertising? They do not advertise, I think, cigarettes on TV now. They do still in magazines. I consider that a rather grim abridgement of freedom of press, freedom of speech. Why aren't you allowed to advertise cigarettes[4]?

---

[4] This restriction was actually requested by the tobacco industry. It eliminated the equal airtime that was required be given to antismoking advertising. Only broadcasted content required equal time be given so they shifted their focus to printed magazines, billboards, and sponsorships. None of which required equal time be given for antismoking ads.

THE NATURE OF DRUGS

I consider that to be slightly improper. Yes?

STUDENT: They also don't advertise hard liquor anymore.

SASHA: Not on TV. I've seen ads in magazines. I consider that to be an abridgment of freedom of the press. I think if a person is taking responsibility for what they want to sell, they should be free to talk about it. And this is back to my very first lecture. I am very much against the abridgment of freedom. Personal freedoms, I think, should be protected to the last degree. We are run by a superb constitution, not by a presidency and a Congress, but by a constitution which assures that freedom—freedom of speech. And I think that if a person wants to advertise the sale of handguns or cigarettes or hard liquor or whatever, kits for contraceptives or what all, and will take the responsibility of what is going on there, they should have that freedom.

That had been long avoided in much of the early advertising of drugs and of drug use. People were advertising with unfair, dishonest, improper, and irresponsible claims. I think honesty should be there. Do you enforce honesty by blocking advertising? I don't know. There is an immense amount of energy put out by the American Tobacco Council, which is a funder of grants and is a supporter of information on smoking. Their contention is you should be free to choose. And I go along with that. One will say, "Well gee, that's said by the side who wants to promote smoking." I go along with it. The freedom of choice should be yours. If you choose to smoke, that should be your prerogative. I don't think there should be restrictions on what you do as long as it doesn't screw up other people too badly. Really, nothing you can do will not have some influence on others, and you can't be totally black and white in the matter. But to a large measure that should be your freedom. Be aware of what you go along with. The passive smoking argument? Be guarded. "I don't think people should smoke because I don't like it. I'm nearby and I'm getting smoke as a consequence of it." "Please go to the other end of the plane" or, "No smoking on planes" or, "Let's have part of the restaurant over here where there should be no smoking."

What is the true problem with passive smoke? The term "passing smoking" is when you didn't intend, or you didn't take an active role, in doing it. You're in the area where the people smoke, you do get nicotine in you. You

LECTURE 10 ~ *Stimulants II*

do get cotinine in you. You do get tar in you. There's no question on that. The amount is very small. We did one study at the hospital in which we took a total of five or seven stewardesses who did not smoke, or had not smoked at least for a year. They were, in essence, nonsmokers. How long does it take a person who once smoked and then stopped to become a nonsmoker? How long does it take the person who decides to raise a beard to become a person who is not unshaven, but a person with a beard? I don't know, there is a sort of a feel in there. I would say probably a year or thereabouts would be the point after which the chance of slipping back to smoking habits becomes quite unlikely.

But the stewardesses had not smoked for a year or had never smoked, and they were on a flight from San Francisco to Hawaii, and we went along with them. We took blood levels and urine levels at San Francisco. They worked the tail end of the plane, the smoking section of the plane, for the five hours or so to Hawaii, at which point we took blood levels and urine levels again, and determined the amount of nicotine and cotinine in them. They had something like a 36-hour layover; we bled them and took urine again, and then came back to San Francisco and they did it again, to find out the amount of nicotine and cotinine. Tar is difficult to measure in humans. There is no good measurement of tar directly. You have to go indirect for that. I'll talk about that in a moment. But nicotine you can. You draw a blood sample, there is nicotine. A smoker will have 20, 30, 40 nanograms per milliliter of nicotine in them. A smoker will have maybe 300 or 400 nanograms per milliliter of cotinine if they are a chronic smoker. The stewardesses would have maybe five or six nanograms per milliliter of cotinine and maybe two or three nanograms per milliliter of nicotine. There is real nicotine—not nearly as much as a smoker would have, but more than you would expect in a person without any exposure.

How much does a person have who has no exposure? Let's take a person who has spent the last 35 years on the top of a mountain in Himalayas, with nothing but goat meat, the goats having been nowhere outside of Himalayas. No tobacco, no cigarettes, nothing. Avoiding all exposure. The ideal you can't get, but a person who makes his rule of avoiding any contact with tobacco has nicotine in him. You get nicotine from the environment. Nicotine, I think like cocaine and DDT, now are throughout the environment. If I were to take a swipe, if I were to wipe this bench off where no one presumably had spilled tobacco, I could probably, from a swipe on that, come up with maybe a sizeable fraction, if not several micrograms of nicotine. It's in the environment. It's on the windows. It's on your hands. It's in your clothes. We run, at the hospital, assays for nicotine. You cannot take brand new glassware that comes out of the manufacturer, out of the glass manufacturing company, packed in cardboard boxes and in essence is sterile, it has never really seen any use. You can't run nicotine levels in them. The nicotine background in that brand new glassware is too high. You have to take solvents that have been treated to be free of nicotine, wash the solvents, then take the solvents and wash the glassware to get glassware that brings the level down. The sensitivity of picking up nicotine far exceeds the background level of nicotine. It is throughout the environment. You have it in you, as you have in you . . . I don't know about cocaine, but you certainly have DDT and nicotine in you.

What is the actual time? I don't know what the minutes time is. I'm running over.

ANN: Seven after 12.

SASHA: Seven after, thereabouts? Okay.

Additives, passive smoking. Oh, an interesting thing about additives; I was picking up some things for today and unexpectedly came across this musk. People, I think, are familiar with the concept of musk as a flavoring agent. It comes from some embarrassing place in the deer, I think. Anyway, it's a specific thing from the musk deer, but it's a general term for things that have an appealing scent, not particularly an immediately recognized flavor, used as a flavor agent. And I found specifically "musk ambrette." I don't know if it rings a bell. Musk xylene is the straightforward name. Musk ambrette. It's kind of an interesting compound for those who like organic structures.

Purely synthetic, a dinitro anisole. I discovered that musk ambrette is added to what's called perfumed chewing tobacco, and a lot of additives are put in there, to over 2 percent. Musk xylene is added to the extent of 1 percent. So this is actually added, no one gave permission. It is mutagenic. It is known to be a potent mutagen, which means it probably is associated

LECTURE 10 ~ *Stimulants II*

with the potential, at least, of causing cancer. It is known and potentially added as a commercial advantage. I consider it to be totally irresponsible and actionable on the part of the company. But who is to make a comment? There is no government agency that regulates it.

What else have I got in here...? Cigars, pipes, snuff, chewing. I've mentioned tobacco already in conjunction with betel nut. Who actually has used betel nut? Someone raised their hand in that very first class, they may not be here now. I mentioned in the very first lecture the use of betel nut throughout the world is probably just behind nicotine in general usage. The fact is that the two are used together and this originally began in India. It is common now. I was talking to a good friend of mine. In fact, I asked him to get some betel nut for me if he could. He did. He is from Thailand. He says it comes in from Thailand. The FDA says, "Thou shalt not import betel nut, because it is a—" I don't know why they say you can't use betel nut, it has some health problems associated with it, but I imagine anything does. It is certainly used by a minority in this country, but it's by a large minority. The samples I got through the Indian community were kind of old. I mentioned the betel leaf came from Modesto, domestic. I think I'll talk more about this in another lecture on intoxicants.

But what I did find out is it's imported frozen from Thailand. And he brought in a great big package of frozen betel nut. Little individual nuts. He said they're absolutely fresh, fresh frozen, and took one out and cut through it as much as you can through a fresh nutmeg. Inside was the actual arecoline-containing content. This is what's chewed. And it's available in San Francisco, fresh frozen. It comes in routinely from Thailand. And so the betel nut is used broadly in this country. It is used in Guam, Indonesia, Micronesia, Asia, Korea, in China, I don't know about Japan but certainly in China, southern China through India, Pakistan and up to the point that it begins competing with hashish, which is in the Pakistan-to-Afghanistan area. It's used by this entire portion of the world.

Now, spreading through there is a mixture of it and tobacco. And I did mention in either the first or second lecture, that combination is murderous. And that combination now is really promoting mouth cancer, like smoking promotes lung cancer. Cancer in and around the mouth, not specifically at the mouth, but mouth in the sense of jaw, gum, and larynx, for some reason, cancer in this area is now the principal form of cancer in India. It exceeds lung, it exceeds bladder, and exceeds liver. Okay, that is my lecture on "don't mix betel nut with tobacco."

Addicting snuff, chewing, cigarette smoking—yes?

STUDENT: Are betel nuts sold in the United States?

SASHA: Under the counter, or in ethnic areas where they're under some other name. You look at the package, it doesn't say "betel nut." It says everything in Thai. And if you can read Thai, then that's fine. If not . . . [Laughter.] The betel nut is sold in Berkeley, distinctly under the counter at the store all the Indians usually shop in, roughly Ninth and University, and the leaf, I mentioned, comes in on Wednesdays from Modesto, and is gone by Thursday. It comes in by the hundreds of pounds and is spoken for ahead of time. So it is sold out usually within a couple of days. Are there any questions in general? I don't think I have much more to say about betel. Pardon?

STUDENT: How do you take betel nut?

SASHA: How do you take betel nut? A number of ways. In a general sense, you use the inside. Get the husks off, they're not edible, they're harsh, and you take the inner portion that has kind of the texture of an almost dried walnut. If you put your thumbnail into it, you make a notch. The old stuff is kind of hard. You have to cut it with a knife, or you have to shave it. Some people in India will roast it, or not roast it. The people in Indonesia will take it as it is and wrap it in what's called a "betel leaf." The nut comes from a palm tree. The leaf comes from a pepper plant. They're different plants; absolutely unrelated. Often, they will add a little lime to it or something that contains a slight basic change that will raise the pH, and put it in the mouth and suck on it, or put it in the mouth and chew on it, or work on it as a quid—as a little bolus of material and keep it in there. In India typically it'll stay in there, be replaced every few hours, stay overnight, it will be there 24 hours a day.

STUDENT: Have you tasted it?

SASHA: No, I have not tried it. I'm kind of curious to. I will, now that I have fresh supply. Yes?

LECTURE 10 ~ *Stimulants II*

ANN: Is that the thing that turns the teeth red?

SASHA: It is a thing that stains the teeth red or black, typically through Micronesia, but it's not as effective a stainer as you would think. The spitting of it is a brilliant red, that comes from the colors. If you do not use the dye, you do not get the red color. But the red color would tend to stain the gums and the teeth, and the darkening of the teeth has been a mark of age and wisdom in these areas of the world. So much of the darkening is done artificially just so you look wiser, and it's more than the betel nut would do. But there is a darkening of the teeth. It's considered a marker of prestige, and not a marker of cosmetic beauty.

I'm going to try to do the same sort of rambling thing on cocaine next Thursday. That is another major drug that I think merits a whole hour.

And here are some show-and-tells. Oh, I forgot to talk entirely about Nicorette gum. Here's gum that is now being used to get you off nicotine. The example, this is four-milligram gum. In this country you need a prescription for reasons I do not understand. In England it is available but specifically over the counter, not prescription. Their argument is that if people want to conduct themselves in a lousy personal habit, the government's not about to pay for it. It's socialized medicine. You're on your own. So here it's prescription; in England it's not prescription. Four milligrams. It doesn't, apparently, quite do the job for a good heavy smoker, but it does provide nicotine to the body; it is nicotine. Nicorette gum, it's called.

STUDENT: What is this?

SASHA: That is a dipping snuff. Fine snuff is the white powder here. That's ground-up tobacco. I think most of those are open. Take a look. Open them if you want. The chewing tobacco is very moist and a much larger thing. That's pure nicotine; do not get it on your hands. Smell it if you want, but don't get it on your nose. And this is the one that my friend who is a chronic tobacco-chewer tried one time and it was almost too much. He said it was just about the fourth time chewing it. He may not have even realized it was too much. He spat it out and then he got lightheaded. He got nauseous and he had to sit down. He didn't know how far he was going to go. He went through about half an hour of agony. Very, very potent that

THE NATURE OF DRUGS

PROPOSED ROUTES FOR THE METABOLISM OF NICOTINE

LECTURE 10 ~ *Stimulants II*

pure chew. Very dense. Feel the density of that; you could drive nails with it.

STUDENT: And this one is different?

SASHA: Yes, it's even denser. He did not dare try that one. But this is the one he used, something of this ilk. It's got sort of a looseness, a loose, leafy, moist character to it.

> **LECTURE 11**
>
> March 5, 1987

Stimulants III

SASHA: Two, 4, 6, 10, 12, 14 . . . [Around two minutes elapse with background noise of students and occasional chalkboard sounds.]

How far is the minute hand off of the right minute? What is it? About the hour now?

STUDENT: A little past.

SASHA: Past the hour, okay. By a couple of minutes?

ANN: Yeah. It's a couple of minutes slow.

SASHA: Strange. This way it will never be right, haha.

Okay. I want to spend the hour today on the area of cocaine. How many people have tried cocaine? Weeee! Boy, that's over half. Up the nose? Down the mouth? Intravenously?

STUDENT: Not me.

SASHA: Not I. [Laughter.] One.

Okay, it's a stimulant, and if you get it in fast enough, as you know, if you can get a real good shot going in there quickly, it's a lovely stimulant. In fact, it's so lovely that it's one reason there's an awful lot of use of it, and a lot of compulsive use of it. I brought in some samples to show-and-tell. The two bottles with the red-and-white labels are cocaine hydrochloride and cocaine freebase, both USP. The little vial, which I put together yesterday—you'll find in it little whitish beads and little crystals and so forth—is what is called "crack." I'll talk about how it's made and what it's like. And in the little plastic container are dried coca leaves. So, feel free to take a look at them if you like. And because someone had asked after the class last hour, how do you keep bringing in illegal drugs and not find yourself in problems? I also brought

THE NATURE OF DRUGS

along a slip of paper which is a Bureau of Narcotics drug license, Schedule I through V, and my name is on it, which gives me a very nice situation of being able to bring in little vials of LSD and crack and what have you, and not have any problems.

Yes?

STUDENT: Where do you get it?

SASHA: That was a cute story. It's worth a little bit of background. I believe at this point, I'm probably the only person, or one of two people alive in the country, who has one as an individual. There were three. One dear friend of mine in Los Angeles got one. He is now dead. It came out about 15 years ago. I wanted a drug license so I could do work in my lab—analyzing, making, evaluating, judging, taking care of drugs. Doing things with drugs. And I approached the federal government. They said, "No problem. Your institution has a license. You can work under their auspices."

I said, "I don't have an institution. I mean, I have my own lab. I live all by myself out there somewhere." They said, "You've got to work in an institution." So, I delved in to look at the law, and it didn't say you have to work in an institution. So, I went back and said, "Please, I would like to have a license."

They said, "Well, we can't give you one until you get approval from the state for whatever research you want to do." So, I went to the state and said, "Please give me approval for whatever research I want to do." They said, "We can't give you approval until you have a federal license."

Well, there's a term for this that was a bestselling book for about 20 years, known as "Catch-22." So back and forth, back and forth. No one had gotten an individual license. Don't need one. Institutes, administrations, all government things are exempt. Industries have theirs, academics have theirs, and institutions have theirs. So, I finally said this is ridiculous. I can keep going back and forth for a long period of time.

So, I happen to be a member of a rather conservative place downtown. That's why I wear a tie on Thursdays. It's called "The Bohemian Club," which is an unusual place in its own right. But being there is a useful thing, and since you have to be a little conservative and wear a tie, if you also wear sandals and a beard and long hair, you aren't too conspicuous. So, I said, "Gentlemen, why don't we have lunch at my place downtown and talk it

LECTURE 11 ~ *Stimulants III*

over?" And so, I got kind of a bigwig in the state thing and kind of a bigwig in the narcotics federal thing down there for lunch, had a delightful couple of glasses or couple bottles of wine, and it was a very good thing. I said, "Gentlemen, we have a trivial problem which we can just resolve right now. Who goes first?" And the state said to the feds, "Why don't you go first?" They said, "Okay."

So, I got a federal license, then I got state permission, and that was it. And I kept it renewed for paying initially $5, inflation made it $20, every year, and the licenses are renewed every year for possession, for analytical purposes, Schedule I through V drugs. I personally believe that anyone who wishes to could apply and get one if they are not part of an institution, which can apply for one and get one. It is a doable thing. All you have to do is write an application and pay $20. They'll give you a lot of hassle, but of course, persistence. I think it can be gotten if you have a reason to do analytical work on scheduled drugs, which is the story of why the slip of paper is up there.

Another thing I want to talk about, and cocaine I think brings it into good focus, is a matter of governmental honesty. There's an awful lot of palaver, which is a nice word for bull, in the whole area of drugs and what is wanting to be done, and I like to try to bring it into some sort of evaluation so it can be judged in its own right. A good example of this: when I was coming over in first gear on the bridge this morning with the traffic, I was listening to the radio and I heard an announcement that apparently some company is manufacturing a vitamin B12 preparation that you spray up the nose that gives you energy. Anybody hear of this?

STUDENT: Yeah.

SASHA: Okay, what do you know about it?

STUDENT: Well, it was in *People* magazine, and it was—

SASHA: Oh, really?

STUDENT: It's a fad and it's selling really well.

SASHA: Yep! Anyone else know about vitamin B12 up the nose for energy? I don't see why B12 up there should do it—you can't even absorb

THE NATURE OF DRUGS

B12 without going through intrinsic factors and a lot of biochemical nonsense in the gut. By the way, B12 will come in later in the discussion on nitrous oxide. Anybody familiar with nitrous oxide, laughing gas? Hey, we got a lot of hands! Nitrous oxide, which they say does relatively little long-term damage, does some short-term mischief in the sense that it depletes the body of B12 and it depletes it for a good number of hours, if not several days. We'll get into that later. There's a connection between B12 and an altered state. But stimulation from B12? I don't know.

I have a clinical syndrome, what is called "hypomanic." Anybody familiar with the term "hypomanic?" It's a nice one. "Manic" is a pure pathology. You know, you go go go, and you crash and you creep around and you're obsessed and depressed and have paranoia for a while. Then you go again, and it's this sort of looping up and down, it's called "manic depressive." If a person is kind of manic but not quite pathologically out of control all the time, you don't want to say he's manic because manic is a pathological thing, so you say he's "just under manic." "Hypomanic," which is what my syndrome is. So the result is I'm hard put to tell if I'm taking a stimulant. I can take 20 or 30 milligrams of methamphetamine and not be aware I've taken any drug whatsoever because it kind of goes all the time.

So, I don't know if B12 up the nose will just smell good or if it will make things move. But what I enjoyed thoroughly was the immediate statement of the FDA, the Food and Drug Administration, which said this claim is nonsense because anyone knows you cannot get energy without burning calories. You have to burn calories to get energy, and there are no calories in a vitamin. Well . . . I mean, you could inject a few milligrams of speed and get a lot of firing around in places, climbing up the walls, moving fast, and not going to sleep, and there are no calories in speed. And there are calories in a hot cup of tea, but not the kind of calories you're going to burn for energy. What you're doing, you are releasing, you're triggering, you're mobilizing the body's energies. You've got a lot of energy in there. If you burn the fuel fast, you may run out of fuel and have to restoke.

I'm going to touch back on amphetamine and methamphetamine again, because in a sense, cocaine is a problem today in part because of the problem we had with methamphetamine and amphetamine in the '60s. And that was closed off, restrictions were made against the importation, against

LECTURE 11 ~ *Stimulants III*

manufacture, care was taken against illicit distribution. Companies that were making it were told not to make it and they obeyed. And the result was that amphetamine became more and more underground, scarce, and expensive. In compensation, cocaine, which had always been very cheap and reasonably unknown, became a very popular drug. It is a speed drug very much, and in many ways, like methamphetamine. How many people have used methamphetamine intravenously? Okay, of those, would you have a vigorous preference of one over the other? I don't want people to necessarily expose themselves, but there is to me not that much difference between the two.

Okay. *Erythroxylon*, spelled that way, *coca*. A plant grows 5 to 10 feet high. It has pointed leaves pointing at both ends, an inch and a half to two inches long. Origin, probably northern Peru. Let's see if I can do this upside down... It's a real challenge to draw South America upside down and still put the countries in. And this is a poor way I have of dividing in my mind's eye where countries in South America are. You have three little buttons up here. The Guianas are this... [Drawing.] I went into some harangue on the Baltic states being alphabetical from north to south. The Guianas are alphabetical from west to east, which coincide with north to south. This has nothing to do with cocaine. The old days they were B, D, and F, the three Guianas; British, Dutch, and French. Now the British is called... Help me...

STUDENT: Guyana?

SASHA: Guyana? The Dutch is called...

STUDENT: Suriname.

SASHA: Suriname, and the French is called...

STUDENT: French Guiana.

SASHA: French Guiana. So anyway, B, D, F. And what I like is the equator comes right around through here. [Drawing.] This is the equator. So you obviously have Ecuador sitting up against the equator. And so this whole thing divides above and divides below, which gives you sort of four countries running down the west coast. You have Venezuela, you have Colombia, you have Peru, and you have Chile, which really makes the whole west coast

— 71 —

of South America straightforward. Then from Peru, we could argue, from Peru, you sort of cut a line down that way over to the other coast, and you have Bolivia. Of course, here's Argentina, Uruguay... Uh, no. Paraguay sits in here. Uruguay sits in here, and of course the monster of Brazil.

So this is, in essence, a poor man's guide of South America, and the origin of cocaine is probably in the northern part of Peru—the whole area between Peru, Chile, and Bolivia is the area that was known as the area of the Inca civilization hundreds of years ago. The Spanish put an end to it and put cocaine into a very interesting light. But this is more or less the origin of the plant. The plant has been moved around the world. It has been moved

LECTURE 11 ~ *Stimulants III*

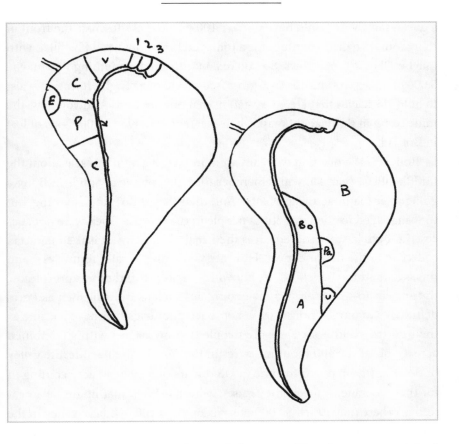

into the lowlands and it has changed character. The plant in its original form contained about a percent of cocaine, as best they can deduce what the original plant was, maybe between 1 and 2 percent.

Anyway... today, Bolivia, Chile, Peru, into Colombia, into Venezuela, into Brazil, down into Argentina, and into Paraguay, and into Ecuador. Every country thus far... I don't know an exception, though I don't know specifically of coca being raised in Chile. But certainly, all the other countries, the major countries in South America are producers of cocaine. Probably the major producer is Peru, with Bolivia being a close second. The amount of cocaine that is harvested annually in Peru—and this is one I just want you to try to get a feel for—is 100,000 tons of leaf.

Now, I don't know how much leaf it would take to fill a gunnysack. Probably maybe 10, 20, 30 pounds. So, if you to have, say, 20 pounds of leaf in a gunnysack, you would have to have 100 gunnysacks to cross the front of this room, two or three times for a ton. And this room would be filled with maybe 10 or 20 such rows. So you're going to need something like maybe 5,000 or 10,000 rooms this size—this is just kind of a rough projection—just to hold the leaves that are harvested in that one country. And you have the same thing in Bolivia, but maybe half that, about maybe 50,000 tons of leaf in Bolivia.

Bolivia is the one that we really went into with a bit of muscle about the middle of last year. Do you remember the helicopter army raids in Bolivia going after manufacturing plants? And then they didn't find anything, but they managed to alienate a lot of people in the process. They were not successful; Bolivia still proceeds. It is their major economy, it is the major tax source, it is the major export. How are you going to take countries—and this is something which I don't know the answer to and others don't know the answer to—how do you take countries such as Bolivia, such as Peru, definitely Colombia, how do you approach countries that have cocaine as their entire livelihood? Here are people that own acres with a combined acreage of maybe 100 thousand acres in the central, the most fertile valley of Bolivia, who have an income in US dollars of $3,000 an acre coming in for their cocaine? Suggest they raise soybeans? Well, that might be a neat thing. Where do you sell $3,000 of soybeans in a 100,000-acre valley in the middle of Bolivia in a country that doesn't use soybeans? They use cocaine. Half of their entire domestic crop is internal.

What are you going to do, how are you going to approach it? I don't know the answer, as far as getting in the way of producing cocaine in South America. They say, "Well, if we stop using it . . ." Why would we stop using it? We're a big market, we demand it and we'll pay for it. That's been our way. If we want it, we'll pay for it and get it. How are you going to get people to stop using? It is a fun drug and it's got a lot of peer pressure, and it's got a lot of a social thing that's gone upper class. How are you going to do that? The number of people using cocaine in this country in the course of the last year? I don't know how to count them, but other people who have counted said about six million people have used cocaine within the last month in this country.

Over 100 million use alcohol, 100 million use tobacco, probably 18 million use marijuana, and 6 million use cocaine. I don't know how to approach

LECTURE 11 ~ *Stimulants III*

the problem other than say, "The stuff is cheap." It should be cheap. My heavens, it is very cheap in South America. It's that southern market that is the beginning of the drug problem because of its availability. I'll talk a little about that in a few minutes. But I got this four ounces . . . It is down now because I used a lot of it in chemistry and a lot of it in analysis. I originally got that four-ounce bottle of cocaine for just under $20. That's right . . . a four-ounce bottle for under $20.

What I did, I put an order into Penick, the company that manufactured it, and I said, "Please, dear Penick, send me four ounces of cocaine. Here's my DEA 222 form and here's a check for $18.23." And they sent back a very nice letter saying, "We're sorry, our minimum order is $100." So, I turned around and I said, "Please, Penick, send me four ounces of cocaine . . ." And I went out of their catalogue and got four ounces of morphine, four ounces of codeine, four ounces of thebaine, and four ounces of meperidine, in addition to four ounces of cocaine, which put the bill just over $100 . . . $103. I sent a check for $103. Back by UPS came five bottles containing amongst them four ounces of cocaine. The intrinsic value is very, very small of all these drugs. The street value is monstrous.

And what you get on the street is unbelievable. In fact, I saw an article in the *JAMA* discussing drug screening—something we're going to talk about with some vigor and some vehemence later on in the course. Loxley was talking about drug screening and how necessary it is, how important it is, and how it's going to bring our drug problem to a stop. Crack is a subject we're going to talk quite a bit about. Of cocaine crack, they said, "It's a mixture of freebase cocaine along with talcum powder and a number of other things that runs now as high as 70 percent pure and it is a mush that is made, or a solid that is made, on the addition of bicarbonate to cocaine as a solid mix." It's nonsense. It's utter nonsense.

Cocaine crack is exactly that. It's freebase cocaine. It's made in a very straightforward way from cocaine. You take cocaine as a salt in water with some bicarb added to convert it to the base, or cocaine as a freebase suspended in water, heat the water up to a boil. Just short of the boil, the cocaine melts. Its melting point by the Merck index is 98 degrees Celsius. Water boils at 100 degrees Celsius. So, just short of the boiling—because of the water, it tends to hydrate and go a little bit lower, but just short of the boiling, it all melts, you have a clear oil, a little film at the top, and the bulk of the oil settles to the bottom of the beaker. Take it off the flame or off the heat, and let it stand and just a little bit below the boiling point, it crystallizes. And the stuff that's in the bottom crystallizes as little balls, the stuff that's at the top crystallizes as a fragile crystalline sheet. And as it cools down, the little stuff crystallizes out from the matrix inside, and that's what's called "crack." Freebase and crack are identical. They're both cocaine. Cocaine hydrochloride is cocaine. All of these are the same drug. I'll be damned if I know why they are considered so evilly different.

Yes?

STUDENT: Why is it called "freebase?"

SASHA: Because there is no salt present, no component that is salt. Most bases—

STUDENT: Okay—

SASHA: Have you taken much chemistry? This new territory? I probably haven't gone into it. Most materials are either acidic or basic. This means they are acidic—they turn litmus red. They have a sour taste. The word in English, which is "acid," in German is "säure," from which comes the English word "sour." Like lemon. The taste is not the bitter of quinine. It's the lemon; the sourness of lemon. That is the general feeling of acid. You urp [makes sounds] and feel that rubby taste on your teeth, that taste is sour. This is the hydrochloric acid from the stomach.

The other side of this balanced coin are things that are basic. These are things that are caustic. I think everyone at one time or another has tried to put lye down the toilet when it was all stoppered up, and you dump in this lye and a little bit of water and so forth, and finally something starts fuming, and maybe it clears. And if you've gotten it on your hand, it has this oily feeling. This is caustic. Any sort of an ammonia will have it. Household ammonia. Who's ever gotten household ammonia on the hand? It's got that sort of almost lubricated, soapy feel to it. This is caustic. This is the basic side. Alkaloids, most things from plants and 90 percent of the drugs we talk about are bases. They are on the caustic side of this arrangement. They are not water soluble. That's why they work in the body. They go in

LECTURE 11 ~ *Stimulants III*

the body, they go into tissue, they go to the brain. The body can't get rid of them because they are not particularly water soluble.

So, things that are acid and things that are base tend to neutralize one another, and the neutral product of acid and base is called a salt. So we have a base and an acid in the same amount, from the point of view of its molecular weight, that combine to form a salt which usually is water soluble and has neither the baseness or the acidness of the components. These salts almost universally taste bitter. Bitter is a property of the alkaloid or the alkaloid salt in the mouth. These salts are water soluble. Now if you want to put cocaine up the nose, for example, you'll put it in the form of a salt. If you want to inject it in the arm, you want it in water solution, you want it in the form of a salt. But you take a salt and you try to put a flame to the salt, it won't volatilize, because it is an ionic water-soluble thing that has very little vapor pressure. These are generalities, but they're generally true.

If you want to volatilize something that is a salt, you take away the acid and leave only the base. And hence, from cocaine hydrochloride or salt or crystal or whatever it's called, you make freebase, which is a component without the acid half. This, then, is volatile. It's organic soluble; it is not water soluble; it could be volatilized and inhaled. So, if you are using cocaine through a water medium such as in the nose or injection, or orally, the best way would be as the salt. The salt is a loose white powder; this is the actual cocaine hydrochloride, as a white salt which is water soluble.

In making the freebase you have to somehow get rid of that acid. This is where a lot of the original mystique came in. Basically, in the household we have several very common bases that can be used for neutralizing an acid. We have bicarb in the kitchen, which is sodium bicarbonate, baking soda, also sodium carbonate, which is used for washing powder. Bicarb is a much weaker base but adequate. Stronger bases: washing soda is a strong base, lye is an extremely strong base, ammonia is an intermediate strength.

They take these and then they extract them with some organic solvent. An organic solvent that gets used is ether. And that, of course, came into the news. Many tragedies and fires stemmed from the fact that ether was used to extract the cocaine from the base, and the ether had to be evaporated off. Ether is an extremely flammable material. In fact, when we get into volatiles and toxins, I'll bring in the samples of ether and chloroform and nitrous

THE NATURE OF DRUGS

oxide so you can actually get a tangible feel for what these materials are like. Somehow people didn't think that putting a beakerful of ether to be boiled down over an open flame would constitute a fire, so there were fires and immense damage.

But the freebase then becomes a volatile form that can be smoked, not the acid form, the salt form, only cocaine base. Cocaine is the only material that I know which has this particular property of crystallizing out quickly in this very ball-like, granular form. This is one of the reasons that smack—not smack, crack has its reputation of high potency and high abuse potential— the fact that it is highly pure. Because if you have a mixture of half cocaine and, say, sugar, the mixture goes into hot water, the sugar will stay in the water, and the cocaine alone will come out. If you have other bases, it will come out such as, oh let's say, procaine or benzocaine or xylocaine—procaine is Novocaine. The reason they're being used is the fact they are numbing agents, they are local anesthetics. If you have any of these that are sometimes used as a diluent, they will not crystallize out in the same way, but the next day you'll find that these too will crystallize out, and the first material out is the cocaine.

And so this is a device by which you can go from 20 or 30 percent pure, to a material that is substantially totally pure. And you have a material that is not diluted or, as they say in the drug traffic world, "stepped on" or "buffered" or "cut." You have material that is of full potency. The abuse potential is, of course, exactly the same as freebase, exactly as it is of cocaine, milligram for milligram. You may be taking it by the faster route, and you may be taking more than you would otherwise, but it is not a different drug, and it is not a different mystery. The whole idea of writing an entire law about cocaine, then a whole second law about freebase, and now a whole third law about crack doesn't have scientific reality. Freebase and crack are pharmacologically identical. They are chemically identical. They're the same thing. They may look different, but they are chemically and pharmacologically the same. I saw a hand in back. Someone had a question?

STUDENT: Well, we started all the talk about crack. It was a term that was made up.

SASHA: It is largely a press, political, promotional thing in government.

LECTURE 11 ~ *Stimulants III*

There is a beautiful article . . . I don't know if it's appropriate to copy and have it here as a handout, but it was, "Did the Narcs Invent Crack?" It was an essay in, I think it was the *Atlantic Monthly*,[5] a surprisingly aggressive essay for a relatively conservative journal in which they say, "Look at what the value has done."

I have a personal suspicion that there is a small but real possibility that Len Bias, the fellow from Maryland, was murdered. I'll say that just straight out, because we have politics coming up. Maryland happens to be the very sport-oriented center for Washington, which happens to have Maryland as their identity for things. Here is a very young, upstanding, and very much respected young fellow who had no history in drugs whatsoever, who suddenly died of an overdose of cocaine. How many people remember the lurid drama of the inquest and the coroner's report? I've got a copy, but I may even have some of it here. I may have it in the file; I certainly have it at home. "Here's a person," said the coroner, or at least said the newspapers who quoted the coroner, "who had never used drugs before, and with a one-time using of cocaine of high purity went into a cardiovascular spasm and died of a coronary-whatever-it-was, of heart failure."

I have spent a lot of time in criminalistics, in toxicology, dealing with coroner's reports, dealing with autopsy reports. I will make the following two blanket statements and there's no one, *no one* who can call me a liar for it: There is no way of telling from a coroner's report the purity of the drug that was used. You can't even tell the purity when you look at the drug half the time. Sitting there in the body with ten billion biochemicals and god-knows-what-all, you can't tell the purity. All you can tell is how it went in, and they didn't say how it went in. The newspaper said he was snorting cocaine at a party. You look at the medical report, the evidence says not snorting. The evidence says it went in orally. But I'll get into that in a moment. You cannot tell the purity. You cannot tell from an autopsy report whether a person has used the drug before or not, or if it's the first use. What can you look for that tells you the history of drug use? There's nothing you can look at. You know the tattoos on the arm. You don't know. Nothing you can look at. So the coroner could not deduce that the drug was high purity. He could not deduce that it was the first time. You look at the

---

[5] "How the Narcs Created Crack," *National Review*, December 5, 1986.

coroner's report and neither comment was there. The comments came from the newspapers and from the press coverage and from the news release and the promotion from the government: "This proves that crack is bad." Then a report—and I have the actual report. In fact, I have brought some aspects of it. I can give you names.

STUDENT: Was he using crack or cocaine?

SASHA: Who knows? Cocaine was there, it was cocaine. [Shuffling papers.] Oh, oosh ... I'm organized. I'm always organized. Everyone said he never used drugs. They found quite a bit in the stomach, by the way. There can always be drainage from the nose. Buried in the sports section of *SF Chronicle* on 9-14-86, three staff members of Prince George General Hospital and Medical Center analyzed the autopsy report, concluded Bias had ingested about five grams orally, and volunteered (these are the people in the medical group who did the autopsy) that it could have easily been dissolved in a soft drink or a beer. They found no alcohol in him, so he had not been drinking.

This, in essence, ignited the fuse that never let up until the elections, and culminated in the passage of a $1.7 billion anti-drug bill that contained some of the most outrageous losses of freedom you'll ever see. I will talk more about that at length. I believe there is a microscopic chance that this is criminal mischief on someone's part.

Yes?

STUDENT: Five grams is a lot of cocaine.

SASHA: Sure it's a lot of cocaine! The lethal dose orally is usually considered between a gram or two grams to a relatively naive person. That, I would say, would be very blatantly a cause of death.

STUDENT: So they found five grams residual in his stomach?

SASHA: No, they found evidence that five ... No, they found in the stomach ... I had the autopsy report ... Something like 20 to 30 milligrams with a great deal of very, very thick yellow viscous fluid. Suggestion is he had vomited because he did not have a normal inventory of stomach contents. But vomiting could have been totally incidental to the whole thing, just the

LECTURE 11 ~ *Stimulants III*

trauma of a seizure could cause vomiting. They found it in various organs. They found a monstrous amount in blood. The autopsy report, in the eyes of the people who looked at it, suggests it was used orally.
Yes?

STUDENT: I guess you would have to ask, what would a college student, athlete . . . How would he have enough money to buy five grams of cocaine to get into his system? That would cost a lot of money. I mean, why would he take five? That doesn't make any sense. I mean it doesn't add up.

SASHA: I mean, basically, if you trust the picture painted of the fellow, he was a very respected, very clean, very ambitious, and very much like the typical, you know, the image of a person who had not had any drug experience. In fact, the people said there's no way he had taken drugs. I mean, it was that kind of thing. People who knew him just said no. Okay, I don't know. I'm just saying there is a microscopic possibility of criminal mischief. And I'm going to keep very much eyes-open to that possibility. Because, as a result of that, we had another one in California about a week or two later. Someone who was visiting from outside the States and gotten into a cocaine death. I forget the name—Washington, something Washington or Washington Someone.

STUDENT: Rogers.

SASHA: Rogers. Okay. I don't know any of the details. I do have the autopsy report, but I have not gone over it.
Yes?

STUDENT: Why would somebody want to kill him?

SASHA: Beyond that it would galvanize the public opinion toward anti-drug, and allow the passage . . . ? Okay, have I talked about this bill that passed last November, October? I haven't much. It has a very elusive title, I don't know the final title.[6] It had a lot of little components. This was the one that avowed $1.7 billion to bring the drug problem to a stop, and a lot of the original proposals were actually thrown out. There was some semblance of

---

[6] The Anti-Drug Abuse Act of 1986.

moderation, but not a great deal. Among its original proposals were some such as, "Require that the military bring an end to all importation within 90 days."

I mean, stop and think. You're charging the military with all of its force to totally involve itself in a civil criminal action and use. They can't even keep people from smuggling god-knows-what-all in and out of this country. I mean, there's no way that could be done, but that was going to be a requirement. In the law, there is the—this has nothing to with cocaine, but here I go—in the law, there was the exclusion from responsibility of responding to the Freedom of Information Act by four government agencies, including the DEA, the FBI, the CIA, and I forget which the fourth was. They do not have to respond to freedom of information requests anymore. They've been exempted by law. You have, and I have mentioned—

STUDENT: That passed as part of the—

SASHA: That was in the law. Yes. You have $1.7 billion being devoted toward the battle against drugs, largely federal judges, prisons, jails, of which I think something like $20 or $30 million dollars is toward education, but the bulk of it is restrictive energy of some kind. A lot of that money by the way, some of the money at least, has been quietly withdrawn. It's just not available or they're deferring the use of that money, which I consider to be extraordinary. The bill also has a law against what they call "designer drugs," which means anything that resembles, that is substantially similar to a scheduled material and is a stimulant, depressant, or hallucinogenic as intended to be given to a human, who, for the possession of it, the manufacturer of it, the giving of it, the importing of it, the exporting of it, anything doing with it, can be charged as if that material were a Schedule I or II drug. So in essence, you're putting a completely open category on all chemicals that do not have to be named, but can be named at the time of the indictment.

One of the difficult things for me to handle with this bill is that it was enacted in the latter part of October last year, part of it was to be enforced immediately and the rest to be enforced by the first of November, but the bill is not obtainable. I cannot get a copy. The Federal Department, Bureau of Narcotics, the Drug Enforcement Administration cannot get a copy. The

LECTURE 11 ~ *Stimulants III*

Department of Justice cannot get a copy. Copies are not available. It has now been October to November, December to January to February and into March, so the law is now into its fifth month. The bill still cannot be gotten. You can get comments from official bodies as to how the regulation shall be enforced, but the bill cannot be gotten. You can get the full Tower Commission Report out in paperback in 48 hours, but you cannot get a $1.7 billion drug enforcement bill with a lot of very strange things in it out at all. And this is unofficial but I believe it to be so—there's no contradiction to it—the federal government is not enforcing the bill. There is none of that law that has yet been enforced because no one officially knows what is in the law. So, these are the things that came out of that. It was passed, by the way, just before elections.

Yes?

STUDENT: If some of it is enforced, can't then . . . can it go to court as being unconstitutional?

SASHA: Absolutely. Absolutely. This, by the way, applies to a lot of this area. Okay, this is kind of a generality on laws. A law says such-and-such, and the law empowers a group to enforce the law. The group writes regulations saying how they intend to enforce the law. This is more or less the structure. One big law regulation enforcer is the IRS. And these are the IRS regulations. For example, the W-4 Form has been the big palaver. The law doesn't say to have "W-4 Forms." It doesn't say how long or how short or even that you'll have them. These are IRS regulations that enforce the law. The law says, "You shall know." The IRS says, "This is how we shall know." Almost anytime you challenge a regulation that is not rebutting the law, you will win. If you look at these regulations, and you challenge the way things are, or if the law itself were unconstitutional and you can challenge that, you probably will win. At least you'll be able to take it up to the court, to the high court, with appeals up and down the line.

But how much money does it take to defend yourself and go that route? You're dealing with many, many thousands or hundreds of thousands of dollars to address this problem legally. And that poor schlemiel who was walking down the street and got a heavy hand on his shoulder, "Buddy, you're under arrest, because I suspect that . . ." hasn't got the hundreds of

THE NATURE OF DRUGS

thousands of dollars, and he's not about to kindle the interest of the Civil Liberties Union or whoever, the public defenders who might be able to do it. So there will be an immense amount of damage done until someone says, "Hey, uh-uh. Not this time. Boy, you got to take it all the way ... right to the Supreme Court, you know." In truth you can, but there will be damage done before that.

Yes?

STUDENT: So the benefit is that they can get something done before seeing the court?

SASHA: Right. And maybe you'll be able to break the pattern of drug importation, drug use, drug what-have-you. And cocaine is the vehicle for mobilizing the troops. And they take a message home, "Reelect me to Congress because I helped pass the drug bill that's bringing this thing under control."

Yes?

ANN: I'd like to support the suggestion that somebody made that Len Bias may have been murdered.

SASHA: I said, "microscopic suspicion." That's all.

ANN: Yes, I know. But if that happened, it could just as well have been a real jealousy thing, too.

SASHA: Oh, absolutely. It's just the timing, the place. It was just two months before election, and the month before that bill was to be finally put in its final form, and in a very, very sensitive location, right outside of Washington, DC.

STUDENT: And anybody ... I don't think any other fellow student would have enough money to buy five ounces of cocaine to kill someone.

SASHA: Five grams.

Mmm-hmm. So, you're dealing with a lot of money. Sure. Could be a way to get rid of a competitor. I don't know but it just doesn't ring true. I mean, you've got to judge behavior patterns by including what you have personally dealt with when people's behavior patterns suddenly change. For example,

LECTURE 11 ~ *Stimulants III*

this is . . . this is a lecture on cocaine. [Laughter.]

I had a situation just yesterday. I stopped to get a bottle of wine and a bottle of soda water because I was going somewhere out in Marin County for dinner. There were three people standing at the counter, one of them wanting to buy a lottery ticket, another counting out change, it was going to take a while, and two of us waiting to get a bottle of wine and a bottle soda water. A fellow came in, and immediately you had that feeling that what he's doing is strange. You don't know what it is, you can't put your hand on it. I mean, there's just a feeling that something is strange. He came in and he asked the guy behind the counter where the something-or-other was. The guy said, "It's underneath the yellow flowers back there." He was in the middle of dealing with three people, and we, too, were in line. He went over and looked, came back and asked us if the man could show him where it was. "Well hey, hold on buddy. There are three people there and we are the next two. You're number six in line, you know. Hang on!" The guy said, "Give me a moment." Then he went back there and when he returned got another comment that was totally weird, something like, "The flowers were not yellow, or the red-pink flower..." Something about flowers, it was inappropriate. I left. I left the liquor store attendant with this character. But in going out, I just recorded, memorized the license. The behavior pattern didn't follow what you think is a normal behavior.

You always have the far-out nut, you have the guy who's a little bit wonky, you have the person who has an idiosyncratic character, he does strange things but you have a feeling of, "Okay, that's his way." But when you have that sense that things just don't quite jive together . . . And so here I have a feeling that at a party a person taking five grams of cocaine in a soft drink doesn't quite fit the pattern that I have been led to believe that this person had as his self-image and other people's images of him. We have to follow this gut feeling. You don't always follow this when you're dealing with people. You sometimes will argue yourself out of an interesting or a very dangerous situation, or into a very dangerous situation, by intellectualizing. Follow your gut feeling. Suddenly you want to walk on the other side of the street, or suddenly you decide, "No, I'm not going to go there." I don't know why. But somehow I've gotten a lot of confidence in following that kind of a feeling. Okay, okay. [Laughter.]

We had probably 200, maybe 300 thousand people who were involved with cocaine and with heroin at the time of the writing of the narcotics law in 1914. Yeah, "addicts" is a weird word, but it's the best way they can reconstruct the number of people associated with cocaine, by the number of addicts. At that time marijuana was considered not even a drug. It had no legal status at all. But cocaine? Largely heroin got the press, opium, through World War II. And just after World War II the amphetamines became more and more popular. Amphetamine, methamphetamine, the various analogs of this. I think it's probably before your time, but certainly you've read about the Haight-Ashbury area. San Francisco speed freaks who would use amphetamine continuously—they would take whatever supply they had and they would use it up. They would damage themselves financially. And often actually bring harm to themselves.

Many illegal drugs come from supplies that have been stored for use, for example, importation from Mexico. A lot of our drugs currently that are being used and abused come from Mexico, which are there legally because they are exported from the United States to Mexico. You can buy things in Mexico virtually without a prescription. If you have a cold and you think penicillin is the thing, go to the drugstore and order penicillin. You get it. You don't need a script. And I know right now a lot of the ketamine that is in California is made by Parke-Davis, it's shipped in brown kilo bottles in sterile solution to Mexico. It's being bought there, and brought back into the United States and being sold here as vitamin K or whatever they sell ketamine as currently. It's one of the very large, unscheduled, unpopularized drugs.

We may have a ketamine scare one of these days. Currently we're in a cocaine scare. We were in a speed scare before, and these restrictions were put on. The companies were told, "Don't manufacture any more. We're going to move it into Schedule II from III," and it was so moved. And in essence the pipeline of amphetamine and stimulants of that type was closed down. And I think a direct outgrowth of that, cocaine, which is cheap, easily available, came into its place. And now the effort is to close it down. I really think that the trouble is not in our stars, but it's in ourselves.

Why this extraordinary, compulsive urge to use drugs that change states of consciousness, to speed you up, slow you down, and turn you inside out? I think it's in us. I think it's part of our way. And I think if there's going to be a good clamp down on this, then like a piece of mercury you put your

LECTURE 11 ~ *Stimulants III*

thumb on, it'll pop up over here in some other form, in some other guise. And I think a lot of the damage that comes from it is the illegality. A lot of the damage that comes from it is the fact that no one knows for sure what they are putting up their nose or down their throat or in some hole or other. You don't know its identity, its purity, its source, or anything about it. And I think that a very great service would be to let things be and make some sort of information and maybe even quality identity aid available. Maybe this would be an outrageous suggestion. But certainly, I think the suggestion of making 10 years in prison without the possibility of parole for possessing a white powder that intrinsically is not an evil thing is not necessarily a step that is curing the problem.

I was talking about the weird, strange use of drugs Tuesday night. Ha, nothing to do with cocaine. Tuesday night I had the great pleasure of being at the Grateful Dead concert over in Oakland. How many people were at that concert? [Laughter.] You caught that one! That really was a big one. But when they focus the light from the stage out in the audience and you see 5,000 bobbing heads like that sort of thing. [Gesturing.] If you took a collective urine test of that place . . . [Laughter.] Like, you could not walk up the corridor without hearing, "Three hits for a dollar . . . [babbling sounds]."

There was no feeling of violence. That feeling of anger, that feeling of, you know, "Watch out what you're doing," it wasn't there. Maybe it's a bit of a high consciousness there, I don't know. I certainly could kind of feel that. I felt very comfortable with it. I really think you have a good example that you can have at least, let's say conservatively, a third of the people that had consumed something that might be on the edge of being illegal without problems—God, you couldn't walk through that place without smelling the pot, I mean, it's unbelievable—but I think if you could somehow take a hundred anonymous narcotics agents and just let them see the consequence of a thousand people being stoned and having a single thought in mind, I think you could plant a seed that maybe there is not an intrinsic destruction of the human race, the gene pool, that comes along with the use of a drug, because drugs were everywhere there, and I don't . . . Oh I'm sure there are overdoses. I'm sure there are damages. When I went out, there was a fire truck and an ambulance sitting outside. It was not there just because of a place to park. Anyone know, were there problems? I'm sure there were. But certainly, it was not the problem that you would project on that bunch of people using dope or drugs, what you'd expect from what you read in the newspaper that "this is the weed that leads to hell" or "this is cocaine that will drive you into insanity or make you have a heart attack."

Sure, cocaine can do that. More than less of the people on small amounts of cocaine have a drop in pulse rate and have a drop in blood pressure. People say it's a pressor. It is a pressor, and if you take much more, then you get a blood pressure rise and you get a pulse rise. But you don't necessarily go into a convulsion and go to the emergency department.

Modest amount of drug use is not unusual. It's our way of life. Any gross amount, any use beyond what you feel you're comfortable with, or that is taking charge of you, or as I mentioned in the earlier lecture, in which you have a bad relationship with what you're doing—and this could apply equally to cocaine and to chocolate chips or chocolate ice cream—is an abuse situation in which you can do damage, in which you can do yourself harm, and in which you can lose your control.

I'm getting a list of things for my lecture on alcohol. I wish I had it now. What constitutes an alcoholic? The AA, Alcoholics Anonymous, has a list, 10 things. And you can go down through those 10 things not using the word "alcohol" but using the word "cocaine," or using the word "marijuana" or "chocolate chips," "chocolate ice cream," or whatever. Do you tend to hide your use? Do you tend to hide your source? Do you buy your bottles of wine at three different liquor stores so the guy who sells you the wine doesn't recognize how much you're buying? It's your relationship with that drug, with that thing. Are you doing it by choice? Or is there something that's sort of taking over? It is the relationship that I consider is the abuse, and that's a relationship that's damaging. And I agree, I believe that as my personal ethic. I don't know where facts lie, but that's my very, very strong belief. Okay, any harangue? Any questions or comments?

Yes?

Student: I was wondering, if you buy drugs in Mexico that you can't get here, is it illegal to bring them back into the US?

Sasha: Yes. It's known as smuggling. [Laughter.]

LECTURE 11 ~ *Stimulants III*

STUDENT: Well, I mean for medical reasons, like that . . .

SASHA: Mmm-hmm. If they are not available here by prescription, technically it's smuggling.

STUDENT: So, if you got a prescription in Mexico for something . . .

SASHA: If you brought it back by prescription that would be allowed. I think that would be allowed, I don't know where the fine line is drawn, but basically that would be allowed.

Theoretically, anyone who wears a tie and a white shirt and drives a 1985 what's-it can bring back 5,000 tons of cocaine under the trunk and not get stopped, unless he does the basic error of telling people he's driving a 1985 Camaro and he's putting it in the trunk. This is the basic rule, the whole concept of coming across the border. I think everyone who has come across the border at one time or another and has gone through customs? Okay, everyone. It's a matter of . . . again, you're talking about that gut feeling, your body language. These customs people have that gut feeling, notice that body language. They inspect according to how eye contact is made, according to how you carry your body, according to how you respond.

I remember once when I was coming back from Europe, I was extraordinarily weary. I was wearing clothes that, believe me, should have been washed way before they had been washed, because I was going on a shoestring. I was coming back with my mother-in-law, so I was carrying a case with a little shopping bag filled with dirty clothes. I was sure I would trip every alert that could possibly be in the customs man's list.

"How far east did you go?" Typical question.

"Istanbul." I mean, I thought that would do it, being in Turkey or Asia . . . You know, practically in the realms of heroin back there, and of hash.

"Oh fine. Do you have anything to declare?"

"Just what's on my list." And, you know, it's 30 cents worth of this and 10 of those.

"Fine, welcome back." Rolled right through. I could have had a search . . . but I didn't. If you have a suitcase and glance in a suitcase, if a custom man spots in a suitcase two things, just in quick glancing, a bottle of Vaseline and a condom . . .

THE NATURE OF DRUGS

ANN: A bottle of Vaseline?

SASHA: A bottle of Vaseline and a rubber prophylactic . . . full search.

STUDENT: How are they going to notice it?

SASHA: Opening a suitcase. They happen to see Vaseline and a rubber prophylactic, one of the profile combinations for smuggling. Because how do you bring back in large quantities of something? You put it in a rubber prophylactic, put it in the behind, you would need Vaseline for that. And when you're back you go to the men's room and remove it. It's known to be a device for smuggling. And that's the sort of thing they're looking for. Or if your name is on that little magic computer list. The simplest thing is don't get your name on the list.

If you are overseas and you are wanting to buy something, and you really might not choose to declare it, let's say it's a nice fur coat or a nice piece of jewelry, "I will wear it and I'll say I've always had it." Okay, they're familiar with such devices. They don't necessarily think every piece of jewelry or every nice fur coat or every camera coming in from the Orient was bought there. You might have taken it with you. But if you paid for it by traveler's check, you may have gotten a special discount. "We'll give you 30 percent off for a traveler's check." You know why you get 30 percent off for a traveler's check? Because your name is on the traveler's check and your passport verifies your name and your passport has your photograph. And they say, "Aha!" and they inform the customs that they have just sold a $3,000 fur coat for $2,200 in Naples to such and such name and passport number. You're on the computer list within about 12 hours. That's the turnaround time for getting information on the computer list. Every day it's upgraded. And you're coming through they say, "Do you have anything to declare?"

"Three shirts for my little niece."

"Uh, step aside." Because that guy who gives the report gets 25 percent of the price. That is his livelihood. He doesn't only make his money selling coats. He makes his money collecting 25 percents. And this is a way of life. A lot of people do this.

I worked with some people, four of us sort of owned a company that's now a multimillion dollar company, and everyone else has ulcers and I don't,

LECTURE 11 ~ *Stimulants III*

happily. A company known as Bio-Rad, now in Richmond. It was in Berkeley at the time. And the very same thing happened one day. Some people came in and said, "Hey, this is a neat little company. We are finders of people who buy companies, and this is a neat little place and we have venture capital. We think we could make this company really have the potential we think it has. We could make a really good offer." And this very tight arrangement is made with the owner to look and see what the company is really worth—not what they say to the government it's worth, not what is said on the tax report it's worth, but what it's really worth because they're willing to pay a little bit extra. And they get the owner or the manager to reveal, "This is what we told the IRS, but this is what it's really worth." Greed is a driving force for a lot of behavior patterns. At which point they say, "We'll go back and talk to our venture capital owner about the whole thing." They don't at all. They go back to the IRS and say, "Why don't you check this company. They are really earning this amount but they're only paying this amount on their on their return."

"Oh, very nice. Thank you very much. We'll send you your 25 percent." These people make their livelihood collecting 25 percents by pretending to be sources of venture capital, to get someone who's a little bit dishonest to say, "We're really earning this, but we're only reporting that." That's all the information they need. They'll get the 25 percent of whatever fine is made. So yes, customs has not only a good sense of body language, and eye contact language, and a way of talking language, but also it has a lot of information thanks to computers. They have information fast and accurately.

Okay, cocaine, haha. What else have I got to talk about here on cocaine? Cocaine itself is this dirty picture [referring to drawing].

Some points are worth talking about. I'm going to erase this part for the moment . . . This compound is called "tropacocaine" . . . co-ca-ine . . . that

Illustrations of structures of cocaine and ecgonine.

spells it. Tropacocaine is an alkaloid that is present to a few percent in cocaine. Most coca had initially been raised in the 2,000–5,000 foot level in Peru. Then it spread out and there's only so much of that kind of level that you can get, there's much better growth and much richer growth if you move to the lower levels. In fact, cocaine production moved throughout all the countries of the central part of the equatorial portion of South America. It also moved to Java. Java is now one of the major sources of cocaine. And a lot of the material has moved to lusher growth areas, as in Java, and as in the lowlands of Bolivia and in Paraguay, and also Ecuador. Ecuador is also raising a fair amount.

Tropacocaine in cocaine is becoming a bigger component. There's an interesting article in *High Times*. They have a lot of color and light, as well as fact, on tropacocaine as being a source of cocaine overdose. I don't know the answer to that. Tropacocaine is more of an atropine-like compound. It causes a dryness of the throat. It causes a dilation of the pupil that is fixed rather than reactive. And they have found tropacocaine in instances of cocaine overdose. I don't know whether it's going to be a major alkaloid or not, but be aware of the name: "tro-pa-caine" it's often named. That middle "c-o" can drop out. Tro-pa-caine. Tropacocaine.

The cocaine itself—I've brought back the molecule [drawing on chalkboard]—in the body is metabolized almost entirely. People say, "We will run a drug test for cocaine." What they really mean is they'll run a drug test for the use of cocaine. This is worth of moment's mention—the difference in terminology. If you're looking in the urine for a drug that has no business being in the urine, and cocaine is almost in that category, you excrete maybe 1 or 2 or 3 percent of the cocaine you use. So there is a very small amount of cocaine in urine and it's possible there is none in there, because the body has this machinery—we talked about metabolism—that flips off that ester group, that takes off that ester group, in fact, flips off both groups, and that is what's excreted. And so you're not looking in the urine for cocaine. You're looking in the urine for evidence of cocaine use. It is exactly like looking in the urine for THC. You're not looking for THC. You're looking for evidence of THC use. So be very careful of the way you phrase, or the way people phrase to you, the test in urine for drug use. They often say you're testing for a drug, but what you're really doing is testing for evidence of the drug use. The drug may not be there.

LECTURE 11 ~ *Stimulants III*

There is a very, very important name in all of this. I'm going to go right back to the parent compound [drawing on chalkboard], which is an alcohol acid amine, ecgonine. And this is the material, ecgonine is the material that is usually tested for in the body for cocaine use.

Some points in the legal area: In federal law, for years, from the original writing of the Harrison Act up through the 1970 scheduled drug, the Controlled Substances Act, and up until about two years ago, cocaine was not explicitly named in law as being illegal. The law said the coca plant and anything that's found in it, but excluding the coca plant if cocaine is not in it. So, they can have the flavoring of the plant in Coca-Cola, and this type of thing, but they never said cocaine as such. They said "the plant" and components in the plant. Well, in the plant cocaine is the *levo*-isomer, the *l*-isomer. This is an optical isomer. You have these sorts of things, they can be upside down, different ways. These things can be in different geometries and be different compounds. Only one of the number of compounds that can be represented by the structure, cocaine, the *l*-isomer specifically, is the active one in the plant. It is the only one found in nature. And it was the one therefore, by definition, that was illegal.

And of course, then all you have to do is find, "Here's cocaine." You don't worry about it, because only the cocaine you have is from the plant. Then a court case was heard in 1983, someone who kept insisting, "What I had was *d*-cocaine, the other isomer. It wasn't active. It was synthetic, had nothing to do with the plant, and hence it wasn't illegal." And it went up through the court and the judge said, "Hey, you're actually right. The law doesn't say anything about *d*-cocaine. Perfectly legal. Throw the case out." The case went out, which offended enough people who didn't want to see this dealer get off with this kind of a device.

So they passed a law saying now the *d*-cocaine is illegal, which is not active, it's not around, a virtually unknown compound. At this moment I don't know where I would go in the world to get a reference sample of *d*-cocaine. It probably could be gotten, I could make one if I had to, but it's just an unknown thing and it's not an active thing. It's now illegal in the state of California because of this case. So now all the poor overworked criminalists have to assay whether cocaine is *d* or *l*, because if it's *l* it's under this law, if it's *d* it's under that law, and there are different penalties. Then the federal government did the same thing. They said cocaine, but coca leaf has always been said to be the narcotic.

It's ridiculous. We all know that coca is not a narcotic. It's a stimulant. But what they can do is give it legal sanction, so they will say, "As of this law being passed, cocaine is a narcotic." So now cocaine is legally a narcotic. Which is just as silly medically as it has always been. But since you have this *d* and *l* thing, they'll say, "You have cocaine. But if you invert that upper thing, you have something that's called pseudococaine." Greek, "pseudo," "false." "If you invert the lower thing, you have something called 'allococaine.' If you invert them both you have something called allopseudococaine. And so we will make narcotics, not just cocaine, but pseudococaine, allococaine, and allopseudococaine. And each of these four has two optical isomers, so we will make the *d* and the *l* isomer of each of these illegal as narcotics. And by the way, the raw material underneath is ecgonine, so let's make also ecgonine, pseudoecgonine, alloecgonine, and pseudoalloecgonine, and all the *d* and *l* isomers of those compounds for a total of 4, 8, 16 compounds shall all be narcotics." And they passed that law. None of the 15 of the 16 compounds, practically for this purpose, exist. And yet in principle, every criminalist in every crime lab when they get this white crystal material, and it doesn't have to be cocaine, should make sure it's not any of the other 15 materials. And of course, it's not going to be done. Laws that are passed in this way, in which they are not to be enforced become, I consider, bad laws, because there's no way you can run 15 compounds to determine if the injury is not one of those when you do not have reference samples. And it, you know at the bottom of your heart, is not going to be.

How much synthetic cocaine is on the market? None. And that's a pretty dogmatic statement. I'll say none. It may be reconstituted because it may be degraded and reconstituted, but the sense is that cocaine is a pretty rough chemical process. Oh, it may have been made once, and there may be some that I'm not aware of. But I would say virtually all . . . "virtually" here means "as far as you can see," all the cocaine you're going to find is from the plant, and it comes in from other countries. Not all of it. Cocaine is raised domestically, but it's not talked about. Substantially all of it comes in from South America or from abroad.

One point of issue that is a technical thing. Those who like chemistry

LECTURE 11 ~ *Stimulants III*

know that it's cocaine. This is *l*-cocaine. *d*-Cocaine should be the mirror image. In medical language *d*-cocaine is not the mirror image. *d*-Cocaine is the epimer about this position. So *d*-cocaine chemically is *d*-pseudococaine. That's a chemical detail. By the way, I am going to hand all these things out and get this information to you.

Medically, cocaine has two basic uses, two basic medical properties. The one for which it is used very widely is as a stimulant. It is a stimulant that works, as they say in the neurological sense, from the top of the head down. Its very first stimulation is cerebral, the feeling of that rush, the high, the intoxication, the lightheadedness, and then as more and more of the material is used, or used in larger and larger amounts, it begins working its way down the brain stem, leading eventually to tremor, to dizziness, nausea, and eventually to actual respiratory breathing problems and circulatory problems. You get a pallor, and you'll get bad, difficult breathing, short, gasping breathing, usually coupled with circulatory closedown. Death is probably due largely to the circulatory closedown. But that's deep in the brain stem; we're dealing with respiration. Respiration is one of the very last things to go. Respiration and heartbeat are basically built very deep in the computer. They are the last things to be touched by poisons. So it's really from the top, the cerebral on down through the brain stem, down to the lower vital functions as you increase the toxicity of the material.

ANN: What is a toxic level of cocaine, or is that known?

SASHA: Orally, a gram or two. Intravenously, difficult to say. Probably to an inexperienced user a few hundred milligrams; to an experienced user a gram or more. Parenterally in general this would apply, the typical dose is a few tens of milligrams. As a matter of fact, I saw an interesting experiment being done on cocaine at Langley Porter, in the SF medical school, in which they were doing experiments with cocaine users, giving them intravenous cocaine. They couldn't quite bring themselves to let them smoke it because they wanted to study parenterally used cocaine, but they could give it to them in a drip.

These were people who were experienced. You had to come in and give evidence by showing your body fluids that you had been using cocaine. This way they didn't expose a naive person. You'd get paid $75 a day or so to sit, have blood drawn, use cocaine as the person said to use it. They would try to evaluate the kinetics—how it goes in the body, what happens to it, how it goes out. These people were maintained at about 50 or a hundred nanograms per milliliter in blood on cocaine by an intravenous drip. They had been in there for about four hours. They, in essence, were being kept high at levels you'd expect to get in a cocaine surge. Both of them, absolutely fascinating to watch, and to talk to them. They were both asleep, they were dozey. "Hey, hey [blabbering], you know, that's really, really, really fly." They were sound asleep!

That chronic usage actually became a depressant. They could be roused, there was no loss of sensorium. I mean, they're definitely asleep and lost in that sense, but you could touch a person and say, "How is it?" "Hey, it's really neat. Really quiet, and kind of really, really, really love the place I'm in, you know." Asleep. There was none of this stimulation that you expect to see from the stimulant; wired up, hairs hanging on end, eyes dilated, go, go, go. No! They're dozy. And they had been that way for the last couple of hours. Initially, they said it got a little flashy, a little fast, and they sat back, but after about the third or fourth hour, they're asleep. This strange depression comes out of it. It's a stimulant, and yet in larger and chronic amounts, this other side of the coin is shown. You find this in people who are on speed runs, who are injecting methamphetamine at regular chronic levels. They are kind of depressed prior to their next dose, as if the next shot brought them back up to where you think they would have been without any drug.

One of the nice ways of avoiding local entanglements is, at Langley Porter, a lot of the subjects come from Los Angeles at the neurologic clinic at UCLA, and a lot of the people from here go to Los Angeles, so they are being studied outside of their environment. This avoids the contact with people who want to get in on the act, and family arrangements. These are people usually without families, but they're in the drug scene and they're volunteering.

This one woman came in from Los Angeles. She was a chronic cocaine smoker. She was carrying with her a propane torch, and her little glass tube—the mouthpiece—and her supply. We said, "We'll provide the supply." She said, "I need to know my supply." And brought her own supply and was very, very pissed off because she couldn't fire up on the ward. No smoking.

LECTURE 11 ~ *Stimulants III*

She couldn't fire up and she was going from restroom to restroom, so to speak, with her stash and with her torch and with her glass pipe. We could barely wake her up after she got up here on the plane. She was completely depressed and out of it. She was heavily sedated, not by the use of a drug, but by the absence of the drug, and its use brought her right back up to where she felt fine, and within an hour, down again. So, this strange body's response, this reaction to a heavy stimulus and a heavy chronic use of stimulants is something that you see quite regularly.

There is usually an afterglow, in the sense that you're in recovery. The recovery from it can take almost as much time as you had been in the run. People who have crashed from a speed run or crashed from a cocaine run will often have days of getting out of that. Usually, nutrition has gone to pot because you're totally oriented toward the drug and the drug is an appetite suppressant. So you don't get the nutrition. You don't get that feedback, and this is one reason that often, especially in the speed runs of the of the '60s and '70s, you'd have health problems. Hepatitis would be especially bad. Not only because you're exposed to infection in the very act of administering the drug, but you're not taking care of your body's needs, and the little bit of liver poisoning could express itself very badly.

I think we're about at the hour. Anything else I was going to say? Legal? We talked about legal. Chemistry, preparation . . . oh, one thing. One final point on preparation. I think the origin of freebase comes from a misunderstanding of the Spanish. I've got two minutes. When cocaine is prepared, you take a pile of leaves and you make it basic, usually with carbonate. You extract with gasoline or kerosene or some organic material, so you're getting out all the cocaine and all the alkaloids into an organic matrix. Usually it's a white gas that's commonly used—kerosene for example—in which you dump in about a percent of concentrated sulfuric acid, and out comes what's called "basa" or "pasta," but it's basically called "basa" because in Spanish it is the basis of what you're isolating. Not "base" in the chemical sense, meaning acid base, but "base" in a sense of the fundamental thing you're getting out. It has the texture of Portland cement. It is really a good, crude sulphate of cocaine and all the other alkaloids. This then is processed to the hydrochloride. This is smoked more and more in South America. It's mixed with tobacco, or it's mixed with marijuana, and it's smoked.

But it's a salt-based salt. It's terribly inefficient. You put in a couple, 300 milligrams of this sulfate and sell these things. They have their names, I have the names written down. I am not personally familiar with them. But this smoking of basa came into this country as smoking of "base." And I think the idea of making the freebase and smoking as such was a corruption of the translation of the of the Spanish word, "basa." And I think therefore, the origin of that phenomenon was strictly a transcultural misunderstanding. But nonetheless, that which is smoked in South America by those who can barely afford it, but who are using it more and more widely, is the salt, which technically cannot be smoked and hence you use a lot of it. Very inefficient. You lose probably about nine-tenths of it in the course of combustion, but what is left is used as a drug.

Okay, that is the hour.

Anybody want to inspect . . . please don't take them with you. [Laughter.]

LECTURE 12

March 10, 1987

Depressants I

SASHA: We have had three lectures in a row on stimulants. It seems to be appropriate now, in the interests of symmetry, to have three in a row on depressants.

How many people in this country today depend, to some measure, on a depressant drug? Probably the same number that depend, to some measure, on a stimulant drug. Which is, with the exception of a few of the Amish and an occasional Trappist monk, everyone.

The statisticians and epidemiologists would like to be able to say that the heroin-dependent population has decreased by 12 percent last year, but in truth no one knows the numbers for how many heroin users there are, or for how many of them are dependent. And should one add to this number the many people who are routine users of some prescription pain killer, and also those who have a regular pattern of tranquilizer use? And how can we exclude those who go to bed drunk every night?

The most ancient of all depressants is the opium poppy, *Papaver somniferum*. It is a beautiful flower that appears in many colors, and grows easily in warm, dry climates. There are legal restrictions in most countries designed to regulate poppy cultivation, but the direct financial rewards strongly encourage the circumvention of these restrictions. In Turkey (where the poppy is called *hashas*, pronounced hash-hash, not to be confused with *hasis*, which is hashish) it was banned in 1971, but the economic impact was so severe that the ban was lifted three years later. The bulk of the opium crops are to be found in the Golden Triangle and the Golden Crescent. The Triangle countries are Laos, Burma, and Thailand, in the center of Southeast Asia. And the Crescent countries are Afghanistan and Pakistan, in the Near East. There is a sizeable production in India, in Turkey, and in Mexico, and recently Iran has reappeared in the selling market.

THE NATURE OF DRUGS

LECTURE 12 ~ *Depressants I*

The most famous part of the plant, and the part of the greatest value, is the capsule that forms at the base of the blossom after it has been fertilized. The blossoms last only four or five days, and it takes about two weeks for the seed capsule to ripen sufficiently for harvesting. This fleshy part provides the opium gum, and the seeds are used for their oil and food value.

The usual procedure is to cut a score around the capsule of the poppy plant, using a sharp knife, scoring once or occasionally scoring every second or third day for several days. A gummy latex, a liquid, comes out called "chick." It comes out of this area, it's scraped off, put usually into an earthenware bowl. And this liquid then on drying becomes, in essence, crude opium.

The manner of slitting the capsule varies with the geographical area. You can tell by looking whether it's from the east or from the west. In Southeastern Asia, usually the scoring is done as several cuts vertically, usually with a knife with three or four different blades closely spaced. This knife is called a *nushter*. In Afghanistan, a single incision is made horizontally, nearly completely around the capsule. In Turkey, several horizontal slits are made. These scorings are done repeatedly, and again the exudate, the sap that runs out, is gathered and constitutes opium.

Until the advent of smoking as it came from the New World, the world had no opium smoking. It does now, due to the hand of man. Until that advent of smoking, which spread, as I mentioned under tobacco, to Europe, and from Europe on around the world, opium had been eaten and it had been consumed as a tincture. Laudanum is the name for the opium tincture—opium and alcohol, consumed as a tea, consumed as the actual material itself. The properties of opium are in the name of the poppy, "somniferum"—sleepiness. It produces a sleepy state. It is the origin of the term "narcotic," which is a numbing term. It is probably the prototype of all narcotics.

I don't know really how best to describe its properties. How many people have actually used opium or morphine or had medical treatment of Demerol or something of this ilk? Probably most have, usually before surgery. The thing is you may go into surgery with a real anxious, real nervous, uncomfortable antsy way. With a shot of Demerol, once it settles in, it doesn't matter. You get this kind of floating, easy, willing-to-talk-ness, quite amiable to whatever is going on. Everything has lost its anxiety aspect. You have the use of opium in the "opium den." The connotation is, you know, something in the basement of Chinatown with people sneaking in and out doing ugly things. And the truth is, the use of opium smoking is throughout the world. It is still very broadly used in many parts of the world. And at one time at the turn of the century in China, well over 25 percent of all people were chronic opium smokers. Yes?

STUDENT: Is it legal in those countries now or is it illegal in most countries?

SASHA: It is illegal. It is technically illegal in China. It was made illegal at the time of the Communist Revolution. It is technically, I think, illegal in virtually all countries. It is widely practiced in many countries, which is illegal.

A friend of mine went to Bombay, to an international meeting of the Society of Humanistic Psychologists or something, but it was a chance to see Bombay and take it off of his taxes. And he had a couple buddies that had always wanted to try smoking opium. So they managed to go and find some relatively seedy cab driver who was open to bribery and said they wanted to find an opium smoking place. They were taken to a really run-down part of town, and he told them, "Down that alley and the third door on the left" or something. They went there and knocked on the third door. Clearly, they were in alien territory. I mean, they were tourists and they were not welcome—this was not part of the Bombay tour of the night. This was something quite different as it was illegal.

They were allowed in and they went up on the second floor. People passed around the classic opium pipe, which has a bowl about two-thirds of the way to the end where the opium is kept, and it is heated, you draw air over the top of it. They were given a pipe and none of them really felt a great deal of it. They were a little bit anxious, not quite knowing what kind of alien territory they were in. One had the second pipe, and he was at the first pipe, which was really all they were told they could have. He was kind of, as he said, waiting for the dancing girls to come in. He had this story, you know, that you have these visions and people hallucinate; you would hallucinate and see things that weren't there.

Well, he was having kind of a neat little drifting of the mind and a fantasy, but he was not expecting that; he was waiting for the action. And he

LECTURE 12 ~ *Depressants I*

took a second pipe and that was a critical mistake. He took a second pipe, they all took a second pipe, and he did not realize that the end of the first pipe is what you are after. It is that retreat from concern, the replacement of real things with things of the imagination of the mind is almost like that hypnagogic falling-asleepness. You are not dreaming because you're conscious, but you're not really responding in a way to anything outside because you're quite content to retreat within yourself. This is very much in the area of peril; you find much of that today.

Totally subtle! Totally subtle. And so he took the second pipe and he was still waiting for dancing girls, so to speak. And at the end of the second pipe, he realized he shouldn't have. They were kicked out. They found a cab somehow. He was sick. He was nauseous. He was nauseous all the way in the cab back to the hotel, to the disgust of the cab driver. On the elevator, the elevator moved; that was enough. He was nauseous in the elevator. He got to bed. The next morning he got up, he was still nauseous. He had to give a talk the next day. He never gave it. He got some breakfast, it came up. It was about 16 to 18 hours of this absolutely, cannot live with yourself, excessive poisoning. It was a very rough experience. This experience was disappointing, because it wasn't what he expected, but what he got is what is there in opium. It is something that has, very much as tobacco has, a satisfaction that you can't put your hand on. Tobacco you smoke because you want to smoke.

STUDENT: So it's fairly subtle. It's not something that just hits you in the face?

SASHA: Well, it doesn't give you visions, and it doesn't give you a dancing girl and doesn't give you hallucinations and doesn't put you to sleep. But like smoking gives you something that satisfies some need of smoking, opium is very much the same thing. There's more of a fantasy, more of the dropping away instead of the alerting care, a dropping away into not caring that is part of the experience. But that is the experience of opium. It has been used from time immemorial. It is still used, and it will be continued to be used.

The fields of opium growing are many. It has moved around the world. It's now grown in all major countries that have a reasonably warm and dry climate that is amenable to poppy growth. Mexico has moved into large fields of opium. And it is a general static commodity. I wish I had it to show you. Next week I'll bring it in.

The actual tar, the opium tar, the crude opium when it's dry has something of a feeling of asphalt. When it's fresh, it has a gooeyness and stickiness. It originally comes out of the plant as a pale yellow or yellowish-pink latex, and it goes to quite a brown, brown-black when it gets dry.

The largest transactions were the British moving opium into China. The opium was moved around the world, but a lot of what was moved into China came in from the Arab countries, came in from the European and from the western Asian countries where it was grown. It was very heavily grown and it was the British—what was that name of the big company in India at the beginning of last century?

STUDENT: East India?

SASHA: East India, the company. Thank you. They owned the company. That company owned the grounds in and around Bengal where there were literally thousands of acres of opium grown. Bengal was a major source.

In Bengal . . . there are some pictures here you may want to see. The opium came in balls about the size of a child's head, weighing several kilos. They put some 30 or 40 of these balls in a case, and these cases were the commodity of the opium trade. Really, what first brought into Western attention the amount of trade involved in it was the British bringing it into China. They would import it into China; it was not really allowed in. It was against the regulations. And so they would bring it to the Chinese coast and the Chinese smugglers would move it in by literally thousands of cases. Yes?

STUDENT: How widespread was opium smoking in China before the British went in?

SASHA: Oh, quite widespread, but largely limited to the upper class—the Mandarins—and that was in the 1700s. And around the turn of the century, into the 1800s, it began spreading wider and wider and began to be planted. The planting was outlawed, but it became very commonly used probably in the early 19th century. And it got into increasing demand. The voices went against it because it was a depletion of money, of funds. The

LECTURE 12 ~ *Depressants I*

British had a very keen market, primarily for tea and silk; they wanted this from China. But they didn't want to pay gold and silver for it because the amount they needed would deplete these international funds, and so they wanted to supply opium for it. But the Chinese did not want to encourage opium use. In fact, opium was beginning to be recognized as a nonproductive liability in society. And they objected to this. In fact, at one point the Chinese sent officials into Canton to destroy the reserves of opium that were stored there. The British responded by opening up the ports by brute force—1838, I believe, was the beginning of the Opium Wars. They came in and said, "Thou shalt take opium in exchange for silks and tea."

And they quickly opened these ports by brute force, in essence, to make a goal, a target for the opium trade. The Chinese rather objected to it, but there was still a lot of official bribery, so it was being brought in, but not quite openly. The people who were in the smuggling business loved the profits that came from that. And so the British got it to the smugglers in exchange for teas and silks. The Chinese government said, "This is absolutely unallowed." They put an actual ordinance against it being grown, against it being used, against it being imported, and the British and the French added their little muscle to it. In 1842, they really opened up the ports. They invaded, the French and British entered and occupied Peking. They took Hong Kong as part of their retribution for the Chinese having been reluctant to international trade. It was quite a flap at the time. And the ports of China were permanently open largely because of the British action.

It's been said that this was heavy-handed on the part of Britain to get a trade going in opium. I'm sure it was. And yet the rationalists, the defenders of the British action, said this was a move to make international trade as it's understood in the rest of the world, understood throughout the world. Okay, you can swing that any way you wish. The long and short of it is that the opium trade was officially introduced and legally introduced, legally as the Chinese lost the war, and the use of it spread and was only brought under control in its own strange way by two events—one, in fact, in this century.

One was the fact that they began raising their own, which meant that they were independent of external trade. And secondly, the communist regime coming in said that this was clearly a destructive thing in society and made very firm moves to outlaw it. And the moves of controlling drugs by a totalitarian country are never satisfactory, because they don't address the main reasons that the drugs are used in the first place. Same thing you have in the use of opium in Russia. The Russians will say with a great authoritative, firm voice, "There is no drug problem in Russia. There is no this or that or the other problem in Russia because it is not a thing that's needed in the kind of state that Russia is." Well, Russia is, as all countries are, made up of people that have their ways and their needs and their weaknesses, and these ways and needs and weaknesses are met. And there is a trade.

The whole use of hashish, which we'll get back to in some way down the line, I'm sure I have a lecture on marijuana somewhere, if not I'll make one. The whole use of hashish in Russia has become quite, quite broad and quite a serious problem. And there is a small fallout from the Afghanistan adventure, where Afghanistan is one of the major exporters, raisers of hashish as well as opium. These things are coming back with the Russian soldiers who have learned that there are aspects of this they never knew about, which have some value or some mechanism of escape. Hashish is very often for the entertainment-escape. Opium most often for the retreat-escape, the self-indulgent escape, but they're both playing a role of giving this altered state. Yes?

STUDENT: From what I hear there's also the problem with alcoholism in Russia.

SASHA: Woowie! Yes, much more than we have here, and cigarettes more than we have here. China also. Not so much alcohol in China interestingly, but tobacco. Yes, very severe. We have a severe problem with cigarettes and with alcohol here, but it has fallen into that slot of not being a drug. It is not accurate. If you were to take a look at alcohol objectively from a non-alcohol-using point of view, and look at marijuana objectively from a non-marijuana-using point of view—the arguments, the legal approach, the destructiveness—alcohol would way outweigh the marijuana argument. And if you then bring heroin in on that, heroin is not a particularly destructive drug.

With the opium it will lead to morphine, will lead to heroin. There's a connection there. Come on. Certainly. Heroin is not a particularly destructive drug. Opiates, the narcotics, are not particularly destructive drugs. They

LECTURE 12 ~ *Depressants I*

don't kill. They don't cause things like lung damage or cirrhosis. Oh sure, you can overdose. You can get respiratory overdose on any of these drugs. Usually, your respiratory system paralyzes. It ceases. You get what's called "respiratory failure." You stop breathing. If you stop breathing for very long, you run out of air, and that is a cause of death. But respiratory failure is really due to a monstrous overdose, and there are people who use and have used heroin chronically—or methadone, or meperidine, or morphine chronically, daily—for their full, productive life, and have not grown a third horn or done something weird physically. It is not that destructive. It is debilitating socially because it takes you out, if you accede to it, it takes you out of the flow of what's called "the productive society." But so does alcohol in much the same way. You'll find that much of it is used to escape from problems and from complications, and from the flow of things. Yes?

STUDENT: Wasn't Johns Hopkins a heroin addict?

SASHA: I believe he was an addict. But was it heroin or was it morphine?

STUDENT: I don't know. I was just wondering if you knew any other—

SASHA: Oh, De Quincey, the *Confessions of an Opium Eater*. Gosh, a number of poets and writers—

STUDENT: Freud was known to—

SASHA: Freud was into cocaine and into tobacco. Tobacco is what killed Freud. But no, De Quincy comes to mind.

STUDENT: William Burroughs?

SASHA: Burroughs. Oh, gosh, and many other things.

STUDENT: Edgar Allan Poe?

SASHA: Edgar Allan Poe.

STUDENT: That was alcohol.

SASHA: Was it alcohol? I think it was laudanum. Hmm-hmm. Laudanum was considered to be, in fact it was called for a while, in the alchemy days, the "philosopher's stone" because it provided that insight one saw as being authentic. When you go into that, everything is absolutely right, and you're backing away a little bit from the strains of the world. What better picture do you have of nirvana or heaven or what-have-you? And out of these visions, and out of this retreat, a lot of the writers wrote, and were users. They used it as a source of inspiration, a source of ideas.

STUDENT: Oh really?

SASHA: So it's not an all-negative thing. It is something that, however, if you are ever toying with the idea of getting into, be aware of the fact that it is an extraordinarily seductive mistress. Very seductive. And as with many of the habits that provide pleasure, be aware of the price of it. But it is very broadly used; I mean, people use heroin in this country, my lordy. In fact, that brings up an interesting point about heroin use. I'm not following the pattern. Okay, I never do. Often people say, "Why not treat addiction? How do we solve the heroin problem? How are we going to get people off of heroin?" Well, you make it illegal. You make it a crime to use it. You lock people up. But after they've been locked up, they come out, they will still use heroin. It does not really resolve itself using the hard-muscle approach to law enforcement.

One thing I wanted to mention in this area was the so-called British Experiment. I think this is something that should come up periodically on the idea of "why not treat heroin addiction with heroin?" I mean, after all, it's not particularly damaging, and if you use it in a modest amount, you can use it for an indefinite period. But it's not particularly damaging. So why not take people who are addicted to heroin and give them heroin? In essence, you will get them away from buying it on the street, and get them away from using it in a social environment, get them away from becoming dependent on a criminal environment for its use, and they will use it in a medical environment.

Well, this is often called the British Experiment, in which England approached the entire problem of heroin use by making it available by prescription from a physician for treating people who are addicted to heroin. England had much the same legal history that we have. We passed the Harrison Narcotics Act in 1914. England passed the Dangerous Drug Act in 1920. It did the same thing. In fact, it's worded very much identically as far as what defines an addict and how it should be approached. "No physician

LECTURE 12 ~ *Depressants I*

shall treat an addict, except in that . . . No physician shall use or prescribe or make available heroin except in the practice of his medical professional duties." In this country, that went afoul of the state of the law because they defined a heroin addict as being a criminal, and no physician would treat a criminal in a criminal act. That would not be part of his professional duties. So, in essence, heroin usage became a criminal act here.

In England, after seeing the direction that it was going here—this was about 10 years later, and they didn't have very many heroin addicts or opiate addicts in England—they said, "Well, why not?" Since many of them were dependents for medical reasons, rather than from the self-medication, self-serving background. "Why not let the physician help, let that be part of his medical practice?" And indeed, the physicians made heroin available on prescription or on the equivalent of medical call to those who were dependent upon heroin, and it worked out quite well for, oh, for many decades. But it's not really a parallel to this country because here there is a large body of addicts. I mean, you're dealing with hundreds of addicts in England. You're dealing with hundreds of thousands of addicts in this country. So you have a different problem, but it worked there.

And it worked for a while, until about 1960, when there was the beginning of more restrictions. There was a very severe law passed in Canada restricting the use of heroin and making it a more severe crime than before, and quite a few Canadians went to England. Within the Commonwealth they can go from one part to another. A part of the structure of the Commonwealth is this flow of people. There are a number of heroin users in the Jamaica area, in the Dutch East Indies, the British portion of the Indies, which were part of the East and West Indies, and in the Caribbean, who used heroin and went to England. And there was a general movement, in keeping with the move we had in the 1960s of drug orientation, pill using, drug using, pot using. There was an increased amount of opium and heroin use. And this movement invaded England, too. So not only were people coming because they were heroin dependent, but the people were coming because heroin was available in the socialized medicine in England, and it became part of the philosophy.

Heroin use moved from the medical origins to the self-gratification origins. And thus, the volume increased quite rapidly in the 1960s. And about the latter part of the 1960s, a new regulation was passed that was a force of law that only certain physicians, under certain circumstances, in certain clinics or certain hospitals, could prescribe or make available heroin. Then as they began restricting it, about the same time, the concept of methadone came in where methadone could be used in place of heroin.

Methadone is a synthetic narcotic. I'll get to this handout sooner or later that just gives you the pictures of what these things are. It's a synthetic narcotic that has much of the same properties of morphine or heroin, but it's much longer lived. It has the disadvantage that it is still addicting, it's cross-addicting to morphine, but it has the advantage of being longer lived, of being active, and it can be given orally. In fact, it can be doctored in a way that it can only be used orally. So it never had the ugly stamp of being made illegal, of Schedule I, no medical use, as heroin has. So the idea of replacing heroin use with methadone usage came in as being quite a movement in the late '60s and early '70s. It was called "methadone maintenance," in which the replacement of heroin with methadone would be done. You take a person dependent on heroin and transfer the dependency to methadone, which could be administered less frequently and under somewhat controlled conditions. Yes?

STUDENT: As a drug it has a similar effect to heroin?

SASHA: They are very similar in their effect. Taken orally, it does not have the rush and impact that heroin has, but has much the same retreat quality. It was also called "methadone-based maintenance" and "methadone withdrawal," in which you transfer a person from heroin dependency to methadone dependency, and then regulate the amount of methadone and try to wean them off with decreased dosages and some sort of psychological support structure to get them free of any dependence. It has not been terribly successful. I don't know what the bottom line is in the American experiments with the methadone clinics, clinics in general. They still exist, they're still being used. Heroin is still being used. Whether it's added problems, I doubt it. Whether it's solved problems, certainly not very many conspicuously. But in England, the methadone then took the place of heroin. And methadone was used in the clinics and in the hospitals and with the people who had dependency upon heroin. And so the heroin availability, in essence, died by default because methadone took its place.

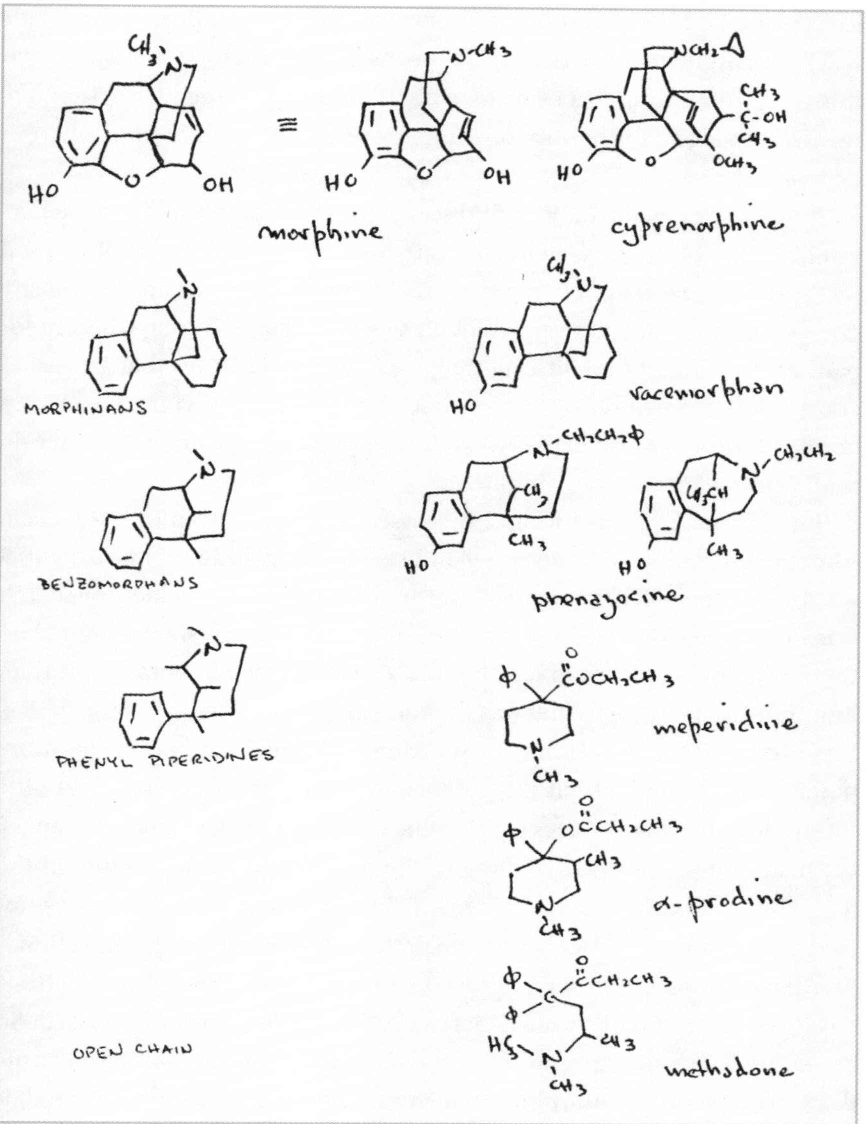

LECTURE 12 ~ Depressants I

[Figure: Structures of morphine, cyprenorphine, racemorphan, phenazocine, meperidine, α-prodine, methadone, and structural classes: MORPHINANS, BENZOMORPHANS, PHENYL PIPERIDINES, OPEN CHAIN]

People will say, "Was the British experiment a success?" I kind of think so, in ways. It certainly solved the problem of addiction on a very small scale, with a very different matter of addiction than in this country, but solved it for a number of decades. Was it a failure? I guess it was a failure, too, because it's certainly no longer being done. They're using methadone in its place, and heroin is not legally available for this purpose. And people will try to argue its success or its failure as being a reason we should try it in this country. The idea of using a drug for a drug dependency—you're using methadone for heroin dependency. Why not use heroin for heroin dependency?

Well, it goes against a lot of ethical principles somehow, because heroin has been branded as being a totally ugly drug, so the idea of using it medically—in fact, even using it as treatment for pain—is not even being done experimentally. It's very difficult to get permission to use heroin for pain control. You'll find it being said, "We have other drugs that are just as good." How the hell do you know that it's just as good if you haven't tried heroin? I mean, run an experiment. Run experiments! Compare it. Heroin is an extremely effective painkiller.

I had made a mistake when I discussed its introduction in 1890. I said something I couldn't document when I said it was introduced as a drug that was free of dependency. That was not so. I cannot find evidence for that. It was introduced as a drug for treating coughs, as a cough suppressant. And there was no argument given that there was less dependency on it than morphine. In fact, the feeling was the equivalent in many ways to morphine. It is preferred for injection over morphine because it is somewhat more potent. And I'll talk about the potency in a moment. But also, morphine intravenous injection causes a histamine release and you get a burningness; the histamines will cause tingling in the fingers and cause tingling in the feet. Heroin does not have it. So it's a slightly smaller dose and does not have that irritation, hence it became preferred to morphine for injection.

All morphine is—by the way, don't worry if I miss a few things. I'll draw out the lecture, it'll be available later and I'll put in things I've missed. So I'm going to bounce around to what occurs to me at the moment. Let me draw out a picture of morphine, which I think is on the handout for today. [See illustration on page 111.]

It's a big complex alkaloid. It has about four rings, one of them sticking out toward you in the molecule. Morphine is the principal alkaloid in opium. It was first isolated—in fact, I brought in this paper—Sertürner, in Germany, roughly 1805, was the first to isolate morphine from the poppy. Opium was known to be a sedating thing, a dreamlike thing, a painkilling thing.

LECTURE 12 ~ *Depressants I*

But put yourself back if you can, this is 150 years ago, 170 years ago. It was not known that plants could be resolved into components. Remember, organic chemistry was just beginning. In fact, organic chemistry, for all intents and purposes, didn't exist yet. The residues were there. They isolated this from air, they generated this gas and knew this gas was different from that gas, but plants were plants, and plants were entities unto themselves. And opium was a part of a plant and opium was an entity unto itself. The idea of something being in there that had such-and-such property was not there. Components of plants, plants being resolved into components, and each component having an action, was totally unknown, totally unsuspected until this person began exploring.

He was really the first one to uncover the concept of alkaloids, the concept of plants having active components. He was diddling around with opium, trying this way, the other, trying to distill it. He boiled it in water, boiled it in this and that, trying to get different things out of it. And he worked in the basement of a pharmacy as a pharmacist's assistant. He had no professional degrees, no medical degree, no scientific degrees of any kind. And by chance, he happened to pour liquid ammonia onto opium and put it aside; the liquid ammonia evaporated. This beautiful white solid was sitting out there. This, he found by tasting and trying, had the active property of opium. He converted opium from a black, gooey mess to a white solid, and the white solid is what we now know as morphine. He had gotten it out just by trying different things. He hauled in mice, and hauled in rats and dogs, and found that given enough of it, it would numb them, put them to sleep, and then kill them. And he began working on the pharmacology of this compound.

STUDENT: Is that why that white powder itself—

SASHA: Oh yeah. He found this much killed a rat and this much killed a dog and so he used less than that amount in himself and knocked himself into strange little places. But it's interesting. In fact, you have insight into this like with Davy. Davy was the one who worked with nitrous oxide. I'll talk about him a bit later. [Reading:] "Sertürner experimented first upon himself, then he found three lads, friends of his who agreed to let him experiment on them." You know, three buddies, "Come on and drop around in the evening."

"When they came silently at the appointed hour, Sertürner admitted them to his lab. Every detail of the experiment had been carefully thought out. The four young fellows seated themselves at a round table and Friedrich Wilhelm allotted the dose, each receiving half a grain." Okay, half a grain, a grain is 60 milligrams, half a grain is 30 milligrams of morphine, and morphine is active orally, by the way. It doesn't have to be injected. That's a pretty good charge. It's a pretty good charge of morphine. "Heroically each swallowed his share and after a little while they all became aware. . . . They felt extraordinarily cheerful. A sensation of warmth became diffused through body and limb. The cheeks were flushed and a general sense of comfort pervaded their frames. They exchanged ideas all day." Oh lordy. This person I was just mentioning who had the little bit of the Demerol going into surgery, always very scared going in, once the Demerol had taken over not only did he relax, he kept talking to the nurse explaining, "This is the kind of work we do in the laboratory. I think this kind of thing . . . blah blah blah." Just on and on. The nurse was apparently quite familiar with this type of response to a little bit of Demerol, she quietly nodded, "Yes, sure," and turned it off.

"For half an hour they carried out a lively conversation luxuriating in the remarkable sensations. Then Friedrich Wilhelm arose, walked around the table, gave each friend a second dose, another half grain. As leader, he had already experience, he took more himself. 'Now fellows,' he said, 'Watch out and tells me what happens, for this is of the utmost importance to the experiment.' The fellows obediently tried to watch out, but their eyelids were heavy. It was hard to keep their eyes open. Their limbs were oppressed by a paralyzing fatigue. All the same, they continued to sit bolt upright. Their friend had impressed upon them the experiment did not concern him alone but had a vast importance in humanity at large. They were determined, therefore, to keep awake as long as possible. With hesitant tongues they gave their reports: fatigue, weight in limbs. They were getting out of touch with reality. They wanted to tell their friends all about it. They wanted to cling to the strange images that were coursing through their minds, but their tongues refused to articulate. Nothing but an incomprehensible murmur emerged. Their eyelids grew heavy, and they were forced open with their fingers, but they closed again. Wilhelm having tried the powder several times before was a little more able to resist its effects. Besides, the experiment was not entirely done yet. To learn the full effect they must each take another half grain."

"You watched how the heads of the three around the table were nodding as if drunken with sleep, but there was no time to lose. He must give them the third dose while they could still swallow. They thrust away the powder but he would not be denied and as soon as they have been persuaded to take it he took his own last dose—a little more than gave to the others. Then he tried to take observations and make notes, but everything began to swim before his eyes. Visions mocked him . . ." Romantic, it's a nice story. On and on.

"After long hours the young men woke to vomit. They were suffering intensely from nausea and headache. The three friends staggered home after the experiment, the four bold fellows at the Paderborn Pharmacy had taken each of them twice the amount of morphine that they regarded as a maximum dose. The second experiment had been planned for the next evening, but Sertürner could not persuade his friends to it." [Laughter.]

Yes?

STUDENT: What is the source of that again?

SASHA: The source . . . I wrote the source down. It is a book written in the 1930s, originally written in French, translated to English, called something like *Triumph Over Pain*.[7] Systematic and beautiful. At the back of the handout I put the title. You can look at this afterwards if you want to; here is the citation.

Yes?

STUDENT: Is there anything to do about the nausea with morphine, things like Compazine or something like that?

SASHA: Yes, you can use materials that are largely like the phenothiazines that will modify the nausea. Nausea comes for a number of reasons. It can come from poisoning. Many materials cause nausea, not due to any burden in the stomach, but to an actual action in the brain that's called the

---

[7] René Fülöp-Miller (translated by Eden and Cedar Paul). *Triumph Over Pain: The Story of Anesthesia*. The Bobbs-Merrill Company, 1938.

"chemical trigger zone." An area in the brain that is affected, and as a result of its being affected you vomit. It's not an upset stomach. It's not due to a metal or due to something churning. It's just a response to the chemical. Peyote causes vomiting quite regularly, much for the same general reason.

Apomorphine is a superb chemical now in the study of dopamine receptors. You treat morphine with a little bit of dilute HCl or fairly concentrated HCl and heat. And that big ring that sticks out toward you goes flat. The whole molecule completely rearranges into a four-ring molecule that is a catechol called apomorphine. But this is a totally synthetic product. It does not occur in nature and is totally artificially made. It was worked out in its structure, but this was totally misleading as far as the structure of morphine, as it was not realized it could undergo such a monstrous rearrangement.

But apomorphine, when injected, produces just that. You inject it and some 15 minutes later, if it's done intramuscularly, you vomit. That is the only real effect it has in humans. And by the way, it is habituating. I mean, you can get very much dependent upon it. It has a property of physical dependency, even though the only goody that you get is vomiting, which is not considered by many people to be that much of a pleasure trip. So, you need not have the pleasure to make the body physically dependent. Psychologically, no. Physically, yes.

STUDENT: Do you know what William Burroughs was saying when he expressed comments about its value in psychiatry?

SASHA: I don't know what he had in mind there, but certainly the drug has been used medically. But in psychiatry, it's been used in, what's it called, it has been very much of a fad for a while, in conditioned response. You have a person. You want to treat them for some problem. Let's say you have a person who is homosexual in an area where homosexuality is considered a medical problem, you've got to cure it. It is something you've got to cure. So, you give him inputs that reinforce his homosexuality along with a shot of apomorphine. Then you flash the negative picture just before the time that he vomits.

ANN: Aversion therapy?

LECTURE 12 ~ *Depressants I*

SASHA: Aversion therapy. Where you try to connect the physical negative with the visual input to try to get a condition. It's been abandoned as being not terribly successful. But nonetheless, it's one of these fads of psychotherapy that has had its time. They also did it with a material called hexafluorodiethyl ether. It's diethyl ether with fluorines out in the outer part of the molecule. The ethyl groups are flouridated. What it causes is convulsion, so you would get the aversion to something when presented with a negative input you wish to have him averse to at the same time he convulses. The idea is, like rats in a Skinner box, responding to something in place of the actual thing; the thought of whatever it is that is to be purged from your system will cause the convulsion, will cause the vomiting. Not terribly successful but still a valiant try.

Okay, what happened, of course, in this particular case he received all the laudits of the community, except the medical community couldn't stand it. And he had uncovered a neat compound. But it was all fraudulent from the point of view of the medical community because there was no precedent for it and it had no established basis. They said, "It's a charlatan's magic trick." And he was finally drummed out. He had to leave the pharmacy business. He left his entire community. He finally went, as did Daly, into ammunitions and used his imagination for developing better firearms and better powder.

Physicians are no different than the pharmacists, are no different from the educators: they are in a system that works, in their eyes. They don't want innovation because it's change. Something is suddenly rejected as being unproven; well, look at the whole argument of unproven. Sure, the use of morphine as an analgesic is unproven. You have three freaked-out hippie friends of some weird pharmacist somewhere who all turned on and went to sleep and vomited into the basement somewhere. That doesn't prove that it has medical utility. Look how long it took for hypnotism. We mentioned hypnotism as a factor in the medical community. How long it took for that, once it was introduced, to come into actual usage. It was a half a century and even now you'll find people who will say, "Oh, hypnotism . . . I guess it has its role, but it's not really the practice of medicine." Look at acupuncture. I mean, good lordy!

One of the first times I was in Paris back in the early '50s, I came across a bookstore right outside of Sorbonne that was filled from floor to ceiling—it was a high ceiling, about a 12-foot ceiling—completely filled with books on acupuncture. Every single book in there was on acupuncture. They were in Greek and they were in Latin and they were in French and they were in Chinese. They were in all languages, but virtually none in English. Acupuncture was not a part of medical practice. You couldn't explain it, hence it didn't work, and hence it was not valid. How many people know physicians who will say today that acupuncture is not valid? I certainly know them. And yet it's demonstrated it works.

STUDENT: It's like chiropractic.

SASHA: Chiropractic, absolutely. So it usually takes an overflow of several generations to bring these changes in. They may be valid, but they are not proven. You cannot explain why they work. You cannot explain how they work and hence they're not acceptable.

Yes?

STUDENT: There are a few different theories.

SASHA: There are different theories of operation; nerve interchange and intersections and all this. They're not acceptable as they do not fit through our narrow filter of scientific inquiry. You will find a strange reluctance to approach the explanation of it because it doesn't fit an explanation, doesn't have a ready explanation. And it really is not completely valid anyway. So why put your energies out and try to explain something like that? It's a strange category.

STUDENT: At some point in the '60s some Korean doctor, I read this somewhere but I don't know if it was ever authenticated, but he found some little ... some unexplained and previously unidentified little fibers or something like that.

SASHA: Oh, there are connections in the body. There's no question about it. I don't think anyone has ever avoided the experience that'll come at strange times, when you're sitting and for some reason you have a sharp sensation hitting you at one point and something elsewhere twitches. Sure, you have a bunch of connections in the brain and everything, if you follow the path long enough, it's connected from somewhere to somewhere. So

LECTURE 12 ~ *Depressants I*

I'm sure those connections are there, maybe not in tangible red strings that go through the muscle, as sometimes you'll find drawn in the acupuncture books. Also a very, very major component, as mentioned when we discussed Mesmer, is the belief that it works. And with that belief, amazing things can work. If you have a pain and someone in the act of pointing at you can cure your pain, that's a pretty impressive thing to explain in any biological terms. I mean, the method is not easily explained. And yet, if you have this overwhelming belief—you've seen it happen, you know what's going to happen, and it does happen—the placebo effect is very real, and it depends on your personal state of mind as well.

Anyway, it was a long time before morphine came into general usage. Same thing with all the anesthetics. We'll discuss some more of those. Okay, we've mentioned about heroin. Heroin—I'll just make a few changes on this molecule as I go—is nothing but diacetylated morphine. Take morphine, you make a diacetyl derivative of it, and the compound is called heroin. There are two acetate groups. The material is a white salt, a white crystalline salt, hydrochloride. Again, I didn't bring any for show-and-tell. It is preferred as a freebase. Morphine as a freebase has some water solubility. Heroin's freebase has very little. It's usually used as a salt. It is preferred, I mentioned, as it's slightly more potent, probably two or three times more potent. And it does not have that tingling aspect that comes with a morphine injection.

Morphine is still the mainstay of emergency medical kits. Injectable morphine sulfate. I forget what they call the little one sterile needle, one dose injecting thing that you hold in your hand and warm up. It had a name.[8] They used it in World War II as a major lifesaver. They're still available in kits you can self-administer. Morphine is still a miracle drug. It is still one of the greatest boons to man. It will allow a person to withstand painful agony for a period of a few hours. As I mentioned, I think in an earlier lecture, you are not numbed to the pain. You're not free from the pain; it's not like a local anesthetic that makes the whole tissue numb. The morphine causes an indifference to the pain. This is part of the same idea of the general indifference—it doesn't matter, the pain is there, the pain is severe, but the pain does not distress you. It does not agonize you for the period of the

---

[8] A syrette.

morphine's effects. Interestingly enough, probably the reason heroin is more active is that it hydrolyzes to morphine in the body. Heroin is probably more effective, because the acetate groups keep that big phenolic OH group from being ionic, and it will go directly into the brain.

Yes?

STUDENT: Is it possible that if you are in enough pain that you can put, let's say, morphine in you and you won't feel the effects of it?

SASHA: Yes, you can have pain intractable to morphine.

STUDENT: Because I think that's . . . my father was on the army base and they gave him morphine. They did something like they put an X on your forehead so that they'll know that you were given it, so they won't give it to you again and you will die. But it didn't have any effect . . .

SASHA: Many times internal pains—this is one of the problems you have with instances of certain cancers in their terminal stage—are extremely painful and some of them do not respond that well to morphine. And hence the search for some way of medicating. It's one of the arguments being brought forth with heroin. Is it not possible that heroin might work in certain areas where morphine doesn't? It might be. Heroin has a little bit more impact to it primarily, as I mentioned, because in the body heroin goes zango, back to this OH group very quickly. Half-life—remember this discussion of half-life? In fact, I think half-life is in this handout. I did finally write it out.

The half-life of heroin in the body is about three minutes, one of the shortest half-lives known. Hence, heroin can't act over the course of three or four hours—which it does—because it doesn't exist for more than a few minutes. But for those few minutes, the diacetate goes right into the brain and forms active metabolites. Morphine itself—just to diverge a little bit into chemistry—morphine has in its own structure, it's both an acid and a base. It's a base because it has a big basic nitrogen up there [gesturing] and it forms sulfate salt, hydrochloride salts normally. But it's also an acid because as a phenol, an acidic function in the molecule and being an acid as well as a base, it is its own salt in a sense, and being its own salt, it's ionic, and being ionic it's difficult to get into the brain. As a diacetate with its two acetyl

LECTURE 12 ~ *Depressants I*

groups, it goes in much more readily. Hence, it's faster acting and somewhat more potent. And then people say, "We have drugs that are just as good," I think the drug has value and I think in a sense it's a shame that's blocked for purely legal and bookkeeping reasons. But much of our drug laws are there for bookkeeping and legal reasons. Not to instruct the medical research. Yes?

STUDENT: You spoke earlier about the widespread usage in the United States as compared to Great Britain. One of the reasons I have heard why is that morphine was widely used in the Civil War.

SASHA: Morphine was extremely widely used. We had one hundred thousand addicts of morphine at the time of the turn of the century. And I mentioned also, it was used for a so-called women's disease. It was used at the end of the nineteenth century by more women than men. It was suggested for a thousand women's complaints. And it was very readily used over-the-counter. There was nothing but over-the-counter medicine; there was no prescription medicine at the time. It was used for treatment

Some pathways of opiod metabolism; red text indicates prescrption drugs.

of menstrual cramps, used for treatment of being down, being depressed. The art of being depressed and being moody was considered a woman's art at that time very keenly, and hence it was prescribed for women's use.

ANN: By the way, was it Eugene O'Neill's *Long Day's Journey* . . .

SASHA: Right. O'Neill's mother was addicted to morphine in this use. It was very prevalent in this country, much more than in other countries. China would buy it for its volume of use. The Arab countries will buy it for its volume of use. Remember, in the Muslim countries, alcohol is forbidden. Morphine is not. Opium is not. And so, since alcohol could not be used, opium became the counterpart component of usage. So, I think these large cultures, primarily the Muslim culture, the Chinese culture, and the American culture really leaves . . .

ANN: Do they use opium in the Arab countries now?

SASHA: Muslim countries, yes.

ANN: Really? Because they practically execute you just for bringing some marijuana.

SASHA: Hash is used. It may not be legal.

But that doesn't say what the culture of it is. I mean, heroin is illegal in this country. How many hundreds of thousands of people use heroin? It's very easy to look at the law and say, "That's the definition of what the culture is." But the whole purpose of this course is to indicate that the law is not what the culture is. That's what the culture should be from the point of view of the people who dictate what the culture should be. But it's not what it is. Heroin is broadly used in this country.

Okay. This molecule has been a target, and there's a current term that's called "designer drugs," where you modify structures to achieve an end that might not be covered by legal or by procedural restraints. But the whole art of medicinal chemistry, pharmaceutical chemistry, has been the art of designer drugs for decades, ever since it has existed. Here's a good example: you have this molecule and it has two properties. It controls pain centrally. It causes a central analgesia. And it causes the loss of the agony of pain, this sort of retreat and this sort of dreamlike, satisfying escape from problems.

LECTURE 12 ~ *Depressants I*

Two very nice properties in their own way, but medically only one of them is acceptable. The acceptable one is the treatment of pain, and it's not acceptable to have a person retreat into some sort of a satisfied, personal inside world. I mean, that's what we want to get out of the molecule in medical research.

So, they banged and whacked away and hammered, nailed, and caulked, and they did millions of things to this molecule, trying to separate one property from the other. And this led to a number of changes. How many people have this handout? How many don't have it? One, two, three, four, five, six. I gave a few back in that quadrant. Pass them around. If you want to take notes on this aspect, just for the color of it. Ciprenorphine is a whole family of drugs that had been made from the third most plentiful alkaloid in opium. A little bit of chemistry. [Drawing on the chalkboard.] Allow me this time. Opium has morphine—number one alkaloid. Ninety percent of the alkaloid content of opium is morphine. The next major alkaloid—yes?

STUDENT: It says a CH2 and N, on the top.

SASHA: CH2, yes. My angle of CH2 is badly drawn. This is CH2, CH here. Okay, morphine is a principal alkaloid. The second most present alkaloid is a simple ether which, by the way, is less active intrinsically, but will go to the CNS more readily. It is called codeine. Morphine and codeine are the principal alkaloids. A third alkaloid present is one that is of great interest pharmacologically. It has no OH group. It has two double bonds over here. [Referring to drawing on board.] It's called thebaine, and it is the starting material for many, many of these synthetic, so-called opium derivatives.

A moment aside, there are two terms that have never really been defined legally or pharmacologically, but they are used. And I think I can give you a pretty good feeling for their use. One is called "opium derivative." Opium derivative is a thing that somewhere had its origin in the poppy. It could be opium, it could be morphine, it could be codeine, or it could be anything built up from thebaine. I've given one example here, ciprenorphine, these compounds are called "oripavines." Etorphine is another commonly used drug. Extremely potent. These are made by performing a Diels-Alder reaction, generating a whole new ring system.

These are very potent. They are amongst those that are used in bringing down wild animals, long-range dart narcotics, knocking the animal down

THE NATURE OF DRUGS

without having the risk of respiratory paralysis. They are primarily intensely rapid-acting narcotics. They are used some in man; the acetate derivatives are very much like heroin. All of these are used, by the way, one way or the other, as narcotics.

The second term is "opiates," a term again, not accurately defined, but it is a general term meaning "compounds that do not come necessarily from the opium plant but have actions akin to those in the opium plant." So your things that are totally synthetic are often called "opiates." Things that are maybe totally synthetic, but have stemmed from a material that occurs in nature are called "opium derivatives." Get rid of this [in reference to his drawing on the chalkboard].

One of the first people to take the molecule and bang away, taking off different portions of it to try to find where this central analgesia resides, one of the first things to go was this oxygen ring, not necessary for the analgesic properties. On the second line of the handout you have the morphinans, which is the same molecular structure but without that oxygen. Racemorphan. Dextromethorphan. How many people have heard of dextromethorphan?

STUDENT: Cough syrup?

SASHA: Cough syrup. Yep, Romular is its common name. A good example of a compound that has an interesting, checkered history. I remember back about 15 years ago, 20 years ago, Romular was available as Romular in the drugstore. And although one tablet pretty effectively kills a cough reflex, about 10 or 12 tablets puts you in a very strange place. It's sort of a deathlike place; it's not a nice one. It's sort of a morbid, death, anxiety kind of a spot. It's not a particularly good place, but it's a different place. People sometimes don't look for good places, they look for different places. The drive to the altered state is quite a motivation. Anyway, the word passed around, this is back in the House in 1960s and Romular was having quite a run, so pretty soon they decided this had been enough, something was amiss with it. They withdrew it. It was withdrawn from availability over-the-counter entirely and was locked into a prescription status for a while. Then after some time it sort of sneaked back in and I was quite amazed. A couple of years ago I saw one of these sample handouts—I live out in the

LECTURE 12 ~ *Depressants I*

country and they have mailboxes down the line, and they pop things in the mailbox, "For the Occupant" type of thing—and there was a little package of Vicks 44. How many people have used Vicks 44? Got an orange, goopy yuck kind of flavor. It's hard to get a bottle of it down.

You use a teaspoon. It says use a teaspoon, not more than once or twice a day or something. But the bottle was Romular right back again, but in an orange syrup you can't take down because the syrup will make you urp. But it's there and you still find that strange, strange place. And I really had a morbid feeling, a cynical feeling I should say, about it being put in every mailbox, because that means the kids are going to go down hauling the sweet syrup out of every mailbox and they're going to have some medical problems as an outgrowth. I have the same feeling when I find Ex-Lax made up in chocolate and being given as handouts in the mailboxes; kids are going to want that chocolate, they're going to get themselves into medical problems because of the Ex-Lax.

Do people know what's in Ex-Lax? It's kind of a neat thing. It's not commonly known. Ever hear of the idea of litmus paper, acid-base titrimetry? You know, red is acid, blue is basic; sort of plugs into the mind. One of the major indicators that's used in organic chemistry is a material that's white and it turns red in base, it's called "phenolphthalein." I think that's kind of a familiar sound. Phenolphthalein is the active component of Ex-Lax. For some reason—this is back in the grandfather days of drug regulation—for some reason the yellow phenolphthalein, which is the impure phenolphthalein is used. They don't know what the impurities are, they just know that the next to the last stage of purification is two or three times as good a laxative as the pure stuff. And so actually Ex-Lax is impure phenolphthalein. Yellow phenolphthalein. Okay.

ANN: Are people still using Ex-Lax these days?

SASHA: Is Ex-Lax still commonly used or is it available? People have heard of it?

STUDENT: Yeah.

SASHA: It's a laxative.

THE NATURE OF DRUGS

STUDENT: I have a question. Does it really taste like chocolate?

SASHA: Well, they add chocolate flavor to it. [Laughter.] Yep.

STUDENT: Because I remember a lot of people in high school used to sneak that to people and make cookies with that—

STUDENT: Is it a coating?

SASHA: No, it's put into it as a flavor.

Okay, back to analgesics. [Drawing on chalkboard.] Having got rid of that portion of the molecule, it was found that this whole portion of the molecule could be gotten rid of, and what is left is a tricyclic compound in the general family of the benzomorphans. Here are the so-called mild analgesics, Talwin—pentazocine. I don't know really what they mean by mild and strong or hard and weak in drugs, but they are classified as the milder analgesics.

Oh, by the way, someone asked about Darvon and I gave a harangue about it being not particularly effective. Who was the person who asked about it? Here's an actual article that tells just how ineffective Darvon is. They recommend aspirin in place of it if you have a little bit of a pain. It is not a very complimentary review, but it's documented.

STUDENT: I was in England and the British doctors use Darvon.

SASHA: Didn't do much?

STUDENT: Well, if you take about five or six of them.

SASHA: Okay. But as prescribed it's not particularly effective. Another molecule to take apart. [Referring to chalkboard.] By dropping out this portion of the molecule you come down to strictly a two-ring system. And that two-ring system is, in essence, what's called a "phenylpiperidine." And this phenylpiperidine system itself is the nucleus of a lot of the drugs that we are currently using. Demerol which we've mentioned, is an ester on this phenylpiperidine ring system, this phenyl ring and this piperidine ring is . . . This phenyl ring and this piperidine ring is apparent morphine if you can sort of see that . . . [drawing]. Let it pop out into a six-membered ring. The position opposite to nitrogen is the position where the phenyl on the piperidine is

LECTURE 12 ~ *Depressants I*

occurring. So, Demerol really had the carbon skeleton of morphine. It is not a terribly potent compound, but it is one of the most favorite now used in medicine. And interestingly, in the medical community, it's one of the major ones that medical professionals, physicians, nurses, other medical professionals are dependent upon because of easy availability.

STUDENT: Fentanyl?

SASHA: No, Demerol, meperidine. Fentanyl is catching up, but Demerol, probably because of its availability, is still one of the major ones. A very interesting outgrowth of this particular one is the chemical requirement of this so-called ester group. That ester group can be as a little COO, reading out that way. It can be replaced with anything that contains carbon and oxygen. If you put it the other way, you'll have material that in essence is a propionate ester of a piperidine alcohol instead of a piperidine carboxylic acid with an ester group ethyl. It's the other way about. And this material actually had gotten on the street. Its common name now is MPPP. It had gotten on the street in Los Angeles about 15, 20 years ago, and it was called "RE." It was the trade name, sold as as heroin. It was called "RE heroin." And "RE," once you stumbled on the code was "Reverse Ester." They had no idea what ester meant. But they knew it was called "RE heroin" or "RE smack." This reverse ester appeared, then it disappeared. And then it reemerged in the San Jose area about four or five years ago with tragic consequences, called MPPP, which is methyl phenylpiperidyl propionate. I'll draw it in here. I'll get into fentanyl next hour. But this MPP is being made from the alcohol, the alcohol itself, if you treat it too vigorously in trying to make your reverse ester that can be sold as heroin, dehydrates. You end up with a compound that contains no oxygen at all. And this compound is called MPTP—methyl phenyltetrahydropyridine. And this is a compound that has received a great deal of notoriety in its criminal ability, its tragic ability to cause parkinsonism, cause the symptoms of parkinsonism in humans.

This came out I think—some of you may have seen the TV thing, *The Frozen Addict*. It's very real. I mean that's not antidrug propaganda. That's a very real statement. I know the people who have covered it, and I've seen a couple of people who have been involved in it. MPTP, in humans and in primates, no lower animal, it has to be a monkey or a human, will in essence goof up the substantia nigra, the area in the callosum where you actually have changes over time that are associated with parkinsonism. Parkinsonism—an inability to move your muscles in a controlled way, including the muscles of the mouth, the talking, the throat, the arms, the legs, a staggering situation. It can be relieved with certain medication. The medication will relieve the people who have gotten into this. But this is a permanent input. And it's a consequence of an illicitly manufactured drug they manufactured badly. The person who had manufactured it is known. He was, at the time, not selling anything that was illegal. They knew who he was, he moved to Texas. A rather interesting twist is that he happens to have parkinsonism. So, he is a victim of his own mistakes. But nonetheless, as a bright side of the picture, it has become a tool for the study of parkinsonism because there had never been one before. So, in essence, it allows an experimental entry into a serious medical syndrome. But again, as I say, it's only in humans or monkey and no lower animals.

Yes?

STUDENT: Do you know if much of that is still on the street these days?

SASHA: I haven't seen any record of it in the last year. The hour is over. I never got to where I wanted to get. I want to get into fentanyl and that whole argument next time.

LECTURE 13

March 12, 1987

Depressants II

SASHA: What sounds like a reasonable time to wait for the lecture? Some people tear out of their classes at the hour and make it over here out of breath at five after, so I don't want to get too far in. What seems like a reasonable time to wait? I'll wait until no one comes in the door for a minute. [Laughter.]

More show-and-tell. I found my bag that I had not brought in last Tuesday with the opiates. It was on the back seat in the car. So I brought in morphine, codeine, and two forms of opium: an opium gum and black opium. I've also brought in thebaine, which is the third alkaloid, the principal alkaloid of the opium. Morphine by far is the major alkaloid, then codeine and thebaine. Thebaine is of importance because it is the alkaloid that is used synthetically in making all the opium derivatives.

Just in passing, thebaine is an interesting material. There is a variant of poppy that is known and has been proposed to be raised in this country as a source of thebaine. It contains very little morphine, very little codeine, but produces large amounts of thebaine. The proposal was made to raise it in this country as a source of thebaine so you could have it as a source of the opium derivatives that are major narcotics. Everyone, I think, has heard of Dilaudid and oxycodone and oxymorphone and dihydromorphinone, on and on and on. There are literally dozens of commercial opium derivatives, and all of them are made from thebaine. So the idea was to raise thebaine in this country. And since thebaine is from this poppy, it would be under the control of the DEA and the FDA; they made requests and they went back and forth in hearings for a number of years and finally decided no. It's just too much risk. If it were raised, it might go afield and literally become a field. It might be raised without adequate control, and someone could get at it and convert thebaine into something terrible, and it's not worth the

THE NATURE OF DRUGS

risk, which makes no sense, but it takes away one more possible risk in the area of narcotics.

I also brought in the meperidine, which is probably the principal surgical narcotic. Now that's being displaced with fentanyl. But meperidine is a major synthetic narcotic. I mentioned it briefly on Tuesday. And now I want to mess up the entire program. I was going to finish up a little bit on the narcotics that I didn't have the time to do on Tuesday, and then get into sedatives: barbiturates, tranquilizers, and then on next Tuesday cover alcohol. I think I'm going to do the alcohol today. So I'm going to transfer the sedative hypnotics and the tranquilizers and the barbiturates to next Tuesday. I'm going to reverse the order of those two lectures.

Oh, also show-and-tell. I brought in, no stranger to most of you, a bottle of alcohol. It's just to give you the feeling it doesn't have to come in a beer bottle with a twist cap or in a booze bottle. It is a chemical that's used in the laboratory, along with its use as a drug.

Okay, I want to spend a little bit more time on the subject that I finished up on in the last hour. I had a question afterwards, and then a couple of misgivings. There is a lot of concern, currently a lot of publicity, a lot of newspaper coverage on the area of parkinsonism and connecting it to drug use, drug misuse, and what have you. And I touched lightly on how it was tied in. And I think it's of enough importance because I heard just a couple days ago from another source, "Is it true that using illegal drugs can destroy your nervous system, give you parkinsonism?" It's a connection that's been easily made. It's a facile connection. You can very easily take the whole chapter of illegal drugs, take the chapter of neurological damage and parkinsonism and teratogenicity and malformed offspring, and find enough of an overlap between the two. You can zero in on the overlap and say, "Look, this shows the problems associated with drugs." It's not a fair thing to do and that's why I want to spend the time just to try to open it up a little bit more. I want to go back to where I left off and give it a little bit more detail. I'm going use a dirty picture again. Let's see how I can best do this. This is a benzene ring. Methyl group. I'm going to go right back to the starting point in chemistry.

This is meperidine. I'll draw it out only because I want to look at this one little portion. This very similar compound, similar chemically, similar pharmacologically, is MPTP. I'm going to use this, just don't worry about

LECTURE 13 ~ *Depressants II*

the details of it, just get the feel of it. These compounds at first glance are identical. They're both white solids. They both cause a heroinlike narcotic loss of pain, loss of concern, loss of anxiety, and eventually, an overdose will cause breathing problems and will cause a narcosis, a coma narcosis. They are identical in their structures, except for this one little thing. In one case a grouping has gone this way, and in other case a grouping has been reversed. This, as I mentioned, has been called a reversed ester. Neither compound is dangerous, any more dangerous than heroin or morphine or meperidine or fentanyl or any other narcotic. To emphasize again, narcotics are not particularly damaging to the body. Despite what they say, you know, "He wasted his life being hooked on morphine, opium, heroin, whatever . . ." The hazards that come from it are sociological hazards. You're into legal problems because of it. You have nutritional problems. You have hepatitis problems because of lifestyles that are changed because of it. You definitely have an increasing commitment and attention to the use of the drug because of the rewards, and you have this psychological investment in it. These are all potentially very negative but it doesn't make lumps grow on the liver. That's my point. It is not physically particularly damaging as opposed to alcohol, as opposed to tobacco, those two come to mind. I was going to say caffeine. Caffeine does not have any particular damage. Marijuana I don't think has any particular damage.

Yes?

STUDENT: You can have a heart attack.

SASHA: Oh, yes. Oh, absolutely. You can have cardiovascular accidents. You can have cerebrovascular accidents. You can have heart attacks in the sense of having a strain on the heart that the plumbing can't handle.

There is the acute overdose. You can be killed by an overdose. You can be killed by a drug exposure. This is not what I mean by damage to the body. I mean damage to the body as chronic drug use erodes, damages progressively, some part of the body. That just isn't there with the narcotics. You can have a clean bill of health. You can overdose on narcotics and stop breathing and be dead. There's no argument on that. Yes?

STUDENT: The risk of hepatitis is just from shooting it?

THE NATURE OF DRUGS

Sasha: Primarily, yes. Entering the body.

Student: Another way, you wouldn't have that effect?

Sasha: Orally? Nope, no hepatitis argument. Strictly, hepatitis is transferred. Occasionally you'll find some bizarre exception, but in general it is transferred by blood contact, one way or another, by the use of the needle in the drug community. Okay, I'll get back to this. I'm off on a tangent again, something I just read in the paper this morning. When you have a controversial point and you have two opposite sides, two sides that are validly in conflict, there is no way of really resolving it.

What I read in this morning's paper is the argument of making rubber prophylactics, condoms or "condos," if you read the comic strip, available to male prisoners in prison. It's a beautiful example. There is a lot of AIDS in prison. There's a lot of homosexual congress in prison. The use of rubber prophylactics would minimize the transfer of AIDS. But making rubber prophylactics available would say we are condoning or acknowledging the fact there is homosexual behavior going on in prisons. And there shouldn't be. Well, of course there shouldn't be. It's not allowed. Of course it's not allowed. It really shouldn't occur. But it does occur. Where do you sit and balance a thing such as that? I don't know. I don't know a fair way of approaching the answer. Basically in the heart, honesty would probably reveal the answer. Honestly, of course it is going on. Of course, if it is, they should be made available. There's no reason to enforce any behavior patterns. But I think honesty should probably be the thing to tilt the balance in this kind of thing. Yes?

Student: Same thing happening with the needles and the distribution of free needles.

Sasha: That's exactly what I'm leaning toward. Should free sterile needles be made available to drug addicts? It's exactly the same question. How many people think clean, sterile needles should be made available to drug addicts? How many people think they should not be? One, two. Okay, basically, I'll take your side first. What's the objection to it?

Student: How does a drug addict become a drug addict?

Sasha: That's a very interesting question. Does it have a bearing?

LECTURE 13 ~ *Depressants II*

Student: Well, the fact that if you make available, let's say for instance, the effect in Britain. If you have teenagers who look towards narcotics as an escape, you know, if you make available to them this option, this opportunity to go that route, then they might be persuaded—

Sasha: You will have more people using than you would have otherwise because of availability, because of lack of expense.

Student: You might have the same amount of drug addicts, but you won't decrease the amount of disease because of, you know, the mentality of the drug addict, using and sharing needles is a phenomenon besides the drug.

Sasha: Okay, fair enough, then I will accept that. But then I want to go over to this side. What about the educational process—getting to the people that this is a means of transferring disease? I think that would probably be the most valuable single thing to do. Australia has made needles available. I don't know if they're free or not. They're certainly not very expensive. I don't know what the effect has been in the hepatitis argument. But the AIDS has been apparently—

Student: What about someone who's not using needles, but is experimenting around with opiates and then all of a sudden decides, "Well, I can get a free needle down . . ." I don't use needles. Now I'm using a needle.

Sasha: That's exactly how the argument is presented. I don't know the answer.

Student: So once he gets hooked on the needle, the needle is what he's hooked on. He's not hooked on drugs. The drug can be secondary anyway.

Sasha: I don't want to single you out because you're in the minority. But I want to address that point of view because that is the argument that is made: "The availability will get people into patterns that are more destructive." I don't know. If a person is going to get into a drug-using pattern, I have heard the argument made, why not have at least some availability of information as to the purity and the correctness of the drug. Are you going to stop drug usage by stopping drug availability? It's not worked out. Maybe it will. Yes?

THE NATURE OF DRUGS

STUDENT: Generally, when somebody starts using a drug, though, they're doing it with another person who introduces them to it. Therefore, they're probably going to get a needle from them. They're not as likely in their first use to go—

SASHA: Go to the drug store and say, "Yeah, I want a free needle. I'm gonna go home and by myself shoot up." Certainly, there is that social interaction. I'll come back to that in a moment.

Yes?

STUDENT: I agree with your point entirely. I don't think very many people are going to start; if they never used a needle before, it is a scary experience. Along that line, I also question whether or not they're going to be that effective in actually stopping AIDS because I think that drug users especially and methamphetamine injectors, it's been a real ritual kind of thing to share needles—

SASHA: Oh, absolutely.

STUDENT: I don't think it's going to change over, you know, or even over a period of time, if it's going to change or not.

SASHA: Mmm hmm. Yes.

STUDENT: I don't think that's true. But I think that's one way to make it effective because some people would not want to show themselves in public to get the needle. They wouldn't want to be known. It's a stigma.

SASHA: Like a 15-year-old going to the drugstore and asking the female clerk for a rubber. Yeah. Okay, the argument was "free and easily available," but that is a good point. Because you are, in essence, showing your hand, you're branding yourself as being a person oriented that way. I don't know the answer, but yes.

STUDENT: If you experience risk, are you going to change your drug habit or risk possibly getting AIDS?

SASHA: Let's go to hepatitis. I think hepatitis is much more severe. No, AIDS is more severe. Hepatitis has more common association with needle

LECTURE 13 ~ *Depressants II*

usage. If you were into using needles, and you had a very supporting habit, or you had the choice of risking hepatitis, I don't know. I'm not in that role and I don't know how to judge that.

STUDENT: I've found the people that I work with . . . I work in drug rehabilitation. A lot of them, their mentality is "yeah, people get hepatitis, but it's really not a concern." I find, you know, it's not an issue with them. AIDS is becoming more so. But generally, the attitude is "it won't happen to me."

STUDENT: Let's go back to the fundamental point. Where does the needle begin? When does it start? Now we're talking about after someone's already taking it by the needle, but the idea is that we have morals against the needle. The needle is bad. Taking drugs with a needle is not good. You have these pros and cons. If you start to make needles available, then you start to break down the moral walls you have against the social point of view about needles. And if you can keep hammering out that moral wall before you know it, you're going to have the needle in your hand. And that's where—

SASHA: I like your argument. I'm going to hold my vote in advance because I like that argument. I don't know where it's gonna go, but valid, yes.

STUDENT: We have a sort of moral wall against needles that may not be as effective when the person finds out that a needle isn't as scary and as dangerous as they thought. People could be informed as to the danger of needles, et cetera, et cetera, instead of just saying they're bad, they're ugly, drug users use them. They might need a lot more information, a lot more education about it.

SASHA: Legislation geared toward morality, ethics, and education has been notably a failure. Yeah, I saw a hand over here. Yeah?

STUDENT: I think along those lines, better information. I think the posters that are all around about cleaning your needles with Clorox are a pretty good idea. You're not saying go ahead and use needles. You're not saying don't use needles. You're saying, if you use that, if you do it, use a clean one.

THE NATURE OF DRUGS

Sasha: Mmm-hmm. That's kind of the direction has been taken on premarital sex. We're not condoning it. We're not suggesting it. But if you do it, be attentive to what happens and what the consequences are.

Student: If you look at the medical society, needles are good. That's the first thing they do is to shoot you up right away. I mean that's solving a very severe problem. If you have a motorcycle accident or something, they have nothing they can do to treat it right way other than to set a broken leg, but they're just gonna give you a shot of Demerol.

Sasha: The needle is immediately at hand in case—

Student: Immediately at hand in a medical context. As soon as it is taken out of that context, it has a negative connotation.

Sasha: I tend to support your point of view. I think a lot of the people in the medical community who abuse narcotics, who, not abuse—I don't like the term—who have spun narcotics into their lifestyle, the professionals, have probably been aided and abetted by the fact that they're casually using needles all the time and it's an available clean thing. More and more thoughts on this kind of thing. I kind of like the argument. I'm not going to be quite as monolithic on the availability, because it's a good point. Any other thoughts? Yes?

Student: You know, being hooked on a needle, per se. There is a valid point. You can just use the needle and it doesn't matter what drug goes in your system. I've seen people put alcohol into their system just because they get to use the needle.

Sasha: Yeah, the whole procedure involved, the playing with the needle, the sharing of the needle.

Student: You watch the blood go in—

Sasha: Mmm-hmm. And play with it a little bit. Yeah, I've seen that.

Student: I've seen people just sit there and do what they call boost tag, and continue to draw the blood out—

Sasha: The needle talking was the term I heard. Yeah.

LECTURE 13 ~ Depressants II

STUDENT: It was the terminology.

SASHA: Mmm-hmm. I don't know if I had mentioned this story. I had one case where a person had gotten into a heavy experience with heroin and the person who they called—oh, what was the circumstance? I can't remember the circumstances. I think it was Dr. Schoenfeld. It was suggested that just warm milk could quiet this or could quiet that. I forget the ramifications, but somehow the suggestion got warm milk associated with an overdose of narcotics. And the person was so oriented to the needle that the person injected the warm milk.

On morality, behavior, and what-all, on the radio just a couple of days ago coming across the bridge, I heard the argument that apparently there's a law on the way now to outlaw the sale of typewriter correction fluid to minors, like people under 18, because they're using it to breathe the fumes for intoxication. I don't know if that is a success. One way or the other they are breathing cleaning fluids. Glue, airplane cement.

Has that brought that wave of drug abuse to a halt? I don't know anybody who breathes—has anyone here gotten off on breathing airplane glue? I have not. I'd kinda love to try and see what happens.

STUDENT: I was working in an office and was using rubber cement all the time. And the fumes were there all day long. Occasionally I would get a little spacey.

SASHA: Lightheaded. Yeah, there are a lot of solvents that will achieve that, and good heavens, they still use nitrous oxide for foam cans to spray, what do you call the stuff, whipped cream? And nitrous oxide is certainly used.

STUDENT: We could buy those as little kids.

SASHA: Yeah, they sell it in *High Times* for powering your spray machine. I'm not going to get into nitrous oxide—another lecture.

I want to go a little bit more. I want to get that off my mind. I just don't know the answer to it. And it was a little more lopsided than I thought. It was about 10 to 1 in the free needle thing, and I voted in that direction. I like your arguments, and I don't know the answer.

But bear in mind these are Solomon-like questions that don't have easy answers. I mean, every time you turn around you'll see people carrying picket signs objecting to abortions, people carrying picket signs saying "freedom of choice." I mean, it's not a clean, clear answer. You may have very strong opinions, and I hope sincerely that you can grow up with a lot of strong opinions. I have some strong opinions. But there are others who have different opinions and you've got to be able to see their point of view to have your opinion be of your own choosing. That's all. Don't change your opinion necessarily. But make your opinion from information, from both points of view.

Okay, I want to go back into the comparison of the two. Meperidine and MPPP are comparable in potency, comparable in effectiveness, and both reasonably free of physical damage, physical risk. [Facing chalkboard.] The—I'll erase this for a moment. In the course of synthesizing this material ... I'll just draw this in here for the illustration. That adds a group onto this hydroxy group. This material in being driven to this compound can go to MTPP. MP... God, I'm all messed up here. MP... tetrahydro... this is the TP. Methylphenylproprionyloxypiperidine: MPPP. Methylphenyltetrahydropyridine: MPTP.

This is the agent which is not a narcotic, which does not produce a high. It produces a funny feeling. It burns on injection. It produces a weird, almost dizzying feeling when it's injected, but it does not produce a narcotic high or narcotic low, whatever you would call the euphoric state from it. But it does—if used sufficiently or in sufficient amounts or in a person who is reasonably sensitive—it does cause the syndrome of parkinsonism. Parkinsonism is a shuffling, lack of motor response; your musculature does not work correctly. This is the agent from a mis-synthesis, aiming toward a meperidine analog. No other drug in any illicit form anywhere, of any kind, is involved with parkinsonism. This is the only place. You have the general thing claimed... illicit drugs produce parkinsonism. Nonsense! There's not an illicit drug over there. None of those are illicit drugs. This is the material—it was not even illicit drug at the time; it has now been made a Schedule I drug—it is a narcotic. But it is not the thing that produces parkinsonism. Parkinsonism comes from a misdirected synthesis.

STUDENT: Can I have the structure of that?

LECTURE 13 ~ *Depressants II*

SASHA: Oh, yes. Certainly. I have here a very strange PT. [Laughter.] I had for a while the argument—I like mnemonics, I was using a visual picture of a couple of Indians—and I tried it, I found myself getting so tongue-tied trying to use the trick that I abandoned that image, and the result was that I messed up the letters. The chemical structures are much simpler. This is the phenyl ring. It is N-methyl, phenyl, tetrahydropyridine. It is caused by dehydration.

STUDENT: So when you heat it up too high?

SASHA: You heat it up; you form that. Right. If you heat it up in the presence of propionic anhydride, to form this, and you heat it too hot, you form that as well. If you don't have the right reagent, or you're in the wrong order, you may make only this. So some of the things on the street market were only this. There was no narcosis from it, but you had a number of people who were permanently damaged from parkinsonism. The treatment of it with the usual treatment for parkinsonism will relieve the symptoms, but upon abandoning the medication, they go back to parkinsonism. Parkinsonism is a disease that is almost exclusively related with old age. If you know people with parkinsonism, as a rule, you know people in their 60s and 70s. It is not a young person disease.

There is an interesting extension to this. Is it valid? Can't say. You won't know until there's been time to see it occur, but, there is a gradual deterioration of what's called the substantia nigra in the brain. There is a dopamine repository, and a gradual deterioration of it—due to who knows what, ozone, environmental things, pesticides, free radicals from the environment—that has been a contributor to people getting parkinsonism. When you drop down to a certain level, you begin showing parkinsonism, and therefore is it possible that the hundreds of people who are exposed to this—I'll get it right, this MPTP—is it possible that hundreds of people exposed to this have made a progressive step toward parkinsonism and will get it at the age of 40 instead of age 60? Not known. A real possibility. You cannot really determine this without looking at the degree of deterioration in the substantia nigra, and this is a destructive thing to do at this moment. Yes?

STUDENT: Can that be a temporary condition?

THE NATURE OF DRUGS

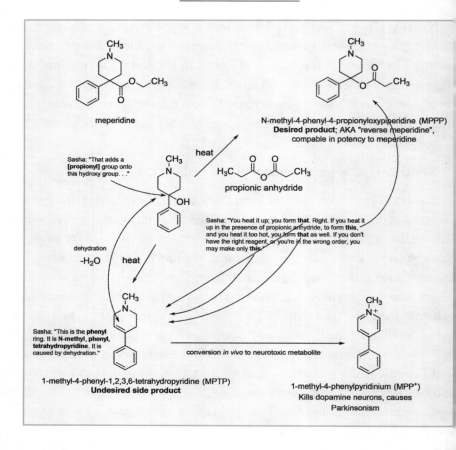

SASHA: No.

STUDENT: Well, I knew someone. He said that half of his face was numb. He went to the doctor and that's what they told him he had, but it was only temporary. It went away after about a month.

SASHA: I'm very suspicious. Parkinsonism is a morphological damage. And it's like nerve damage. It just does not regrow. I'd be very suspicious of that. First of all, it's ipsilateral. Parkinsonism is usually bilateral.

STUDENT: Both hands shake?

SASHA: Yeah, both hands. Both feet shuffle.

STUDENT: Katharine Hepburn...

LECTURE 13 ~ *Depressants II*

SASHA: Katharine Hepburn, good example of parkinsonism. There is a way you can relieve it. There are medications that are dopamine-oriented to relieve it, but they do not cure it. It's cosmetic and it allows a person to walk and to talk to some extent, but he must be able to continue medication, and the medication is in itself somewhat destructive. It's just a nasty, nasty corner.

Okay, I wanted to get to this because I've heard such things about fentanyl, which I want to talk about. I never got to fentanyl yesterday either ... Tuesday. Fentanyl will cause parkinsonism—nonsense. There's not a narcotic known that causes parkinsonism. It is a misdirected synthesis and impurity that does not occur in the pure material.

I should really be a little cautious to say, "It causes parkinsonism." It produces the syndrome of parkinsonism. Parkinsonism is still a spontaneous disease, but on looking at the brain of a person who had exposure to this—there's one example, from an automobile accident, in which a person was in the car and was killed. They ran an autopsy on the brain. Morphologically, pathologically, identical to parkinsonism.

The cells, the pathological cells in the substantia nigra are the same as parkinsonism cells. So it looks like it is somehow getting in there. What has happened on the good side is that now a model is available for testing things in parkinsonism, and you can make animal models that are good displays of parkinsonism. Again, I mentioned, it has to be in primates. But an approach to it they have since the process is intrinsic. I don't know if I used the word before, "intrinsic." It's a good word. I know I put it in one of the handouts. "Intrinsic" means a drug in its own right does what the drug appears to do, as opposed to conversion to an active drug, or as an intermediate or something that has to be metabolized to be active. This material is not intrinsically active. If you went into the brain with this material ... they've done it in experimental monkeys. They put this directly in the brain. It does not lead to this type of motor problem. What it has to do is to go to a totally aromatic system by being dehydrogenated, by being oxygenated, and this conversion occurs in the body. And the conversion is done in part by a material called monoamine oxidase, which I have mentioned briefly, which is the same enzyme that decomposes and destroys many of the neurotransmitters desirably. It's a natural body process of getting rid of neurotransmitters. By inhibiting this enzyme you protect against the damage from MPTP. You minimize the damage from this in experimental animals by giving an enzyme that interferes with the oxidation of it to the active component. But it is not a therapeutic tool.

Yes?

STUDENT: Which one is MPTP? [Referring to drawings on the board.]

SASHA: MPTP is the compound with a double bond. The mono-ene in this compound. In the body it is oxidized to the totally aromatic material, which is the actual destructive agent.

STUDENT: In a show on Channel Nine they showed a company that was making this found out that activity. They did drug studies on rats. It didn't show up. They were making this compound.

SASHA: Oh yeah. And they did not show toxicity to rats.

STUDENT: But then when they started shooting into primates. . . . So they knew about this compound—

SASHA: Oh, the compound was known well before it was commercially available. Aldrich Chemical Company I think sold it for a few dollars for a hundred grams.

STUDENT: I think if people actually were getting this stuff . . . but they couldn't prove it because they were doing it in a rat and wasn't showing anything.

SASHA: It does not show. This effect does not show. This shows the limitation of cross-species studies in animals. I mentioned this some hours ago in the study of teratogenicity, where you're looking at the misformed uterus in an animal to determine if it might cause birth damage to developing children in humans. There are species differences; you cannot just go from one species to another. One of the most toxic materials that's around is a material called dioxin. It is an outgrowth of some of the herbicides that were used in Vietnam, and it is known to be an exceptional carcinogen, exceptionally dangerous, especially in the guinea pig. The rabbit has quite a bit of sensitivity. Fortunately, you might say, in humans it does not appear to have a very high toxicity. The species differences are very real. Here, rats

LECTURE 13 ~ *Depressants II*

mice do not show toxicity. They show this conversion, but it does not lead to the syndrome of parkinsonism. That is a primate or a human thing. You would not have found this with all the animal testing in the world if you had not used a primate. And primates are desirable to use, but primates are exceptionally expensive and are used only with great need.

STUDENT: So what was the new law that was passed, that now any experiment involving primates is restricted?

SASHA: I don't know with primates and the law. It may be. I don't know what the FDA regulations are on that.

STUDENT: Some particular incident occurred.

SASHA: I doubt it because this happened too recently. But it might be. I don't know.

Drugs require very extensive testing for teratogenicity. But there's no reason that a primate should be closer to humans than a mouse is closer to humans in some of these toxicity studies, because primates can have extreme individuality. Humans and humans can have extreme individuality, individual differences. There are groups that are called "fast acetylators" and groups that are called "slow acetylators" in human-to-human populations. There was very much damage from the use of some of the compounds for treating tuberculosis when they were first given to Eskimos. Eskimos were keeling over dead from a normal treatment of a material treating tuberculosis because they happen to be rapid acetylators. Either that or slow—I forget which way it was, but as a group they metabolize things differently than other groups. So even a human testing against a human is not always a valid thing.

You can have, I don't know how valid it is, but certainly they have made studies on, for example, determining the migrational course of people on the basis of the texture of the earwax, strange as it may sound. You'll find people from Asia and the Aleuts and down the Indian groups of the West Coast all have flaky earwax, whereas people from Africa or Europeans have gummy earwax, have gooey earwax. I mean, it's just a different way, person to person. And there are many little markers that have been used genetically to try to trace out these courses of migrational movements in history.

And amongst these are markers in the enzyme system such as things that metabolize drugs, and you'll find that some groups will metabolize drugs dramatically differently from other groups.

MAO is the enzyme. MAO is the monoamine oxidase—it's a good enzyme. You'll find the term in the literature, MAOI, for the monoamine oxidase inhibitor. That is the enzyme that that gets rid of unwanted neurotransmitters.

Hand, yes?

STUDENT: Is MPTP what is called—

SASHA: Reverse ester. Yes. Okay, now this is going to get into a little ... this is as good a place as any to get into some arguments of designer drugs. I'll take this off. I want one more structure up here and then I'll stop drawing structures.

Again, all of these are outgrowths of the study of morphine's breakdown products to try to maintain the activity of a compound as a narcotic, as an analgesic, without the euphoria. This compound is fentanyl.

Extremely potent. A hundred times more potent than meperidine. Relatively easily synthesized. I draw these structures up here because they are probably representative of the type of things that are the major worries of the authorities in the area of designer drugs. Fentanyl is exceptionally potent. One of the first things that started this whole thing going was a material that was called "China White." It was on the street in San Francisco, maybe five, seven years ago. It was a heroinlike material. It was sold as heroin. It was used as heroin. People looked at it—there was no heroin.

People would come in with heroin- or morphine-type overdosages or difficulties, and they would look at the blood and look at the urine—no morphine. Remember, I did mention that heroin has a very short half-life in the body. When you look in the urine for heroin, you will not find heroin—you'll find morphine. Morphine is evidence of heroin usage. It's also evidence of morphine usage. There is no heroin in urine, just as with cocaine, there is almost no cocaine in urine. You find metabolites of cocaine but not cocaine.

Okay, this material ... A material came in called China White. They got hold of a sample. They isolated this material from it—this was done

LECTURE 13 ~ *Depressants II*

back in Washington—and they came up with this new variant of fentanyl, they said it was alpha-methyl-fentanyl. And so you had a big news release about alpha-methyl-fentanyl. *Newsweek* had an article about alpha-methyl-fentanyl, *Chemical & Engineering News* talked about it. "They've got to do something." They start proposing alpha-methyl-fentanyl for emergency scheduling. Then they backed off a little bit and said, "Well, we misinterpreted the NMR," and so forth, and really it wasn't alpha-methyl-fentanyl. It was 3-methylfentanyl. The material is about 20 times more potent. It is of a potency that a few micrograms is an effective dosage. There are analogs in here in which less than a microgram is an effective dosage. You're dealing with hyper, super, super, hyperpotent materials.

Yes?

STUDENT: So if you synthesize different ones, and went from the three to the five, would that be ten times more potent?

SASHA: The five is the same position. Three and five would be the same, but not necessarily. If you get into two position the activity drops off.

STUDENT: I mean, if they can reduce it and add to the three position, it might be very easy to then add to the ring?

SASHA: Oh yeah! I mean, it's true you can modify it and you probably have activity, and you probably have potency. Oh, this is what's scaring the bedickens out of people. Someone over on the third floor of a place in Noe Valley happens to have running tap water and a hot plate in his place, he knows a little chemistry, he'll turn up 4-fluoro-butyl-butoxy-alpha-methyl. And what bothers them is they might have an idea as to where it's going on, but they have not really gotten to what's going on, and they don't know if there are 50 places where it's going on—there probably are. A DuPont chemist was recently arrested. He, in his chemical enthusiasm, made about 20 or 30 grams of para-fluorophenyl fentanyl, and in his naiveness tried to sell it for a lot of money to an undercover agent from the Bureau of Narcotics. That was the end of his professional career as a chemist for DuPont, who quickly disavowed both his employment and any knowledge of it.

There is a fundamental drive that influences many people in this society of ours, and that is making money. And if you could, for example, spend a little time in a laboratory—or I know some people who prefer to even spend four years in college to get the information to spend a little time in a laboratory—to put together a little acrylic acrolein, and phenethylamine [drawing structures on the board] and to the carbonyl, from the carbonyl, go out with aniline in a reductive amination, and hit it with propionyl anhydride. There you go. There you have fentanyl. And they'll put this in there or that in there, make 20 or 30 grams, which is not that much . . . you know, 20 or 30 grams—here's 100 grams of material in that bottle—mix it up with 1,000 times or 10,000 times its weight in lactose or mannose, put it in a shoe box, and get rid of the laboratory. You don't ever have to go near a laboratory again. You have a lifetime income.

Someone has built up a supply of a dozen different analogs. When one becomes illegal, he'll provide another, and he's living a very comfortable life on his paid-for yacht . . . who knows? And every now and then he comes out and sells another $500,000, a million dollars' worth of stuff. How are you going to approach this from the legal point of view? It is a scary thing from the point of view of the law enforcement who says, "We want to stop this." They're invested in stopping the drug trade. But they don't. They don't have a legal handle on the drug. Hence, you have this argument of the Analog Drug Bill that was just recently passed, that says, "Anything that quacks like a duck and walks like a duck and smells like a duck will be a duck" in the point of view of law. Anything that has an action of fentanyl and has a structure that resembles fentanyl should be treated as if it were fentanyl, from the point of view of law.

It is a dangerous law in the sense that it is absolutely without boundaries. It has no restraints, it is an unlimited law, it's an open-ended thing. In it they have named 200-and-some-odd compounds as being drugs and that's it. If you have another one out here we'll discuss it, and if it looks like it's going to be a hazard, we'll put it in the law, but each one is handled separately Here instead is an open, unsigned check. And I don't know what's going to come out of it. It has not been enforced yet. I have heard through the legal grapevine that the first challenge of it is going to come up in Texas very shortly. So we'll see how it's handled in court.

So let's go back to what can be done. I'm not going to go into much chemistry, but I will say for example: If I were to put a chain similar to this

LECTURE 13 ~ *Depressants II*

on this compound. Let me just draw a make-believe compound. [Drawing on board.] I'm going to keep this proprionyl, I'm going to put a methylene, I'm going to put an OH. Not much of a change. I'm going to put that on the bottom of this. Sort of a homolog of ephedrine. Don't worry about the chemistry. This compound that I've drawn is almost 1,000 times more potent than fentanyl. It's in the literature. A 150 nanograms is a full dose that intravenously will knock you out.

A 150 nanograms. Nanograms. Not micrograms. Not milligrams, not grams. Grams, milligrams. How much is a milligram? I put this in a previous lecture. I had a quite a bit of struggle in the handout. I think it's one you just got, or the one I'm just now writing. What I said, "What is a milligram per kilo?" What's a kilo? I suggested that I have a couple pounds of coffee. There's a kilo. You can visualize a couple pounds of unground coffee. But if you're not a laboratory-oriented person, how do you visualize the milligram? Well, how about a coffee bean? A coffee bean is 150 milligrams. You don't visualize 150. It's in a coffee bean somehow. And so what I did on an inspiration, I was trimming my fingernails and looked at a fingernail clipping. It weighed about five milligrams. A milligram would be a fifth of that. Some of the fentanyl derivatives are potent and even lethal at a thousandth of this amount. A fifth of a fingernail clipping in two tons of coffee beans.

A nanogram is miniscule (1,000 nanograms is one microgram; 1,000 micrograms is one milligram). An active amount of that drug is 1/1,000th of the weight of the period after the B of a Susan B. Anthony three cent stamp. You cannot see it, you cannot taste it. And yet that amount in a solution will knock you out. These are things that are being used for drugs. I mentioned fentanyl. Let me give you two precursors—"al" and "lo." Both of these drugs, alfentanil and lofentanil, are used medically along with fentanyl. The "car" fentanyl is not. Carfentanil is something that is used to drop rhinoceri, whatever the plural of it is, in their tracks, if you want to go and do something with an unconscious rhinoceros. You go like that, the rhinoceros falls over and does not affect the breathing too much. The first discovery of carfentanil's potency was in the laboratory of the Janssen laboratories in Belgium when a person was taking a melting point of the material. You take a melting point by putting little solid down in a glass tube, and the glass tube's open at the top. When you get the solid in, you

THE NATURE OF DRUGS

put it in front of a thing, you turn on a heater, the temperature goes up, and pretty soon it melts. And in the course of melting, eight people got affected by the fumes of what was coming out at top of the melting point capillary tube. The guy who was melting passed out, and the seven people who came over to help him with what was going on were affected.

You're dealing with things that have nanogram potencies. Well, they're marvelous for dropping rhinoceroses. But they're also superbly good as narcotics and they will not be spotted in the body. How do you pick up a picogram of stuff coursing through five gallons of blood? You don't. And hence you have people coming into an emergency ward, they look like narcotic overdosages. You don't know what they look like. They're coming in, they can't walk, they can't breathe, there is something very much wrong. Run a blood on this. Send up to the clinic and get a blood and urine to see if you pick up any drug. No drug.

STUDENT: Will Naloxone reverse these symptoms?

SASHA: Yes. That's one of the standard things you do. Shoot them up with Naloxone. Very good. Thank you for bringing that point up. There are drug antagonists that will block narcotic effects regardless of what the narcotic effect is. It's called an antagonist. Naloxone is a good one. There are a number of these materials. Routinely you'll give this if a person might be a narcotic overdose. You can tell by pinpointing pupils. The rule is pupils tend to go to a slightly constricted form, given a narcotic. And if that reverses it, you know you're onto something. But what are you going to do about the legal point of view?

This is a real, scary thing. So the answer is not at hand. The answer I still think must be along the line of: You've got to somehow address the whole drug-use problem rather than writing laws to restrict the availability of chemicals you can't find or can't identify. They had one hell of a time with this 3-methylfentanyl. They finally found a sample. Some poor guy down in Mississippi had published a paper on it about three or four years earlier, how to make it and what it was like. And they descended upon him, "Where are your supplies? Have any graduate students walked off any chemicals lately?" You know, just really trying to find . . . I guess some compounds were made and they looked on the shelf. They were there. Everything's fine. Everything

LECTURE 13 ~ *Depressants II*

was intact. Back over to Janssen: "Dear Janssen laboratories . . ." They got the Belgian contact, a Belgian agent of the DEA in Belgium, to go to Janssen and ask them if they can get reference samples of 3-methylfentanyl.

"Sorry," said Janssen, "We don't have any reference samples. We didn't keep any of those samples. We made them years ago—we didn't keep them." Or, "We never had them." I don't know what they said. The thing is, they did publish on them. And I'm sure industrial companies do have records, but they couldn't find them. There's not been the strongest cooperation with people such as the Janssen pharmaceutical company in Belgium. Companies try to market drugs in this country—it is a perfectly legitimate activity, fentanyl is a drug from Janssen—only to find out, just as they're getting ready to market them, they've been put in Schedule I, because we have heard that these materials are highly hazardous. They have a high abuse potential. And clearly, if they're not marketed in the US, since they have no medical use, and they're put in Schedule I, suddenly Janssen has a lot of politics to get it out of Schedule I and get it looked at, to get permission to be used as a narcotic.

Alfentanil is now available as a narcotic but was placed in Schedule I for about two or three years. So there's this lack of full cooperation, which is almost understandable.

Okay, here's an example. This compound, I drew it up because this is in the literature. It was published by, of all people, the Janssen laboratories, about 20 years ago. In fact, Janssen wrote the review article. That review article is in a very neat little book on drugs affecting the central nervous system. This review article covers, I would say comfortably, 200 drugs. Almost every one of them is more potent than morphine because that's what they're looking at. They're looking at families that are potent, and they publish the animal levels of their activities, things such as this are in there. Once someone is going to make these materials, if he is geared toward making a synthetic narcotic and getting away from an opium source, there's no way you're going to turn him around from making the material available for financial purposes.

There's a little sinister thread on all of this, the idea of chemical warfare, which I will get into later, but I want to touch on it here because about half of the published literature now that is appearing in the area of fentanyl analogs, extremely potent narcotics in this area, is being published in China.

And I wondered, the last time I saw that type of thing was the publishing of quinuclidine chemistry coming out of Russia, and it turned out that it was in a declassified chemical warfare chapter. So I wonder if this may well be an area that's being explored as chemical warfare research in China.

STUDENT: How would they be administered?

SASHA: Aerosols. These things can be achieved by aerosols. A lot of work has gone into developing explosives that in the course of exploding cause a fog, and the particulate fog carries the drug that is being used.

ANN: Are there more people developing this stuff?

SASHA: They haven't shared that information with me. That's a whole chapter I would like to maybe get into some day on the virtues and the disadvantages of security clearance. But that's another chapter. Okay. So this is an area that you can modify these things right and left and maintain activity. A little scary is the use of carfentanil in darts, in the forest. They're using this to control wild animals and bring them in for tagging purposes, and for identification and survey purposes. Not all the darts strike their target. And the darts that don't strike the target are still sitting around in the forest containing probably dozens of active dosages per dart tip. I think if kids on their Sunday outing were to come across these darts, you could have tragedies from it. It's a release of the chemical in the environment in ways that I don't know the consequences of. But carfentanil does not have a medical use. It is used in wild animal control.

That is more or less what I wanted to get into in the last hour. Okay, we have 20 minutes left. I'm not going to be able to handle 20 minutes' worth of alcohol. I'm going to start alcohol and see where it goes. So this is really the beginning of next Tuesday's lecture, which I should have started at the beginning of this hour. This is going to be one messed-up week, but it's already been a messed-up week. Okay! Alcohol. How many here have not used alcohol? Okay, that sort of confirms what is generally in our society.

Probably 10 million people are dependent upon alcohol in the country. I would say in this country five million people are alcoholics. It is a very hard drug to get good statistics on because everyone has alcohol or has had alcohol in them. You say, "Does alcohol correlate with crime?" Well, you go

LECTURE 13 ~ *Depressants II*

and grab a lot of people who are involved in crime and smell their breath or run their bloods and yes, there's alcohol in a lot of them. And you get a lot of people who are coming out of the movie theater very quietly without having done anything and there's alcohol in a lot of them also. So it's a facile correlation to say it correlates with this, or correlates with that. One thing that's interesting: the number of people who are involved in the act of pulling the trigger, so to speak, of homicides—the number of people involved with alcohol is about 40 percent. Forty percent of people who are implicated in the pair—a homicide requires two people, one who does and one who gets done in—40 percent of the people who are the doers in homicides have been drinking or are legally involved with alcohol, if not drunk. Seventy percent of the victims are. It's some kind of a mystery, but more frequently it is the victim rather than the perpetrator with alcohol use. I don't know the meaning of it, but that's what has come out of the statistics.

What is alcoholism? Damn if I know. There are lots of requirements. I meant to get it for the lecture. I did not. But generally I think you have a gut feeling of what it is. It's denial at one level of the degree of dependence you have for the drug. That's probably the bottom line of what dependency is on a drug, is a denial of your dependency, denial of that degree of involvement.

Everyone has seen in the movies, or has read in books, the hiding of the extra bottle, the chandelier reserve in case of emergency, the putting it in the trunk of the car so that you know where there's an extra one, buying it in three different liquor stores so the clerks don't realize you drink quite as much as you do. All these are forms of denial.

Let's take a look at the alcohol getting into you. This is where we can use our numbers of human figures. But let's start a little bit back behind that—200 proof. Everyone I think has heard the term "proof." You know why booze is 86 proof? It used to be always 86 proof. Whiskey, scotch: 86 proof, never below that. Now, for your convenience, you can get it lower than that. But it used to be that 86 proof was the bottom proof. Why 86? What was meant by proof?

STUDENT: Flammability?

SASHA: Flammability is 86.

THE NATURE OF DRUGS

STUDENT: That's when you could tell if it was 86 proof, was when—

SASHA: On powder.

STUDENT: On powder.

SASHA: Yeah, that's very good. That's a neat little bit of old-English history. Pure alcohol is 200 proof. It's twice the number of the percentage of alcohol that's in it. Eighty-six proof is 43 percent alcohol, is the wettest that alcohol can be. You can put it on powder and the powder will still burn. And you're not apt to go into the local liquor store with gun powder and put it on the counter and say, "Let's see if the alcohol is up to specs."

But back in the days when you had powder and you had to keep it dry, and you didn't have hygrometers and analytical instruments, and someone sold you a bottle of booze, you would put some on the powder and strike it with a match, and if it burned it was good booze and if it didn't burn was not. Eighty-six proof or 43 percent was the break point for igniting powder. Hence, all that is called "hard liquor" had that. How much alcohol is in wine? Any feel for this? Twelve percent up to 20 percent. You can't get much over 20; yeasts just don't like growing. In fact, you have to even fortify very often to get there. But under 12 percent—I think there are probably wines that are 11s. But generally red wines 11, 12, 13 . . . In Algeria they produced—because of the hot climate—14 percent red wine; it almost upset the French economy. The French, as you may well gather, enjoy their wine. It's easy to point to them and say, "They've got an alcohol problem." And we'll go down to the bar here and discuss how much of an alcohol problem the French have.

All drinking countries have alcohol problems. Almost all countries are drinking countries. It is a very truly international difficulty. But the French usually have 12 and 12.5 percent red wine. And in Algeria there's a 14 percent wine. They could make 14 percent wine from the red grapes they grew there and it was harsh wine. It was a little bit corrosive, a little harder to drink, but it had 2 percent more alcohol and it swept the French economy and all the supermarkets—*les marchés, supermarchés*—sold bulk one-liter plastic bottles of Algerian wine, and the internal consumption of domestic wine dropped off very strongly. It was one of the things that was very difficult. The OAS, in the days of Algerian independence—

LECTURE 13 ~ *Depressants II*

Anyway, beer has a maximum limit. I don't know what it is, but it's around 6 percent. Did I mention about running a blood alcohol with a very dear clinical friend one time, and it never came in and we got drunk? No? It's kind of a nice little story. We got a what's called a "stat" emergency for blood alcohol. In a clinical laboratory, a stat is something that when it comes in you drop everything and do it right away. It's an emergency—you do it. And we were told we were getting a stat alcohol. And so we cleaned up, it was on a Saturday, no one was around, all machines ready. All primed, all standards, standard curves and everything, ready to go. We were waiting for the blood sample, and it never came. We were standing there waiting. Well, I was told to keep the thing operating until it came in at about two in the afternoon. We had everything set up for it.

So, I went out got a six-pack of beer. Instead of getting a six-pack of beer with six different cans of beer, we said we want to know how much alcohol is in the beer. So we pretended it was blood, made dilutions, and began shooting it into the GC, and we began assaying. Actually, it's hard to find out how much alcohol is in the beer. You ask the company and they will hang up on you. They're not going to tell you; it's a competitive thing, industrial secrets and all that sort of thing. Then we began getting other things—ales and stouts and malts. All kinds of things. We were really rolling by the end of that afternoon. We probably would have been in a very awkward position to defend the quality of our work. But nonetheless, we measured the contents of the can—how much did it weigh? Was it 12 ounces? Almost all of them were within 1 percent of 12 ounces. You're getting an honest weight in a beer can. How much alcohol? My golly, that ranged. And it turned out . . . we got, for example, two cans side by side that both said "Colt 45." Am I close to it? Okay. Colt 45, one of them was 1.5 percent and the other was almost 3 percent. What is going on? We looked at it very closely underneath. Code. No code. And apparently, if you sell it within a mile of a school or something or other, you've got to sell it with this code on it, and if you don't, then you can sell it with that code, or in a bar with that code and not this code . . . And the alcohol content varies by a factor of two. But who has ever looked at the bottom of a bottle coming out of a six-pack? No one. But that's where the information is tucked away on some of these things. And somebody found one that ran over 8 percent. [Laughter.] Olde English 800. [Laughter.]

THE NATURE OF DRUGS

But not a beer! What is it called? 800 Malt? Ale? Stout? I don't know what these things are. But the variation on that is really big. I phoned up one of the companies and asked about their alcohol quality control, and I got totally turned off. They would not talk to me at all.

But how much alcohol does it take in you to make you, say, a bad driver? I mean, over half of the arrests in this country deal with alcohol. And usually behavior with alcohol, or driving with alcohol, or public nuisance with alcohol, or misbehavior of some kind with alcohol. You're talking about the drugs? This is a major drug problem dealing with the law.

How much alcohol can you drink? How many here have actually driven at a time when in true, rational knowledge they probably shouldn't have? I think most at one point. I certainly have. I mean, this is the problem. How do you know how much alcohol you can have in you and be a good driver? Well, the Swedish say none—two hundredths. And these are numbers we're going to begin talking about. Let's put the number up here: 0.10. I'm talking now in grams per hundred grams. One hundred milligrams in 100 grams. That's a tenth of 1 percent of alcohol in your blood system. This is in California. You are not legally allowed to drive a car. Eight hundredths—Canada. Less than two hundredths—Sweden. Different numbers.

You can run a 0.01 on a sloppy analysis. Cadavers are allowed to have two hundredths and be called legally none because you do have a noise level. How much does it actually have? I'm going to slop alcohol over the next hour or two. How much alcohol does it take to get up to 0.10? Don't memorize numbers, get a feel for it. How much? Take an 80-kilo person. An adult male, maybe a 170 pounds, say call him 80 kilos. Let's say for fun that this person is half water—you consider tissue, consider blood, so I'll say he's 40 kilograms of water.

Let's go into hard liquor. Let's take four ounces of hard liquor. An ounce is 28 grams. So four ounces would be let's say 100 grams of booze. And let's say it's 80-proof, which means it's 40 percent, which means that if you've taken down four ounces of liquor, you've taken in 40 grams of ethanol. Okay, so in essence, we're getting about the same numbers. If you were to take four, or three, or whatever, I mean, you go to a bar, "I'll have a shot of hoozie." No, you don't get an ounce of hoozie. You get an ounce of hoozie from an ounce and a half shot-glass, and glug-glug-glug on the side

LECTURE 13 ~ *Depressants II*

because they want to keep you there for another drink. So sometimes . . . "I've only had one." I don't know how many times in court, I've heard the expression, "I only had one or two." Which means "I've had quite a few, but I better make it sound smaller." A few is monstrous. One or two is quite a few. Maybe one means three. There is this self-denial of how much you've actually drunk that comes out in spades when you're inhaling, trying to talk to a policeman who just pulled you over. Have you ever tried carrying on a conversation while inhaling? Give it a try sometime. It is amazing how hard it is. [Laughter.]

So, 40 grams of ethanol in a rubber bag that is filled with bones and tissue, namely you, and you have 40 kilograms of water in you. You're going to have one, one tenth g%. One part per thousand. One gram in 1,000 grams. You're dealing with a tenth of a percent. And that is about the feel for what you have. If you have two or three or four drinks, as an average-weight person, you're going to be up in a position where legally you're not allowed to drive. How fast does it get up there? Well, it depends how fast you drink. It gets burned off, out of the body, oh, I'm going to make a stab at it, keep it to round numbers, let's say two hundredths g% gets chopped out per hour. Again, some people will be faster, some people will be slower. But this is not exactly; this is for the feel of it.

In a few hours, you're out of it. In a very few hours, couple or three hours, you're probably down to where you can legally drive. Five hundredths is an area above which, but below this, between these two numbers, you're in a gray area. You're not disallowed driving, you're not illegal driving, but you're not sober. It's an area in which other factors must come to play. If you're stopped at eight hundredths and that's the level you made, that is not enough to prosecute you for drunk driving. Driving under the influence. Driving under the influence, I think I mentioned before, is not a felony. It's not a misdemeanor. It lies in between as peculiar thing all its own. It has its own regulations, its own rules. There are a lot of efforts made in the area of just over this, but not much over that, to say there are mitigating factors, and indeed there are.

You have three methods primarily used for taking alcohol levels in the body; there's blood, there's urine, and there's breath. These are the three primary vehicles. The law talks about blood. This is a blood level—one 10th

THE NATURE OF DRUGS

is level in the blood. If you huff and puff and blow in the machine, and the machine gathers the air and measures the air by some sort of titration and says, "Aha, we found this much alcohol in this much air," this is not a blood level. There is no machine from the breath or from the urine that can tell you what's in the blood. So we have to use an equation, and indeed there is an equation. Twenty-one hundred, you multiply the breath level x 2,100 to get to a blood equivalent. You take the urine level and divide by one and three tenths to get to a blood equivalent. But these are numbers that come from the masses, or they may not be your specific number.

I am not a 2,100 person. I happen to be a 2,400 person. I happen to be an 1,800 person. You may well be, but they've taken these things, they've taken the chisel and they've taken a chunk of granite, and they would hammer and hammer these numbers into the granite. These are the numbers that are going to be used toward you. They're in there. Make your arguments. You will probably not get away with it unless you are very close to the break point, at which point you say, "Well, I had just recently had a drink and probably it was still in my stomach, and I burped just before I was pulled over."

"I have a set of very loose dentures up here. And I think I got part of my dinner in there, but I had a little bit of wine with dinner. I'm sure some wine was up in my dentures. And when I gave my breath sample it was enriched, it was not a fair measure of my blood." You get away with it up to a certain point. The requirement is you've got to have a blood level that is in this illegal area, 0.10 and higher. And you have to have a certain amount of absence of sobriety in what's called a "field sobriety" test in which your behavior, your handling of yourself, your appearance, your smell, your physical coordination, coupled with the fact that the police officer saw it in the first place to pull you over, those things all taken together, plus a level of one-tenth or higher, is virtually unbeatable.

There are some people who will guarantee to beat it, but they are very expensive. There are a lot of little traps, but there are a lot of little tricks in the maintaining of a blood specimen. Little problems such as, if you give a breath sample, who holds a breath sample for two months until you get to trial so you can get independent analysis? You don't keep breath samples for two months. There's no balloon that will hold a breath sample for two

LECTURE 13 ~ *Depressants II*

months satisfactorily. So you have no possibility of getting a confirming analysis. This is a weakness of breath. It's valid for urine. You put a urine sample on the shelf, it may mold but the alcohol will stay there. And you can check alcohol in urine. You can check alcohol in blood months after the event. You cannot check it in breath.

There are other problems associated with alcohol analysis, such as the integrity of the analysis, the chain of custody, the chain of command of how the sample got from your arm to the laboratory to the clinical laboratory to the analysis, and did it get mixed up with Mike Jones, who probably had a sample coming through at the same time? I'm going to get into a lot of this quality control argument when we get into urine analyses for drugs, because that is a major weakness in the urine analysis. But it also applies in alcohol analysis. So I'm launched into alcohol; I probably will do a little bit more of that next Tuesday. Read Andy Weil's things on narcotics and on alcohol. And I'll finish up the other depressants on Tuesday.

LECTURE 14

March 17, 1987

Depressants III

SASHA: Oh good heavens, show-and-tell. [Crumpling sounds.]
Two, 4, 6, 8, 10, 12, 14, 16, 18, 19. I'm dumbfounded. You get 19 or 22 or something. I have a list with 32 . . . What is "open university?"

STUDENT: That's when you're not enrolled in university, but you can enroll in a class without being registered.

SASHA: That's kinda nice. Why isn't everyone in open university? You have to pay for it, though? Okay.
What's "rad-tel?" Radio, television, I presume? There are two people in "rad-tel." I've been fascinated by what people have as majors in here.

STUDENT: Radio and television.

SASHA: Rad-tel, is it producing, acting, electronic soldering irons?

STUDENT: Maybe it's all the above.

SASHA: All the above, okay. I love "undeclared." That was the history of my life, being undeclared. "Int Rel." International relations, I guess. Art, film, rad-tel, fantastic. "Lib studs," is that "library" or is that "liberal" . . . Library? Liberal Studies, not library studies. Okay!
People are still arriving.

ANN: Take a roll?

SASHA: Yes, I think I will, just to find out how many people . . . Obviously people who aren't here can't respond saying they should have been here. But people who are here who are not on the list are potentially in trouble, that's what I wanted to say. So what I'll do is, without any effort of pronunciation and with a tallying of hands, I'm going to read 32 names and ask people whose names I don't call to wiggle their hand. [Takes roll call.] How many people did I not call the names of?

Good. Okay, so no one is here who's not on this list. The other way around I can do nothing about. Okay, good.

I want to finish up today. Today is the last day of the first half of the course, and there's some sort of an ominous ring to that. What is it, "the first day of the last half of your life?" That was the phrase out of the '60s—

ANN: The first day of the rest of your life.

SASHA: "The first day of the rest of your life." Thank you. The last day of the first half of the course. We'll have our anonymous midterm Thursday, which is going to be fun for me. I don't know if you're going to enjoy it or not. I'm going to love it. It's a case where I will ask questions and just go around the class. If someone knows the answer for someone else, fine. This way I'll find out what I want to go back over. Sometimes the question will lead to 10 minutes of lecture, which is a nice way of getting out of an exam. I'll just lecture for 10 minutes, if necessary, because something didn't make sense, or I didn't understand it well enough to explain it, or I explained it and you didn't understand. So I will go into that on Thursday.

Today, I'd like to finish up the depressants. I got into alcohol in the last lecture. I want to do a little bit more with alcohol, but I want to get into the other things that have often been called in the medical literature, "solid alcohols." Solids that are alcohol; they have the same effect as alcohol but are solids. I brought in two of them for show-and-tell. The first of the barbiturates made commercially and used in man was Veronal, or barbital, and the other is phenobarbital, which is probably the best studied and the most broadly used. But back into alcohol itself. If I recall, I got into how it is in the body, the amounts you find in the body, the degree of impediment to motor coordination, to intellectual integrity, to consciousness.

And the fact also, I don't know if I mentioned or not in the last hour, the margin between alcohol level and life-threatening level is very small. Remember, oh about three, four, five weeks ago, I mentioned the term "therapeutic index." It was a measure of the amount of material, the so-called effective dose, that causes an effect, and the amount of material that causes a life-threatening situation—a potentially lethal level. And the difference

LECTURE 14 ~ *Depressants III*

between the two, the quotient really of the two, the one divided by the other, is a measure of the safety of a drug. The effective level can be very low, but if the toxic level is not much above it, it has a small margin of safety. If the effective level is very high and the toxic level is somewhat above it, it can have the same therapeutic index, the same measure of safety. So there's nothing intrinsic in the toxicity, the lethality, the poisonousness, the potency, of a drug that has anything to do with the therapeutic index. Using a very small amount of a very effective drug can have the same safety or risk as using a large amount of a relatively ineffective drug.

Alcohol is relatively ineffective. It takes handfuls of it to get effects. But it doesn't take many more handfuls of it to really induce a life-threatening situation. I think the best way of measuring that is what we have done before, using the blood level. And I put on the board the following code as the blood level of alcohol in a person which makes it illegal. And that part of the vehicle driving code, the driving laws, says it's illegal for him to drive a car if he has one tenth of a gram percent. I have a few comments around this term.

If you have that level in you, you legally may not drive a car. It does not say you're drunk. It does not say you're a bad driver. I have a mother-in-law who is an atrocious driver. She does not drink alcohol, she does not have this level. So this does not define a bad driver, exclusively from the point of view of a bad driver. [Addressing student.] Well, you just came in. It was you I was wanting to talk to about the 2-methylcyclopentanone?

STUDENT: Yeah.

SASHA: Okay, I have the answer to your question. Hang around for a few minutes afterwards.

This level, this is in blood. By law, this is the number of grams percent, the term "percent"—"per hundred." I'll comment on that. This term people have seen everywhere, in compound interest, what have you. It is the amount of thing per hundred of those things—per centum. A phrase you're going to run into once in a while and be mystified about—who's ever seen that one? [Referring to the chalkboard.] One. That's it. Per mille. Not per hundred; per thousand. It's a weird little thing, but you'll find it tucked away in the literature in strange places. In essence, you have one part per thousand, or one thousand parts per million.

You're now into terminology that is very often seen. Parts per thousand, parts per million. One loves orders of magnitude. How many people looked at the handout they got last week with about 15 or 20 pages in it? Good. In the back of it there was a whole series of ones followed by zeros and zeros followed by ones. Prefixes for three orders of magnitude. Who knows what an order of magnitude is? Good. That's a question for Thursday on the midterm. What's an order of magnitude?

STUDENT: A base of 10?

SASHA: Ten is the heart of it. A factor of 10. Tenfold. A factor of 10 is an order of magnitude. It's not a lot more, it's not several times more. It is 10 times difference. So a half an order of magnitude is an exact number. It's halfway between one and ten on a logarithmic scale, or about three. So, "He was off by several orders of magnitude" means he is off by several digits, integers up there in the exponent. And one number in the exponent is one order of magnitude. All these prefixes outside of the middle clutch are three orders of magnitude things. You go from whatever unity is to a milli, which is a thousandth, to a micro, which is a millionth. And then you get into my famous alphabetical orders—nano, pico, and then you get into really funny things down below. Getting up, you get into kilo, you get into mega, you get into—what's a thousand megas?

STUDENT: Giga.

SASHA: Giga. What's a thousand gigas?

STUDENT: Tera.

SASHA: Tera. What's a thousand teras? Hah! It's in your notes from last week. Know it for Thursday. Don't know it—know where to find it, that's all. That's the whole art of the whole course. It's not what you know, but knowing that you can find what you need to know, and that you don't have to clutter your mind up with things you memorize. The idea of people having all the declensions of the Greek irregular verbs written on their cuff coming into an exam is the most frightening thing.

I remember, I took a very strange disciplinary course in conjugation

LECTURE 14 ~ *Depressants III*

of Russian verbs one time, when I was quite small. And I had this urge to write everything, you know, on cuffs, and you run out of cuffs pretty soon. That is so ridiculous because you have a book in the library that has all that. Don't clutter up memorizing the names of the bones of the right wrist, or the cranial nerves, if you have access to it. Learn how to find it and use it. Learn what it means. But don't memorize it. I mean, how many people know the integral tables? Who can really can spiel off 312 basic integrals? No one. I certainly can't. But I know where in my library at home I can get that book that has it if I need it. That's the heart of learning; get the feel of how it fits together.

Why am I doing this? Alcohol, okay. By state law, this is the form it's got to have. You don't talk about parts per million. Don't talk about nanograms, percent nanograms, per milliliter or micrograms per liter. You talk in terms of hundreds. It is the only place in law, that I'm aware of, that concentrations or numbers are based on hundreds and not on thousands. Per hundred. Grams per hundred milliliters or grams of blood. By law there is nothing to the right of two decimal points. Zero point digit digit. This is completely aside from anything dealing with the history of drugs, but it has to do with the mechanics of blood alcohol, which is a very important thing, because we took a tally last Thursday and it turned out everyone in the class at one time or another had dipped into alcohol and tasted alcohol.

So, it's a drug you're familiar with. It's a drug you will probably be exposed to again at some time in the future. It's a drug that you may very well be exposed to while sitting behind the wheel of a car in the future. Maybe not, but maybe, and as such, know what the state says about driving with alcohol in you. If you have this level in your blood, you are not a legal driver. And there are ways of getting around it, but that becomes a legal strategy. This is not a course of law. It's a course in pharmacology. Okay. Grams. Per hundred grams. There is no other digit. There is no such thing as .107g%. By law, the thing that lies to the right of the second place is discarded. You do not round up, you round down.

This is by law. There is, for example, no 0.0997. You're a 0.09, and you are not under the influence of alcohol from the point of view of not being allowed to drive. Did I tell the story about the public health and taking alcohol and watching people get drunk while eating chips and crackers at state expense? Oh, neat. Okay, I will. And this is something that really offended my first wife. She said the idea that the state would pay for all this booze and have you go in there and have a cocktail party at our expense was absolutely outrageous. But it's a very neat little thing.

They still hold it about every couple or three weeks over in the Department of Public Health, which the main lab for this area is over in Berkeley. They will get maybe 5, maybe 10 people who are licensed to involve themselves somehow with working with people who are under the influence of alcohol. They may be highway patrolmen. They may be people who are in clinical laboratories who run alcohol analyses. Something to do with the officialdom of alcohol analysis. And then they'll get 8 or 10 people who are in that area who, pardon, 16 to 20 people, all of them working in this area. Those who are drinkers will be in the half that are chosen to be the volunteers. The other half are the observers. The only requirement is you have a one-on-one with a person you know pretty well, so you know their behavior pattern, you know how they respond. You know how they interact. You can talk to them. You can pick up little subtle changes in their behavior.

And you pair them off, and they go down to the public health thing and get some booze or beer or whatever it is. Whatever your booze is, they'll bring in that booze. And they will, at the given start of the experiment, give you precisely—I think it was 30–75 milliliters, or maybe it's 50 grams, whatever. Enough of an alcohol to bring you up about a third of the way or halfway to being legally drunk.

And you sip it. You know, glass, mix, stir, ice cubes, dips, pastrami, whatever you have in the way of a conversation, imitating a cocktail party in which everyone stands around, talks, and has booze. The innocent observer—the one-on-one who doesn't drink, does all the talking and eating but none of the booze thing. Then you take a break, and you go and get bled, and you blow down this little snorkel into a machine, you go through the whole rigmarole of things that determine how much alcohol is in you. So you have a person who has taken a known amount of alcohol, titrated and exactly known, who is being measured by all the machinery that the state uses to determine how much alcohol is in you. So the Public Health Department, by the way, is able to say, "Oh, yes, we assayed the breathalyzer, the intoxicator, or drunk-o-meter." I cannot even begin to tell you how many

LECTURE 14 ~ *Depressants III*

prefixes and suffixes have been jammed together with hyphens to boost machinery that's being sold to the state to determine how drunk a driver is.

You take all the prefixes, all the suffixes—they all fit. Then you blow in, and the needle goes over and says, ".17—take them to jail!" That kind of thing. All these have been used. They all have to be assayed by the state to see if they're valid. Their validity is determined by how they compare to blood alcohol. You do not determine blood alcohol by blowing in a machine. You determine breath alcohol by blowing in the machine and then that is converted to blood alcohol, but with a slide rule. So in essence, you're measuring indirectly the amount of alcohol that's in you. But the law does not say how much breath alcohol, how much urine alcohol, is considered to be driving under the influence. It says how much blood alcohol. So each of these things has to be equated somehow.

So you have the alcohol going in, a social situation, bleeding, blowing, talking, misbehaving, the whole thing that goes on at a cocktail party, and then 45 minutes or 40 minutes later, another round of booze, more food, more talk, more bleeding. And the whole argument, with you being observer or being observed as a relationship—I happened to be the observer, a very good friend of mine was being observed, I know him very well—is to determine at what point you pick up the first signs of intoxication. You often do not see it, if you are in those first signs of intoxication. You know this from your own experience in parties. "Gee, how did I get this far along? I better be a little cautious. I have to get up early tomorrow" or "I have to drive home" or something. You get this little phraseology in your head.

You can hide your intoxication. This is something that's very common between alcohol and barbs that we're going to talk about. You tend to hide the degree to which you're intoxicated, and you tend to not see it in others. You cannot see everyone who's getting a little bit woozy and wozzy. You don't see those who are *not* woozy and wozzy. You're not observant to that. Those who are not woozy and wozzy and have been not drinking can pick right up on it and say, "God, he's had his third beer," or "He's had his fifth Old Fashioned." You can see it from that point of view—you don't see it the other way. But watching from the sober point of view of a person taking his second and then his third drink . . . I watched Bill Annolson—an alcohol technologist from a large clinical lab on the other side of the bay—at the end of the second drink go in and be bled. And he was absolutely not feeling his alcohol at all. But I could tell it. I could tell it because I know him. He didn't normally smile quite the way he smiled when he was being bled by this very comely lady who was taking out the blood sample, leaning just slightly, you know, like that. It's not his character; he was just showing that little change. He did not acknowledge it. He did not acknowledge it until the third drink, then he said, "I'm feeling the drink." He showed it in the second drink. You can see it and, by the way, the second one the machine said his blood level was 0.08g%.

He legally was not barred from driving. But I could see the behavior character change in him. I think at this level or around this level, if you know a person well, you can see the behavior characteristics. It's that laughing, saying more easily the one thing that would have normally been preceded with caution, instead of that, bloop, out it comes. That little loosening of inhibition. You get it with alcohol, you get it with what we're going to talk about today—the solid alcohols: barbiturates, Valium, Librium, Doriden, phenobarb, secobarb, right across the board. All of these things are intoxicants. All of them are sedative-hypnotics. All of them will lead to a relaxation, to a loss of motor coordination, to a loss of intellectual integrity, and eventually to a pattern of sleep or of comalike loss of consciousness. They are all disinhibiting to begin with, and then they all become, what would be the term for that? Motor uncoordinating and mind uncoordinating, at a later point. It's a transition. That transition, I mean, that's why people drink.

The idea of taking something that's neurologically infusing—it gets into the myelin sheath, it causes nerve disruption, it causes a certain amount of damage. It has a caloric value, which you can burn but not get much substance out of it. People say "empty calories." Hell, calories are calories. You burn calories, you get heat. I mean, it is a food, you can get energy from it. It supplies energy. Why do you drink when it has all these negatives? Because it puts you in a neat little place; that's why you drink. You go down to the bar to have a couple of beers and talk to people. Well, you're going to say you're going to have a couple of beers to relax. The truth is you'll have a couple beers to talk to people. You'll be able to talk a little bit more easily after a couple of beers. Somehow, that pattern has been instilled into what

LECTURE 14 ~ *Depressants III*

you're doing. It is a disinhibitor.

You're holding your drink and you're disinhibited and you're talking more easily. About the third drink you find it slips out of your hand, if the motor discoordination gets in there, or you find, "I'm going to walk over to the table ...Whoa!" You're a little unstable. "How did I get so unstable? Am I tired?" You don't think, "I've had a little too much to drink." That doesn't come until later. It is a blend. You go through this little maximum of disinhibition and the virtues of alcohol, and some people have a very narrow range. Some people have wide range. Some people I don't think have any range. But in any event, you get through a range of disinhibition, of ease, ease of talking, like that famous old cartoon, "The three most marvelous things in the world are a martini before and a cigarette after." You know, that kind of thing. There's that disinhibition that comes in interaction with people. It is a disinhibitor, but it is also a robber of physical performance, a robber of motor coordination, a robber of intellectual coordination. It produces a state of amnesia.

I wonder if—none of you, I'm sure—but perhaps you know people with whom this is true, in which they have responded to something after about the seventh drink, or the end of the six-pack of beer, and do not remember the next day? You phone a person at 10 o'clock at night, "Sure. I'll see you in the morning." You phone up the next morning. "Where were you?"

"What do you mean, where was I?"

"Well, you said you'd see me at 10."

"I didn't say that. Did I say that?"

"You know, you sounded a little wooshy on the phone." That amnesia.

Somehow, at some level, you've seen it, and somehow, at some level, you know that is a property that comes along with alcohol. Okay, I mentioned .10. I gave numbers. They are, as I say, they're hewn in granite, there's 2,100, 2,300. I've forgotten the number. Lordy, I already can't do that. I'm an expert in the area. It's either 2,100 or 2,300, is the times more dilute than alcohol is in the breath, 2,300. Mainly if you breathe, you don't breathe grams of alcohol —you breathe milligrams of alcohol. By the way, how many people think that vodka gives you an intoxication which you cannot smell on the breath? How many people really think that? Good. Yeah?

STUDENT: But they add something to it to make it—

Sasha: No! Alcohol just reeks. Alcohol smells. I brought in alcohol and I can smell it. That's the smell. And alcohol inside you smells, your breath smells of alcohol. The lungs are one of the major ways of getting rid of alcohol—breathing out through the lungs. You smell. Try it some time. Give the best vodka you can buy, or whatever label they're pushing at the moment, to a friend and have him drink a couple glasses at a party and you just don't drink. Keep yourself clear of anything that would interfere with your sense of smell; and smell your friend.

Student: Don't they add some sort of perfume ... ?

Sasha: Pure alcohol leaves a smell on your breath.

Student: What do you associate with the smell of heavy drinking, you know, from scotch or whiskey and stuff like that, what is that?

Sasha: Oh, there are flavoring things. There are flavors, there are esters, there are hosts of beautiful compounds that are in there for the flavor. Wines have the smell of wine. Booze and whiskey, the smell of whiskey. But alcohol has a smell, and you can smell alcohol on the person's breath. Absolutely. There is no way to hide it, and you can put whatever these little things are that have sharp smells to hide the smell of alcohol, and you can smell alcohol behind those if you learn to recognize the smell of alcohol. You do not hide that. Anyway, the breath carries milligrams of alcohol. You'll multiply the amount in breath times a multiplier to get to the blood equivalency. This is a multiplier. You have 1.3 times as much alcohol in the urine, as a rule, as you have in the blood. And so you divide by 1.3 the urine. This is urine and this is breath.

If you want to take spinal taps it's 1.2. If you want to take a saliva sample it's 01.1. These are things that are not in the law. But these are the approximate numbers.

Yes?

Student: We used to do blood alcohol tests with GC, and we used to compare to MEK, methyl ethyl ketone. How is that relationship?

Sasha: Okay. It's a bit of a technical thing. Let me take a detour because it's a good one. In fact, I'll address it in a little different way. How do you

LECTURE 14 ~ *Depressants III*

determine what is in the breath? How do you determine what's in the blood? How to determine what's in urine? I will get to your point. It's a good point, thank you for bringing it up. I'll approach it in this way. The law changes each year, and it changes slightly in its wording. I don't know, I cannot give you off the top of my head what this year's law says on this. But, in essence, "If you don't cooperate, you are violating the law." You are required to testify against yourself. But the Constitution says, "Thou shalt not be required to testify against yourself." Therefore, if you don't wish to cooperate, you need not. But since they have given you the privilege of driving, they'll take the privilege away from you. In the old days, you had the right to drive. Now you have the privilege of driving, and that's known as a driver's license. And if you don't cooperate that can be forfeited for a period of time.

In Canada, I discovered if you have an alcohol level that is below .08 but above .05, they can suspend your driver's license for 12 hours, I think it is. Which is kind of an interesting gimmick. Probably totally in violation of the Charter of Canada. It has not been tested in court. But the idea is *don't drive for 12 hours*. That solves the immediate problem. It's a social virtue in the sense that you're off the road, and you're out of the driving possibility. But it doesn't go through court and you don't have a record. You just don't technically have the license for that 12 hours; then you have your license back again. So it's a nice way of achieving an end. I mentioned in Sweden, you don't drive at the .02. It's illegal.

Yes?

STUDENT: The other day I was down on Haight Street, and the police officers pulled over this guy. I was walking to The Ivy. And this guy was smashed, totally smashed. And I guess the computer just let him go. They said, I guess, "You're on your own." I can't see how they would, you know—

SASHA: Driving?

STUDENT: Yes, he was driving his car. If the police officers can't get a hold of, let's say, through the computer, the downtown station, then they can't hold a person—

SASHA: The process is known as putting under arrest. They can. There's a fine point known as being put under arrest. Again, this is an aside, but it brings up this point. Prior to being put under arrest, you have a hell of a lot of rights. You have a lot of freedoms and things. You can decline such-and-such, you can cooperate, not cooperate, that sort of thing, until the point you are under arrest. Once you're under arrest, a lot of your rights have changed. And a lot of your prerogatives have changed. You have, for example, no need of a lawyer nor any right to contact a lawyer before you're put under arrest. There's no reason. You've not been stopped—arrest, stop—you've not been stopped legally. Once you've been put under arrest, you need not provide any information but can call upon a lawyer and that can be your only vehicle of communication.

So there's something that goes on at this point of arrest that starts the criminal process in the criminal record. Now, this person is mentioning a person who is clearly smashed, stopped in the car, was allowed to continue going. I consider that to be irresponsible on the part of the policeman, unless there was a reason that they felt was overwhelming to allow a person who, as description was given, was clearly smashed to continue. Yes? I see a hand waving.

STUDENT: Perhaps they let him go because they don't want to deal with the booking and the arrest of that suspect. They may not want to do the hours of paperwork.

SASHA: Right. There must be, to them, some sufficient overpowering reason. It may be the person clearly had no alcohol in him.

STUDENT: Might be on drugs.

SASHA: Might be on drugs. "So do I want to get into that hassle?" The hassle of alcohol. "Is this the hassle of drugs? Is it that?" And the guy may say, "Look, I only live a block..." wherever he lives, two blocks away. And it's late and for some reason they don't want to get into the hassle. Or I can be cynical and say he has a cousin who happens to be a DA of Menlo Park or something. I mean, there are all these kinds of reasons. I don't know. I believe that letting a person who is not able to drive, in your opinion, continue driving then you have to share the responsibility for that person's behavior.

I really, really believe, for example, there is value in the approach toward

LECTURE 14 ~ *Depressants III*

a person who supplies drinks to a drunken person and then allows that person to drive, as sharing the responsibility for that person's behavior.

But I don't know the answer to that. It'd be interesting to know. I don't know how you're ever going to find out. There must have been an adequate reason because policemen are, by definition, in the role of protecting the public.

Okay, how do you determine the amount of alcohol? There are chemical ways. There are physical ways. The most common one mentioned is made with a material that's used to measure alcohol by using another chemical that has comparable physical properties. I'll briefly touch on a technique known as chromatography. Chroma—for color. Graphy—for drawing. Technically it is "to draw color." Originally, chromatography meant looking and drawing things, or actually "chromatoscopy"—the looking at the spectrum of colors. Years and years ago, a person by the name of Tswett started, I believe, with nasturtium seeds and a column of calcium carbonate, which is nothing but finely ground-up chalk. He took the chalk, ran some of the ground-up nasturtium seeds and the solvent, the green color was at the top as it ran down through the chalk. He saw two green bands, clearly separating something that was green into two things that were green.

This, in the 1920s, was the beginning of the concept of chromatography. It's done now without colors, without chalk, without being able to see what's going on, but extremely high resolution. But if you were to somehow put down a long tube of chalk air that had alcohol in it, so to speak—input here, output there—alcohol would kind of want to absorb on the chalk. Alcohol, after all, has OH groups, chalk has OH groups, but it also wants to move. So whereas the air goes through this fast, alcohol goes through this fast. If you have a recorder at the end that says when things come out, the recorder will say, "Here came the air, here comes the alcohol." So, in essence, you get emergence from a column as a function of time, depending on how fast things go through a column, or how much they want to be retained by the column. The higher and the hotter the column, the less the urge to be retained, the faster something will go out.

Let's take blood—that's the fundamental thing. You take blood and you shoot blood in one end of the column. Water comes through at a given point, alcohol comes through at a given point, other things come through.

Not many things are involved . . . a lot of blood stays clotted at the end of the column, but coming out the other end, that's where the alcohol comes out as a function of time.

So what do I do since I don't know how much alcohol is in here? I'll put in a certain amount of something that shouldn't be in the blood, like MEK. Let's use *t*-butyl alcohol. Let's use propanol. Isopropanol is not good because rubbing alcohol might have been taken. People have heard of rubbing alcohol, or smelled rubbing alcohol? You don't drink rubbing alcohol? Nonsense! I know a lot of people who drink rubbing alcohol. It's an intoxicant. It gives you a fair amount of toxicity. I ran into one case, in a clinical toxicology case about 10 years ago, in which the blood came in from a person who acted as if she were intoxicated, but she had not been drinking alcohol, on doctor's orders. "Stay off the booze." She had a cirrhosis problem that was pretty severe, and she was behaving in an intoxicated way—flushed face, staggering, speech impaired.

And for alcohol? Absolutely no ethanol. But loaded with isopropanol. She had been in the rubbing alcohol. She was told, "Don't use alcohol." Okay, so she used rubbing alcohol—it's not violating the doctor's orders. She got fairly intoxicated on it. Methanol is simple wood alcohol. People say, "Don't drink wood alcohol. It will make you blind." Well, it can mess up the optic tract. It's a human characteristic. It's not shown in animals, by the way. It's a species-specific thing to humans. But a lot of people can drink a lot of methanol without getting blind. It's not very good for you. It's a very bad poison and it will do damage the optic tract, but it's also an intoxicant.

A lot of things that are volatile are intoxicants: chloroform, ether. We're going to talk about a lot of these other solvents later on. The therapeutic index is very narrow. You're in a lot of trouble with a very little bit too much. But they're intoxicants. I don't recommend you using them. It's silly because you're taking a lot of risk, but they are intoxicants. But propanol is hard for the body to handle and is not easily available. So propanol makes a good internal standard. What you do then, you take blood that you don't know how much alcohol is in there, and you put in a known amount of an alcohol that shouldn't be in there—let's say propanol—and then stick it in the beginning of this long tube of adsorbent and see how long it takes. And out the other end comes the water and the air, and then comes the

LECTURE 14 ~ *Depressants III*

ethanol and then comes the propanol. And you can see intuitively, if you know how much propanol you put in and you know how much propanol and ethanol comes out, you can tell somewhat the amount of ethanol that was in there by its relationship to the propanol, based on the ethanol that came out. It's known as an internal standard, and it's the heart of almost all clinical analyses today.

You put a known amount of blood that's clean, you put a known amount of alcohol in, you put a known amount of propanol in. Here's another blood that has alcohol because it's from the arm of a person who had been driving across the bridge three lanes at a time. And you take that blood sample, you don't know how much alcohol is in it, so you put a known amount of propanol in. The one you know has ratios of such and such, and the one you don't know has a ratio of such and such, by comparing ratio to ratio, you can tell the amount of alcohol to alcohol. This is the heart of analytical procedures by chromatography. Almost all analyses use internal standards. The virtue of the internal standard, putting in a known amount of propanol, is once it's in there and mixed, you can spill half the sample, knock half of what's left over, inject a bad injection, and do real sloppy analysis, the ratio will stay the same. If you have a ratio of ethanol and propanol, and you still have the sample, the ratio is unchanged. So in essence, you have locked the analytical value into your system before you've actually started your analysis. The concept of the internal standard is very valuable for this reason. Yes?

STUDENT: How effective would it be if you question their method or whatever, comparing—

SASHA: Very effective. This is an area I want to get into sooner or later, so I'll get into it. Now, how effective is it if you challenge the court's analysis? Remember the court—the judge sitting there. We don't use white wigs in this country, but we do have black robes, same symbology. Court, bailiff, a person taking whoosiewhats. This court does not make alcohol analyses. You're going in there, you've been charged with driving under the influence. The analysis came, it was 0.1. Let's give you a margin. Let's give it was 0.12. You're in a twilight area now. Technically, you are not allowed to drive. But could 0.12 really be .008? I mean, how sober was the person who did the analysis? How motivated was the person who did analysis? Did they not like the fact you had an Armenian name and maybe cheated in the analysis? You don't know. Maybe it wasn't your blood. The court doesn't do the analysis. The laboratory located somewhere in the bowels of City Hall does the analysis.

Look at what is necessary. The blood must get from your arm—if it's urine, the path is even longer—the blood must get from your arm into a tube. The tube must be labeled. It must be transported with lots of other tubes—this will apply to urine analysis and drugs in the urine test. It must be imported to a place where it's stored, from which it will be taken as a 77th sample that day, given a code number, and the code number will then go through the machinery. It will be analyzed by some sort of an operation. Somewhere before and somewhere after, there will be a standard curve. Somewhere else before and somewhere after, there'll be what's called a quality control. That number then has another number attached to it. These two numbers get somehow onto a sheet of paper, into a computer that prints out to a thing that gets into an envelope that gets a stamp attached. It gets mailed to the courthouse. It gets put on the record of the person whose name is equivalent to that number. And there are probably a few more steps that I forgot.

I may say that some of these steps have the possibility of error. A lot of these steps have a possibility of error. The chances of something getting through without error I find to be rather remarkably small. If you take the urine situation, the urine has to go an even further route. If you are stopped while driving, you may take a choice amongst those things are offered. It was up until about three years ago, you had a choice of whatever you wish and they gave you what you chose. You're given now a choice of what is offered.

You may give a urine specimen. Look at the urine argument. You don't want a person entering you. So in essence, you provide the sample by means of a noninvasive device. You pee in a cup. But look at what's missing. What does your urine have to do with the amount of alcohol that's in your blood? Why is your urine a measure of the amount of alcohol? Well, alcohol filters all the time through the kidneys. It gets excreted, it gets reabsorbed, one thing or another, and you assume that 1.3 is going to be the ratio between blood and urine. But it's a ratio between blood and the urine as the urine

LECTURE 14 ~ *Depressants III*

is passing from blood. Fresh urine. And fresh urine is not that which is down in the bladder. Fresh urine is that which is going into the bladder. But you don't get to what's going into the bladder, you get what's already in the bladder. And so, let's say you've been drinking lemonade for three and a half days and nothing but lemonade. You have a bladder this size. And then you suddenly take three quick shots of booze, you're going to clear a little bit of alcohol into this much urine. You might not be able to drive a car and your urine says you're practically sober.

On the other hand, let's say you just had a lot of alcohol going in very quickly and the alcohol cleared into the bladder. And that was some time ago, maybe earlier this morning. You got up in the morning, had three Hail Marys... what do you call them? Bloody Marys. You get a lot of alcohol into the bladder, then you go on through the day, and you begin diluting it with lemonade. Your bladder may say you've had a lot of alcohol, but your blood doesn't have much because it's all been cleared. So the urine can be off both ways, and more much more often it is off by being dilute. So they don't say, "Take a urine specimen, on that thou shall be judged." They say, "Dump the bladder. Empty it entirely. Wait 20 minutes, and then give a urine specimen." The law requires emptying the bladder, a 20-minute wait, and then a urine specimen. And that urine specimen is the one that shall be assayed. It should be broken into at least two parts, one to assay, the other made available to you for a separate analysis if you wish. If you don't produce a urine specimen after 20 minutes, you're not cooperating—and that's the end of it.

So in essence, you can usually ooze out a couple of milliliters, and that's all that's needed for an analysis. But you're given that 20 minutes.

It requires a certain invasion of privacy to watch someone peeing in a cup. And we'll get into this in spades when they get into urine analysis in the workplace. And also, by taking 20 minutes sitting in the clinic waiting for your kidneys to produce a couple of milliliters of urine, that police operation is out of function. They're not out doing the job they are paid to do, which is to get dangerous drivers off the road. They're sitting in the clinic waiting for your kidneys to work. And that doesn't appeal to them. Yes?

STUDENT: We require 15 milliliters of urine. So that shows the dilemma;

if you don't have 15 of urine, we won't—

SASHA: That's your lab requirement.

STUDENT: We won't positively confirm—

SASHA: You won't get state business. The state will say, "Here's your sample. Do it." The vial can be empty or nearly empty.

STUDENT: A lot of lab results are based on residues and things like that.

SASHA: The state requires the 20 minutes, and you don't produce 15 milliliters as a rule in 20 minutes. How much urine do you produce in a day? Any idea? Anyone ever peed in quart bottles for a day? A liter, two liters. Couple liters, maybe give or take a little bit. It's very useful to know because it's the same amount of saliva you produce in the day. You produce a couple of liters of saliva. You swallow it all, so you're not aware of it, but the amount of production of saliva is about the amount of production of urine. Okay. That's known as immaterial trivia.

Okay, so that's how you're going to analyze, that's what's involved. And I want to get on into alcohol-like things. Anesthetics go through the same route. But before I get into that, early in the treatment of anxiety, two of the basic complaints that people have presented for years as disturbing things are intrinsically anxiety and insomnia, things that kind of work hand in hand. There has been always a need of tempering anxiety, some way of permitting sleep. The opiates we've mentioned will produce this dreamlike withdrawal, sleep. It will resolve anxiety and resolve sleep. But there are addicting problems. There are problems of physical dependency.

Chloral hydrate, of the Mickey Finn[9] fame at the turn of the century, this is still used very much with geriatrics, with older people. It has a bad smell; paraldehyde also has a bad smell. Take a half a gram, or three-quarters of a gram of paraldehyde, take a half a gram of chloral, very definitely will cause sedation, will cause sleeping, will cause the loss of anxiety.

Bromides. I think we still have the idiom that something is a bromide? Something that will calm a voice that is spoken in anger or spoken in anxiety. "We wanted answers and we got nothing but a series of bromides." The term is still used, because literally a bromide ion is a sedating ion.

[9] This is where the term "slipping someone a mickey" came from.

LECTURE 14 ~ *Depressants III*

It causes sedation. It causes quieting. Do you remember the W. C. Fields syndrome from the turn of the century—the big bulbous nose with the blue coloration? This is the one of the toxicities of bromides. Bromo-Seltzer was originally there because the bromide was the quieting agent and the calming agent in the material. Ah, the toxicity. These things have dropped out because of what was introduced just after the turn the century. The very first material introduced in the barbiturates was Veronal or barbital. I've brought in a bottle—sodium barbital. Beautiful white solid. It's still used very broadly today—not medically but used in clinical laboratories. It's an excellent buffer. And I always get a chuckle at seeing that sodium barbital is a Schedule III drug. It's got to be recorded one way or the other. There's the little C with a Roman numeral III down on the label.

You go into a clinical laboratory, look under the hood in the solvent room, and there's the 55-gallon drum of sodium veronal being used as a buffer. No one quite puts together that it is a barb that also is quite an intoxicant, quite a stupefier. Relatively fast acting, the very first of the barbs, and it has its world as a material used in the chemical area. It has its world as a drug. It is not used much as a drug, but it was at the turn of the century, and for the first 15 or 20 years of this century up through World War I, Veronal was heavily used. Phenobarbital was introduced about 1910 or 1915. Still used today, not as much as a barb but as an anticonvulsant. It is an excellent anticonvulsive agent.

There's one term I love to use in clinical chemistry. It's a nice thing, which is BAPSP. There's no reason to know it, it's not accepted, it's not pronounceable. The P is merely because I want to make it complete. But BAPSP really represents quite a bit of information. BAPSP represents the five major barbiturates that are still used today. You have butabarb, amobarb, pentobarb, secobarb, and phenobarb. And the reason I put them in this order is, in almost every analytical scheme that you're going to work in, this is the order in which they emerge. Very handy little mnemonic to know. Also, it has a sandwiching aspect. This is sort of a personal mnemonic of my own. Everything is sandwiched. Phenobarb is a long actor. The buta- and the amo- are medium speed. And the pentobarb and secobarb are the short acting. So the short acting are sandwiched between the medium and long acting. And also, from my own point of view, my initials of Alexander T. Shulgin—which I

— 177 —

can remember without even looking at my driver's license now, I remember that A and S are the ones that sandwich my initial T—is a combination of amobarb and secobarb, that defines the mixed barb known as Tuinal.

I had never been able to remember what Tuinal is, but it happens to be my sandwich between these two. So if you remember my name, you remember Tuinal, you remember it's a mixture of amobarb and secobarb. But Tuinal is what the AMA, American Medical Association, calls an irrational mixture, because it's two parts of this and three parts of that combined to produce a mixture of fast acting and a medium acting. So you get a protracted action. The fast acting shows its effects now; as it drops off, the medium acting carries on the effect. So in essence you are mixing two barbs to get a longer duration of action. Rapid onset and yet longer duration. But it's irrational because two to three may be my ratio, but not your ratio. Each person should be arranged on his own. The company that makes Tuinal merely weighs two units of this, three in that, dumps it together, and sells it as such.

Barbs are very much like alcohol. The similarities are dramatic. In 1940, there were a billion barbs that were manufactured and sold. A billion dosages. And that was about the time that barbs were beginning to be advertised as a public problem. They called them the "thrill pills." The AMA wrote an article entitled "1.25 Billion Barbs a Year," and said that we've got to do something about getting this drug problem under control. If I say something that sounds familiar from today, don't let that be a surprise. This record has played every decade since the history of this country. It played in the 1940s; it was barbs. "These things are getting out of control. We've got to do something about it. We will pass a law." "No barbs can be sold without a prescription." The law began painting the evils of barbs. They cause death. They cause intoxication. They gave wild publicity to barbs. In the course of 10 years, the number of pills went from 1.25 billion to 10 billion. They're illegal. They got the promotion of being illegal and they became a major problem. They went into the teenage crowd. And by the end of the 1960s, probably the major drugs used by preteens in school were barbs.

And you don't think of barbs today, unless you're into 'ludes or something, which is the same thing. But it was the major—the major, I can't say. Alcohol was probably way up there. But barbs are solid alcohol; they would cause people to get sloppy. They do exactly what alcohol does—a little loose,

LECTURE 14 ~ *Depressants III*

a little disinhibited, a little dopey. Then you don't walk and think too well. Pretty soon you fall asleep. I mean, what better rock can you have when you're a preteen in school? I mean, it answers all the problems. And they were used extremely broadly. And they are still used. How many people have heard of 'ludes? How many people use 'ludes? It's like a barb. Quaalude—two *a*'s. They really capitalize on that double *a*. Q-U-A-A-L-U-D-E is methaqualone, and methaqualone was a prescription sedative-hypnotic. I think I've defined the term sedative-hypnotic. Sedative—to quiet; hypnotic—to go to sleep. It suddenly quiets you and puts you to sleep. It's a hyphenated thing pharmacologically—sedative-hypnotic. Doriden, Preludin . . . oh, Preludin is a stimulant. Doriden, ethchlorvynol (Placidyl).

A lot of materials very closely related to barbiturates were discovered in India in approximately 1965. This little family of heterocycles, when given to animals, would prolong barbiturate sleep time. This is one of the reasons there are so many barb substitutes and barb-like things; they're easy to assay in animals. What you do, you take a series of rats and you give them phenobarb, and you know that they will lose their righting reflex for a period of 82 minutes on the average, maybe 50, maybe 110, but you have a distribution curve of maybe 80 minutes. You aim for that. Later, you give them phenobarb and another compound. And if the sleep time is extended, it takes maybe 150 minutes for the animal to get its righting reflex, so that you put one on its feet and it doesn't fall over again, you can measure that. Every minute you try to put it on its feet, sooner or later it remains standing. That is the end of the experiment. And you'll find barbs or sedative-hypnotics will extend the sleeping time, stimulants will decrease it. So in essence, you're going into new families of compounds. You can check stimulants versus sedative-hypnotics by just having a series of rats and knocking them out with a predicted amount of phenobarb and seeing if you can exacerbate the effects of the phenobarb or abort the effect of the phenobarb by means of a drug. This family of the methaqualone group was discovered in India. They were found to be effective in man in large dosages of 300 to 500 milligrams, but it constituted a major substitute for barbiturates in the 1960s and early '70s.

Yes?

STUDENT: I thought that methaqualone gives a heroinlike high so—

THE NATURE OF DRUGS

SASHA: So does alcohol in its own way. And it's not a heroinlike high; it gives an intoxication high. You get lightheaded and funny.

STUDENT: In the CSF pharmacology book, it said it gave a heroinlike intoxication. Is that true?

SASHA: I wouldn't class it that way. To me it is much more like a barbiturate.

STUDENT: Like being drunk without drinking.

SASHA: Like being drunk without drinking—I would put it that way. I don't think heroin does that. Heroin makes you kind of withdrawn and seeing into your imagery. No, I would classify them differently. I do not put on the same classification. Have you tried both?

STUDENT: Yeah, I wouldn't say that at all. Heroin is more introspective but methaqualone is more—

SASHA: Giggly.

STUDENT: Giggly.

SASHA: You're out there giggling, interacting.

One reason alcohol is difficult to experiment with is you can't hide the taste or smell of alcohol. You've got to take buckets of it, so you smell it and you taste it and you know what's going in. If you can put alcohol into a nonseeable/tastable/smellable capsule and put methaqualone into a nonseeable, nonsmellable capsule and run a double-blind experiment at a cocktail party, the volume would go up. The noise would go up. The whole activity would go up and you would not be able to distinguish who had which, nor would the people themselves be able to. You're sloppy and giggly, disinhibited, intoxicated. Heroin, you're not sloppy, giggly, intoxicated. You're withdrawn into yourself. You nod out with your heroin. You go out and giggle with your barbs or Doriden or methaqualone.

Yep, someone has a hand?

STUDENT: Yes. What is the decade of the big publicity scare around barbs?

SASHA: It started in the mid-'40s. It really got going by about the mid-'50s,

LECTURE 14 ~ *Depressants III*

and it kind of extinguished itself in the '60s. But it's roughly '40s to '60s.

STUDENT: Quaaludes are still very popular in the Asian community.

SASHA: Oh, yes! Very definitely. I've heard it's very popular in Asia. They were very popular in the East Coast in all communities. In fact, one of the things that gave Quaaludes a bad legal turn were what were called "stress clinics." I don't know if anyone's ever heard of stress clinics in New York? It became big business in which a bunch of physicians would team together, and they would set up a stress clinic for treating people who had stress. And what else are you going to do in a stress clinic? They come in, you've got stress, "I have stress, doc," and they have in hand a script for methaqualone. And they would hand out quantities. I mean, "Here's another 500. I hope it relieves your stress." They were the heart of the drug problem. And the manufacturing volume was unbelievable. The amount they were prescribing was unbelievable.

Several of them were charged with bad practice of medicine, which is an awkward thing. Doctors don't like charging other doctors with bad practices any more than lawyers like finding other lawyers being unethical. I mean, when a bunch of watchdogs watch the watchdogs, you have a little bit of a conflict of interest. It's intrinsic in there. And so, physicians can do an awful lot before they are really called to the carpet. The lawyers, when they're judged by other lawyers, they're judged by their own peers. The peer-judging has its self-serving aspect in every profession. Physicians are not exempt from that. And it's taken a lot of very conspicuous irresponsible practice of medicine to bring these stress clinics under some spotlight. Never stopped them. They may be suspended for six months from practicing medicine. But on the other hand, if you were making 500,000 bucks a year, you might be willing to forego six months of the practice of medicine because you want to take a trip to the Bahamas anyway. I mean, it was absolutely a slap on the wrist. It did not work.

Yes?

STUDENT: I had a doctor give me phenobarbital for stress just recently.

SASHA: Mmm hmm. Phenobarbital is still used largely as an anticonvulsant. But it is also, rather interestingly, still one of the most standard standbys for quieting, for long, protracted, relatively safe, quieting. Yes?

STUDENT: Have Quaaludes been moved by the DEA into Schedule I? Because I hear rumors that it can't be prescribed.

SASHA: No, they cannot be prescribed at all now. They have legally been moved by Congress—not by the DEA. It was done by Congress. Congress said, "Methaqualone shall become Schedule I. The abuse outweighs any bit of utility." Therefore, by law they have no medical utility. Schedule I.

STUDENT: Before, like 10 years ago . . .

SASHA: Schedule III, I believe.

STUDENT: Why did they move it?

SASHA: The abuse was getting out of hand. The prescriptional abuse of it was out of hand. Probably a triplicate Schedule II would have slowed it down. And it was unneeded; methaqualone is not a needed drug. Ninety percent of the barbs that are available are not needed.

STUDENT: Like phenobarbital and prescribed—

SASHA: Take a medium, buta. Take a fast one, take amo. Take a slow one, pheno. Three barbs would serve all the needs of barbs right across the board. A fast, a slow, a medium. What more do you need? Well, you happen to have 35 pharmaceutical houses, hence you have 35 barbiturates or 35 times 3. They put a new group on here to get a new patent, to get a new thing on the market, then you have literally, I would say, 40 barbiturates that can be used in the pharmacopeia.

STUDENT: So are we just subjects of the capitalistic way? I mean, this is a moneymaking scheme to get good drugs on the market and keep making them. You've got to produce these things. You got to produce drugs to use them.

SASHA: Oh, you're touching buttons I don't think we want to respond to. Are we victims of a capitalistic system? This is a course in the history

LECTURE 14 ~ *Depressants III*

of drugs. No. Let's leave that unanswered. Let me only say that if we need barbs at all, which I think there's a sufficient valid medical reason for, we can do it with three barbs. Let's say six, to get to the people who happen to be idiosyncratically sensitive to one can use the other. Methaqualone has no need. It can be handled by a barb.

STUDENT: Are Quaaludes addictive or are they potentially . . .

SASHA: Yes, I would say that they would fit in the classification. And there are not many I would put in that classification of physical dependency. Because they act like barbs and barbs do lead to physical dependency.

I had a question over here. Yes?

STUDENT: Are they made anymore?

SASHA: Yes. Not legally. They're made in underground laboratories. Fairly simple manufacture from ortho toluidine and anthranilic acid and acetic anhydride. You can turn out methaqualone in a couple of days.

STUDENT: So methaqualone is the generic name?

SASHA: Methaqualone is the generic name. Quaalude is the usual trade name. Rorer 714. Am I close?

STUDENT: Yeah.

SASHA: Okay. 714 is the number. And of course, when you make—hmm?

STUDENT: Mandrax.

SASHA: Mandrax—that's a combination, I believe. A question right behind. Yes?

STUDENT: I'm confused about short acting and long acting, fast acting—

SASHA: Fast acting is not the same as short acting, although it very often works out that way. You have a response curve—good and famous graphs—in which you get no response or you get that kind of response from a drug. The question is what's the time scale down below? Fast acting, this starts up early. Long acting, it extends over a long period of time. Short acting, over a short period of time. As a rule, those things which start fast don't last long, and they clear fast. And they're often called "fast acting," but really what they mean is "short acting," and early in the process of having been taken. Short acting—you feel it in a half an hour, 15 or 20 minutes even. You are sedated or very definitely quieted down for an hour. And three or four hours later, no effects whatsoever.

Super short acting are your thiobarbs. Let's define the term "thio" for those who are not into chemistry. I noticed there is one person in the course who is in chemistry, which is interesting. "Thio" means sulfur—the element sulfur. You take the oxygen at the pyrimidine two position—the position that is opposite the place where the aliphatic chains are attached. Take the oxygen off, put a sulfur on. That's called a "thiobarb." Thioamobarbital, thiopentobarbital. How many people have had a thiobarb for tooth removal? I have, for oral surgery, it's very common. In the vein—you're out. Talk about fast acting—you're out there in maybe 5 or 10 seconds; you're unconscious. Now that is fast acting. Putting it into the vein gets it directly into the operating system. And it is short acting and recovery is half an hour. You may be pretty miserable, but you're conscious.

STUDENT: When you talk about long, short, or medium, we're talking about the rate of metabolism?

SASHA: Yes and no. Metabolism is a big contributor; probably the major contributor is lipophilicity.

STUDENT: Solubility?

SASHA: Solubility. Interestingly, it's not quite the direction you would think—it's the opposite direction. Those things that are highly lipophilic, those things that go into fat easily, are sucked into fat then they dribble out metabolically out of the fat and are short acting, are short lived. Your conscious state is not dictated by what's in your fat. So when they go into the blood, they're fast acting and they're fast off. So the more lipophilic, the shorter acting.

And interestingly, with phenobarb, that which is the longest acting is the least stable to base. And that's worth a little comment on the side. In the clinical laboratory, barbs constitute one of the few drugs in which you determine how much of something you have before you determine what you

LECTURE 14 ~ *Depressants III*

have. Most drugs... "What is it? Oh, that's it. I'll send the analysis." "What is it? It's abracadabra." Open the book to the *a*'s. Here's how you determine how much abracadabra you have. Run the analysis. With barbs it's easier to tell how much barb you have, because you can do it spectophotometrically without knowing what barb it is.

Then, having found there's a barb worth looking at, you go back to GC or something that will tell the various barbs apart. Barbs are one of the very few drugs in which you take a large bolus of material in the body, you're taking tens or hundreds of milligrams of a drug. And hence there's a big blood level. You can tell what's in the blood by just putting an extract of the blood directly in the spectrophotometer and seeing how much is there. Then if there's anything to look at, you find out what it is.

Now one of the old, old clinical tricks—it's probably not done much at all anymore—is to take that blood extract and cook it up with base. The faster the barb is decomposed, the longer acting and the less potent the barb is. Phenobarb being one of the longest acting and least potent. The reason you want that information quickly is if you have a lot of a barb, it's often very important not to know specifically what barb it is, but is it a barb that would normally be given in a large amount? And if it's something that is normally felt in a small amount and you've got a large amount of it, you've got an emergency problem, and you've got to get the barb out before it will kill. And you get barb out by pumping the stomach if it's still there, or by dialyzing the blood, which becomes real tricky. Hemodialysis is a really complex thing. But you actually extract material from the blood while the blood is still coursing in the body, and you get the chemical out of the blood. It's dramatic, but it can be lifesaving. It can also be ineffective. It's a very complex operation and it's not done unless there's a good reason it should be done.

If you have a large amount of a very potent barb, you do it. If you have a large amount of a relatively impotent barb, you would not do it. So you can tell the relative potency of the barb and the duration of action to the barb by its fragility to base. It's a clinical test, no reason of getting into it in this course.

But anyway, that is the general feeling of the barbs. We're getting down to the final track. I mentioned thiobarbs. Thiobarbs are extremely fast acting

THE NATURE OF DRUGS

and of relatively short duration. The point you brought up is very good. The barbs are intrinsically active. If you have a metabolic destruction of a barb that's fairly quick, the barb is going to have a fairly short action. Obviously, if it's not there for long, it won't have its action for long.

Urine test for barb; here's a tricky one. You see the junk that's excreted in the urine and you don't see the barb itself. You see residues of the barb's metabolism. Blood is by far the best vehicle—the best tissue—for determining barb levels in the body. Urine, if you know what you're looking for, and you know the metabolites you're looking for, can be used, but blood is by far the preferred tissue.

What other things? Oh. Minor tranquilizers. Okay, minor tranquilizers. Minor and major. I don't like the terms minor and major because often they're used to imply not important or very important; or not very serious or very serious; or not broadly used or very broadly used. The terms have often been worked into usages where they carry messages. Here it's a case of minor tranquilizers, major tranquilizers. The minor tranquilizers are those things that are more of our solid alcohols. They are used for quieting, softening anxiety, de-stressing. You have Librium, Valium. People are familiar with the terms Librium and Valium? Arrange this, *L* and *V*. I love the two terms because chlordiazepoxide and diazepam are also alphabetical—*C* and *D*. Librium and Valium, *L* and *V*. They're in alphabetical order.

I came across another alphabetical order the other day. I think I've gotten into this harangue before; things in alphabetical order. The passes that go across the Sierra into Nevada. The one further South is "Walker"—starts with a W. You get up to the next one, it's Tioga. Then you get to Strawberry, and you have Donner at the top. They're all alphabetical order, north to south. I think Monitor is in the middle somewhere, where it should be. I haven't really gotten to a map—I should run it down. There's probably one that will violate the rule, but it gives you a general feel of Donner down south to Walker, at the curving thing, Lake Isabella. That's kind of a nice alphabetical order. Anyway, the Valium, Librium are sedatives. They are quieting. They may not quite produce the tendency to sleep that you'll get with the barbs, but they are basically the same sort of things. They are de-stressors, they are intoxicants, they are abusable right in the same way that the barbs are abusable and alcohol is abusable.

LECTURE 14 ~ *Depressants III*

Meprobamate I should mention in passing. It was one of the first so-called minor tranquilizers. They're called minor just because they are in this classification. They are not minor or major in the global gestalt sense. They are just classified as minor. Meprobamate—Miltown was probably the first of the tranquilizers in this area—a different type of compound. Meprobamate, Miltown, had a great popularity. Wallace was the manufacturing company, named after Wallace, who was the one who invented the compound, which shows it could be financially successful too. It was one of the first ones introduced as a tranquilizer in place of barbs. In a sense it was almost a paralytic. When one uses the term paralytic, one thinks of being frozen up, not being able to breathe or move. Paralytics need not be like this. They can just quiet the response of muscles, soften the reflexes, soften the tension, which is part of the sedating, part of the de-stressing. Make the person sit quietly. Don't get those twitches. Relax. And things that will turn down that muscular feedback—the continual tension, the tonus—are things that will achieve that sedating quality.

Miltown is a muscular paralytic. Apparently it does not affect the intercostals. A question was asked early on about intercostals; on the handout I talked about anesthetics. Intercostals are the muscles that permit breathing, and paralysis of the intercostals will stop the breathing process and you die of asphyxiation. When you go into surgery, you want muscular relaxation, but you don't want the heart and the breathing to stop. You want the reflexes to quiet down. You don't want a person to suddenly have an uncontrolled twitch when you cut in, so you want that kind of thing quieted.

There are general areas and classifications of anesthesia. It is a kind of referencing to give you. The stages of anesthesia—one, two, three, and four. Four, you've lost your patient. Three is what you want, but you want a narrow level of three. A general thing on anesthesia and I'll drop the whole discussion until the next hour, the following hour.

Anesthetics—anything that's done medically—you want as little as possible. And you'll use as many things as you can to make the actual risk to the body as small as possible, to make the anesthesia as shallow as possible and a minimum threat. Anytime you start interfering with body functions, you're playing a game that is hazardous. You may be able to observe a lot of it, but some of it may be going on that you can't see. So you don't want the person to be totally without reflex, totally without response. You want to be as near consciousness, as near response, as near the proper state of awareness and witness as is possible, as long as you rob them of pain and awareness. The whole art of anesthesia is this finding as shallow a level you can get by with. This gives you a feel. The eye, the pulse, and the blood pressure responses to different levels of anesthesia, that is what's being watched by a hopefully alert attendant, and knowledgeable anesthesiologist.

Ha! Good, we'll have a lot of talk and questions and who-knows-what next Thursday. We'll get back to the rest of this on the following Tuesday.

STUDENT: Is it hard to be an anesthesiologist?

SASHA: Oh god, yes. You *are* that person who's under the mask and under your control. You are running the shots in the room. It's you who is living and feeling what that person is going through. It's really like playing piano. You have to learn the responses.

STUDENT: That's four years after?

SASHA: Oh, you've gone through your interning and then you have another additional couple or three years at least. I know several people who have gone into it. It's murderous getting into it. But some people go on to learn it well and are practiced in the art. It is kind of like becoming a psychiatrist.

LECTURE 15

March 24, 1987

Intoxicants

SASHA: Okay, people have wandered in. That's not bad, not as late as usual. Usually people come in about four minutes after. They must have run.

STUDENT: Are we supposed to be starting at the top of the hour or at 10 after the hour?

SASHA: I understand we are supposed to be starting at the top of the hour. Usually we have about seven people in by then. And the crowd doesn't really kind of congeal for the cocktail party until about 5 or 10 after. I was told it went from the top of the hour to 15 minutes after the next hour.

STUDENT: Most people don't arrive until 10 after.

SASHA: Okay, that's right. And then I was told some other classes stop on the hour, so people have to come in here huffing and puffing. So, I'm sort of getting a sloppy start. I don't know. I'll try to throw away little gems like batting averages. Okay! [Sounds of chalk on chalkboard.]

What do you call things that make you excited? Excited things... genesis of excitement. I don't know...What's excitement in Latin? *Excitantia*? Lewin used to use it that way. That's kind of what I want to talk about today. Which is really, no, not really... everything to do with drugs. I mean, there are things which are completely outside of drugs that are wild. How many people have been on a roller coaster? How many people have not? No—one person hasn't. Have you ever at least watched a roller coaster? When it climbs it goes *jub, jub, jub, jub, jub*. Gets way up to the top. Over the top. What's going through your head? "Oh my Jesus, why did I come on this trip anyway?" And it starts down the other side and begins accelerating. How many people scream? There's this blast of noise that comes from the people. Is it excitement or is it fear? Well, it's got a good charge of both. If it were

THE NATURE OF DRUGS

fear, I mean, if you're pushed off a cliff and suddenly you're heading down toward those rocks below at roughly whatever physics says is the acceleration due to gravity, and you realize there's no net down there, and there's no one to catch you and it isn't a dream, the wind is really going through your hair, you scream too. But it isn't quite the same kind of scream. That's a scream of true terror. I mean, that is the end of it. If you are a screamer of that kind, you've got yourself a big part in a horror movie. And there are not many people who are real, genuine screamers. They're eekers and oohers, but a real blasting scream is quite a technique.

Yes?

STUDENT: There's a rock musician named Lena Lovich who made a living screaming.

SASHA: I saw a very, very off-vocabulary comic on a TV video hoozie a couple of days ago—

ANN: Bob Goldthwait.

SASHA: Wooh! Talk about screaming. Here's a person who's worked it into a million-dollar business. Yeah. So, why would you pay $1.50 to go in a car that goes up to the top of a thing and goes down just so you can scream? There's an excitement in there too. An analogy on that, I watched a documentary on river running in Alaska. And they showed a situation where a canoer—there was an older man, a younger person, and a young girl, kind of—I guess they had multiple canoes. They're on a river and they're coming down to spot where there are very bad rapids. They have a bad name like "Hell Something" or "Somebody's Gorge." And the river goes this way and it hits the cliff and comes back that way. When the river goes down this arrangement, there are two whirlpools that go in opposite directions, and the only way to get down there in a canoe is to know just how to go between the two. Okay, everyone has their own way of turning on and that was their way. Just as they entered that little double whirlpool thing, this guy's scream could be heard from all over the area. That was not fear. That was pure, absolute glee. That's the going-down-the-roller-coaster thing.

Same sort of business. One of my experiences, I was seeing what can bring on an altered state by excitement or some sort of religious fervor. One

LECTURE 15 ~ *Intoxicants*

time I was invited to attend a Baptist revival down in Oakland, 18th and San Pablo. And it was a real operation. I was invited in, I had to stand in the lobby, whatever you call that upper portion of a church.

STUDENT: Choir?

SASHA: Choir? No, there was a choir but that was in the front. But there was a lot of singing, and it was excitement. And it was going great guns for about an hour. And the minister said, "You know, that's it, folks. Those people who came for the entertainment, the TV's off. Show's over. From here on it gets serious." We stayed.

And my god, singing and enthusiasm. People were fainting. They had people with Red Cross badges on them and little stretchers, they would come in and take people out when they collapsed. It was a going thing and you could feel it. There are people who handle snakes in conjunction with religion. I saw a documentary on the cobra in India. They'll go out in the field before the sacred day of the cobra—whoever she is who has protected them or given them fertility or perhaps given them crops, I don't know which. But to them it's very big thing. They go out in the field, haul these cobras out. Always a few dozen people get bitten and a few are killed. They'll have an open line to the fridge with cobra venom sitting next door somewhere. And in they come. They bring in the cobras, they go through this thing and put them in little pots and then they release them and deal with them. And the fervor goes on, this excitement. People end up rolling around and babbling just totally, totally intoxicated with this situation through excitement. It may not be a drug getting in there, but . . . I mean, how many people have ridden a horse? Boy, the first time the horse breaks into a gallop? You know, whooo! What's going on there? Is it adrenaline?

Yes, it is. Your hair is standing up a little bit. You feel that excitement. But that is, I think, a lot of the reason why people use drugs. It causes that "I don't know where it's going; I don't know what it's going to do." People say, "This is a difficult direction and you shouldn't try this and it's going to be bad for you." Well, people don't always listen to, "This might be bad for you." A good example of that is nitrous oxide, which is one of the drugs that has been used in dentistry. How many people have used laughing gas? Good, almost everyone. So, you kind of know what that is. Going in, you don't know where you're going. It's going into a strange spot, and part of it is not where you get, part of it is the excitement getting there—the so-called derivative of the input. I have a bottle of nitrous oxide here. It's an interesting example; there's a story behind it. Part of the show-and-tell. By the way, in the chemical trade, these small ones are called bottles. Not because they're made of glass, but they're small and handleable. Lecture bottles.

Nitrous oxide. I wanted to buy some nitrous oxide and some chlorine for my lab use. So, I went over to Matheson Gas. I have no hesitation saying who it was, a big gas company over there in one of the cities that lies somewhere between Hayward and who knows where. Down in the industrial flats near the bay.

"I'd like to buy a bottle of chlorine and a bottle of nitrous oxide." I'd just walked in.

"Who are you?" Fine.

"We'll sell you the chlorine, but we can't sell you the nitrous oxide."

"Why not?"

"It's prohibited by law."

STUDENT: That's not true is it?

SASHA: Of course it's not true!

"So you mean you will sell me a bottle of a World War I gas that is extraordinarily toxic, with which I can kill the whole neighborhood, but you won't sell me a bottle of nitrous oxide, which is a standard oxidant in atomic absorption?"

"Won't sell it. It's against the law." It's not against the law. "The federal government won't allow it. The FDA won't allow it."

Hand—yes?

STUDENT: Why was he telling you it was against the law?

SASHA: He didn't want to sell it to me. I wasn't Mr. Standard Oil. I'm a walk-in; I just walked into the place.

I said, "But I understand you sell to individuals like the University of California."

"Well yes, but not to you."

"You're in interstate commerce, aren't you?"

LECTURE 15 ~ *Intoxicants*

"Yes."

"You won't be if you don't sell to me."

"The DEA prohibits it."

"Phone the DEA. Here are two numbers. One will get you to the narcotics branch; that's a laboratory. The other gets you to the agents' branch; that's intelligence. Phone them both if you wish. Use my name if you wish. Ask them to document that they won't allow you to sell nitrous oxide."

"Well, I've got to get permission from the boss."

"What's his name?" We had some excitement, tempers were going up. I said, "There will be blood—it won't be mine—on the wall if we don't. Let's get this thing straightened out." And pretty soon—

"Well, we'll let you know."

"No, I'm perfectly willing to wait." They didn't know what to do with me. God, how do you get rid of this kind of a person? So, I'm sitting outside having my lunch, waiting. I gave the afternoon to the project. Pretty soon he came back.

"Well, yes, we will. But you have you sign for it."

"That is fine." I said."Yeah, okay, fine, fine."

So they went out, I don't know what they did back there. They came out with a perfectly fine cylinder of chlorine, which I've used and it is chlorine. But the cylinder of nitrous oxide, I don't know. I just had enough time to work with it and I had to do some tests on it. This is the cylinder of nitrous oxide I got. Notice. The color code is not that of nitrous oxide. Nitrous oxide is on the label, but it is a handwritten thing that's been held down by magic tape. That is not a very professional way for a gas company—who's in the business of making professional gases—to sell a bottle of nitrous oxide.

I have not yet run a test of what's in there. I wouldn't be a bit surprised if at some level it might not be what's represented. In fact, what is represented by nitrous oxide is a tricky thing for those who use nitrous oxide. Nitrous oxide—a weird chemical thing, N_2O—two nitrogens and an oxygen. It's the most reduced nitrogen you can get. Sweet gas, as those who have tried it know. It is available both for medical needs and dental needs as a gas. It's mixed with oxygen. Pure nitrous oxide is extremely hazardous. Let me put that in great big emphatic letters, because you need oxygen for normal life process and pure nitrous oxide has no oxygen. Hence, you go into anoxia and you do not get the fuel you need for your cells in the brain. The brain shuts down in about three minutes. Pure nitrous oxide is deadly.

So, when you get it in a dental chair, you get it mixed half with air. Fine. You have nitrous oxide and the air is there, at which point nitrous oxide becomes far less hazardous. But commercial nitrous oxide is made industrially, made for, as I mentioned, atomic absorption. That is a way of determining the presence of certain elements. And sometimes for the flame, you use nitrous oxide as the oxidant. It's a big machine that can, in a flame, produce the actual elemental form from whatever is present; it absorbs an energy that is produced by the elemental form itself as a spectral line. You can tell how much the element is in there by how much doesn't make it out the other side of this little beam.

Nitrous oxide is commercially made, and sometimes, on certain lethal occasions, has been proven to have in it a few percent of another gas known as nitric oxide, or NO_2.

Yes?

Student: In some cars, it says NO_2. It says "nitrous oxide."

Sasha: It's a hazard with all kinds of words. These names are misused. Nitric oxide is a group that goes into so-called smog. It's the thing that creates oxidation products in the air and makes your eyes run in Los Angeles. It and NO_3 are the oxides that are from exhaust, because exhaust is from an engine. There's a high-performance engine, high temperature, high compression, and the nitrogen that's going into the engine is being oxidized to some extent by that oxygen that's there at that temperature. Under the conditions of the engine to produce these, often they're called NO_x, because they're lumped together in their analysis.

These are nitric oxides. They are extremely damaging to the lungs. This is, in essence, nitrous acid dehydrated into nitric acid—for all intents and purposes dehydrated. You're dealing with nitric acid. You know, nitric acid turns the skin yellow, becomes all kinds of damaging things. This is what it does to the lungs. This stuff is murderous. Nitrous oxide, on the other hand, is quite harmless. Nitric oxide is, however, in industrial manufacture, so there's no reason to be extraordinarily careful. You want it cheap and you want it plentiful. But nitrous oxide may contain a few percent of this. I know

LECTURE 15 ~ *Intoxicants*

two deaths in England that came from commercial nitrous oxide that was contaminated with nitric oxide. Medical and dental material is scrupulously checked to have purity.

The whole origin of nitrous oxide is a good story in its own right because it came into popular usage and was immediately of interest. As with many drugs when they first come in, it took a very strange route. It took 100 years for nitrous oxide to find its medical utility. I dug a little bit in the past. Priestley, Davey, names that are from the turn of the last century, they are well known in the area of gas discoveries. Priestley was the discoverer of oxygen, nitrogen, hydrogen, and he was also the first person to make and generate nitrous oxide and begin characterizing it. And I'll see if I have the—I don't know, I brought in a couple of quotations. Oh well.

Let me go back for a moment. At the time of Priestley, just before the turn of the 19th century, about 1790, perhaps 1800, there was a mythical gas that was known as "contagium." It didn't exist, really. But we have a lot of myths in our society of things that don't exist but are the causes of things. When you can't explain the cause of something, you create something that you can point to and say, "That's the kind of thing you've got to avoid." It may be a devil with horns; it may be some sort of a demon. And in this case, not knowing what caused disease, one of the theories that abounded was the use of contagium.

Contagium was a gas that you couldn't see, and you couldn't smell, but it spread and was the source of disease. Well, nitrous oxide when it was first developed by Priestley was clearly an unusual gas. It's not unknown in nature—nitrous oxide occurs to a very small degree, maybe a part per million, in the atmosphere. So it is a natural material, but a very scarce one. It was generated, and in some of the very early studies it was given to animals and the animals died. Of course, you cut off air from most things, they die. And that brought up the argument, "This is extraordinarily toxic." The same experiment, by the way, done with hydrogen would have shown hydrogen to be just as toxic, or nitrogen for that matter. Pure nitrogen is toxic on exactly the same basis. You must have oxygen in your system, and pure nitrogen has no oxygen. Air has about 20 percent oxygen, and that's what allows you to breathe. If you're in downtown Athens, the air has only 16 percent oxygen, which is an interesting and a sad testament to what smog, pollution, cars,

THE NATURE OF DRUGS

and taking out all the parks to put in buildings can do to a large city. But around 16 to 20 percent oxygen is a very necessary part of life.

Maybe it's worth going a little bit into the mechanisms of gas here. You can live on pure oxygen. In fact, you're required in the military, if you fly above a certain elevation, to put on pure oxygen. Above another elevation, you've got to have oxygen in a pressure system. You've got to have a certain amount of oxygen going in. When you expand your lungs and take in a certain amount of air, it's dependent upon the air pressure around you. If you expand your lungs in a vacuum, so to speak, you're going to expand that same amount of volume but you don't take any gas at all. So in a vacuum, you can't breathe oxygen. There's no oxygen there. If you get to higher and higher elevations, you have less and less air. The density is thinner. My first real firsthand experience with this was driving my car to the top of Pike's Peak in Colorado, I think it was. And I was wondering why all these cars were stopped off to the side and people were walking up to the top. The bigger cars stopped first and then the smaller cars. I had a little Volkswagen, and we went chuggy chuggy to about 200 or 300 feet from the top and I stopped, too. The car wouldn't run. There's just not enough air. The carburetor is the same size; there was the same amount of gas. There just wasn't enough air to make a combustion mixture to make that engine go. So people going above, I think it's 12,000 feet or 14,000 feet, must have oxygen by military command.

Actually, you can make it maybe to the top of Mount Everest without oxygen, maybe you can't. But at that level, you don't get enough air.

Now, the same but opposite is true when you go diving and you get too much air. And you can cut back on the amount of oxygen. But part of the problem there is that nitrogen, which is the other main component of air, tends to dissolve into tissues and into the blood very slowly, but very regularly, very tenaciously. It's held in the blood so when you come up, this material which has been put into the blood under pressure tends to expand. The concept is known as "the bends" in which bubblets of gas occur in the system. Excruciatingly painful and it can be lethal.

Hence the idea of decompression. Coming out of deep-sea diving is a matter of timing and allowing the gas to undissolve. These gases are all soluble in water. The fish don't drink water and get bubbles of air. Fish get

LECTURE 15 ~ *Intoxicants*

oxygen dissolved in the water and hence the aeration of water and getting gas in the solution. And these gases go into your blood in solution. So, what they have done, they have found that if you take out the nitrogen and put in helium, that helium goes into the blood much faster and comes out much faster. And you can decompress at a much more rapid rate. Helium is an inert gas. All these things are considered to be inert gases. Nitrogen is considered to be an inert gas. But none of these gases are inert when you get enough of them.

If you go to very deep levels without helium, but with nitrogen as the other gas, at about 200, 250—in fact, it's 210 feet in diving—you get into a situation known as "nitrogen narcosis," in which the nitrogen that's in there actually serves as a mentally confusing agent. There's a rule that is rigidly followed, that is when you try a new free-dive thing, you always have a line attached. You're free, but you're pull-uppable. And you must go and tie the ribbon on the line to prove the depths to which you were able to go, to establish a depth record or establish your diving record. But the thing is, at about 200 feet or so, it is not unheard of that the person who is diving won't turn, and keep diving down. Clearly a killing move, because they are at the ragged edge of survival, and they'll turn and dive in the wrong way. And that's why they have a line on. It stops them. It's known as nitrogen narcosis. That person is totally without proper judgment.

You take other gases that are so-called noble gases—krypton, xenon, neon, argon. They're not inert. In fact, xenon is a rather good anesthetic. Xenon is bound to the hemoglobin[10], and at sufficient amounts will actually cause a narcosis and a loss of sensibility. It takes a fairly large amount and mechanically it's hard to handle. But these gases do bind in the body, and they bind to hemoglobin, and they cause strange interferences with oxygen and probably with pH balance in the body.

I mentioned that oxygen itself is a strange narcotic. A good example, I don't know if I mentioned this before. How many times have said that phrase? Okay. Take a person who has lost his hemoglobin in its entirety—a

---

[10] Xenon produces an inhibition of *N*-methyl-D-aspartate receptors. Krypton, argon, and presumably radon show anesthetic properties at hyperbaric pressures, like nitrogen, but only xenon is effective at normal atmospheric pressure. Helium and neon are not effective due to producing convulsant effects at the pressures required for anesthesia.

person has blood, the blood is there. But the hemoglobin in the blood is the thing that carries the oxygen; it does not dissolve enough in the liquid of blood. You have to carry it in an associated form of hemoglobin. The oxygen ties onto the hemoglobin molecule. If a person has no hemoglobin, he cannot get enough air, enough oxygen in his brain to live.

How do you get rid of hemoglobin? Say a person is in carbon monoxide, to where all his hemoglobin is tied up with carbon monoxide. That person cannot live because the person has no way of getting air, the oxygen, around. Carbon monoxide and oxygen occupy the same thing. Oxygen goes on and off easily. Carbon monoxide goes on and off very slowly, so a person has no functional hemoglobin. So you put them in a hyperbaric chamber. Go over to Mare Island if you have enough time that they'll survive the trip. Stick them in the high-pressure hyperbaric chamber where you put people who come up with the bends. Turn on the valve. Put them up to about four atmospheres. That means at four times the normal pressure of pure oxygen. So they're getting about 20 times the normal amount of oxygen. Your oxygen is normally 20 percent, so that would be 100 percent—it's fivefold. And four pressures is twentyfold altogether. They'll breathe normally and they'll live normally.

They can get enough oxygen in their brain, and in their tissue, by just driving the oxygen into solution under those extraordinary conditions, so that they can live. And they will live until they generate more hemoglobin or until the carbon monoxide blows off. The physician doesn't dare go in with them because oxygen at that twentyfold increase is right on the edge of a convulsive thing. The person in there may convulse. But the physician is better to stay on the outside and monitor the blood samples—there are ways of doing this—until the person gets to where the physician can begin dropping off the oxygen, because convulsion is a very real thing. Oxygen is not an inert gas from a physiological point of view. It's a convulsant.

But it takes 20 atmospheres, 20 times the amount of oxygen that we normally would get, to convulse. Anyway, this first contagium gas was absolutely—like tomatoes in the 19th century—poisonous. No one dared go near them. Who has ever heard of tomatoes being poisonous? So have I. Where's the documentation of it? What happened? I'm not scared to eat a lettuce and tomato salad. When did tomatoes become nonpoisonous? They

LECTURE 15 ~ *Intoxicants*

were poisonous 100 years ago. I don't know—I can't get at it. There's a good example of something that everyone knows and is promoted. One that came up in a big discussion two days ago is that everyone knows when potatoes turn green, they are poisonous. How many people have heard that? What kind of poison is that? Has anyone eaten a green potato?

No, of course not. [Student raises hand.] Oh, and you're still here. What happened?

STUDENT: Nothing.

SASHA: Okay. Then what's this poisonous business?

STUDENT: I just heard that it's supposed to be poisonous. What happens is it turns green because it's too near to the top of the soil.

SASHA: It gets light? Chlorophyll, very reasonably. But chlorophyll is perfectly fine. I've eaten lettuce salads myself. What's wrong with chlorophyll?

STUDENT: Nothing.

SASHA: Where does this stuff come from about potatoes being poisonous? Are they? They may well be. But you're in that situation of, "I'm not about to go out and try one if they say it's poison." Mushrooms—we're a nation of mycophobes. We're scared to death about mushrooms unless they come from Safeway. "You were going out in the field and were handed mushrooms and you ate them? You're weird."

And yet there are nations in the world—Eastern Europe, Russia—they have as many poisonous mushrooms there as we have here, and they'll go out on a picnic and gather mushrooms and cook them with great glee because they have a marvelous taste. I mean, I'm not going to advocate that you either go out and eat green potatoes or gather mushrooms from the local cow patty. Just because I'm saying that most of them probably aren't poisonous, the emphasis I'm making is these things are known to be poisonous, but you can't pin down the whyness and whereness.

I can't go into the literature and find what the poison is in green potatoes. I can't go and find out, except that everyone knows that tomatoes were deadly poisonous back in the days of New England. Back in the middle of last century. They were called "love apples." "Don't eat the love apples. They are ornamental; they will kill you." And no one ate them, and by golly, no one died. And really, I just don't know. After all, we were talking a few lectures ago—when you write up lectures and you've been giving lectures, I lose track of when it was—I talked about atropine, scopolamine, datura, henbane, belladonna, this whole class of compounds.

They're all Solanaceae and they are very closely allied with tomatoes. In fact, a very dirty trick—you can do it if you want to get at someone someday—is to raise an atropine plant, a datura, and chop it off and splice on a tomato plant. They're close enough that you can graft a tomato onto a datura root. All the alkaloids in the datura are made in the root and they're transported to the aerial portions of the plant. And so up goes this beautiful plant with great big red tomatoes, loaded with atropine. And they are appropriately poisonous. They don't look it, but they taste it. They have scopolamine and atropine in them.

Don't do it. I was making a little silly suggestion. [Laughter.] But it could be done, let's put it that way. Ah, but the point is, could it be that the tomatoes 150 years ago, since they are basically the same plant, actually made atropine and scopolamine? And by golly, they were poisonous, you ate them, you got real funny, your eyes dilated, you went funny and dried up, weird things going on. "Don't eat the tomatoes." And maybe a sprout came along which suddenly evolved and had lost the gene for making the atropine, or it was a little hybrid off to the side or something happened. And the tomatoes kept going. They were known to be poisonous, but they actually weren't. And some kid went out sometime somewhere and ate a bunch of tomatoes and came in and said, "Ma, that's kind of good."

And she shrieked, "You're still alive." She tried it and pretty soon the idea went throughout the Western Hemisphere. I don't know the answer. You can't get a tomato from the 1850s. I don't know where to find one. But I don't know how you can get at this kind of myth. Or is it myth? Is it fact and things have changed? Well, the same thing with nitrous oxide when it was introduced. It was known to be the factor that was the source of disease. Contagium was the name of it. And there were a lot of people who gave very intense medical arguments that it should be absolutely avoided, and could not be dared to experiment with it at all.

LECTURE 15 ~ *Intoxicants*

Until a fellow came along by the name of Humphry Davy, who was an assistant to a physician, who listened to Priestley's lectures and said, "It sounds kind of neat." He said it had a sweet taste—at least a sweet smell. At least he tried smelling it. So, he learned from Priestley how to prepare this dangerous gas. One night while Dr. Borlase—he was the physician—was upstairs sleeping the sleep of the just, Humphry was in the dispensary, preparing nitrous oxide.

By the way, it's made from ammonium nitrate by heating. It's one of the commercial ways of making it. Be careful. Ammonium nitrate carries with it the possibility under the incorrect heating conditions of generating the nitric acid aspects—the nitric oxides. They have to be removed chemically. [Reading:] "Drawing deep breaths of it into his lungs he thought it likely that his last hour had come. He waited. He breathed some more in. Third time, wonderful. Not only was there no sign of sudden death prophesied by Dr. Mitchell (he is the one who was the spokesman of the lethality of this because it caused terrible things) but something else surprising happened. The young Humphry continued with increased boldness, he inhaled more and more nitrous oxide and a strange, agreeable sense of lightness pervaded the body. His muscles relaxed, there was a pleasant sensation in the chest, in the limbs. Then the young man became aware that his hearing was unusually acute, and the agreeable sense increased to become an unspeakable cheerfulness. He wanted to laugh. He had no choice but to laugh. He went on laughing in spite of himself until he put aside an empty flask."

Okay. Yeah, what he didn't notice—one thing that came from further experimentation with others—is he felt no pain. In fact, in early experimentation, they actually found people who would walk into something, would be aware of having brushed something, look down, and they would have cut their leg and the pain was not there. The anesthetic properties of this, of ether, of chloroform, all were being more or less explored at the same time.

The physician got kind of involved, because he noticed that often when he would come down, when emergency things came in the middle of the night, Davy was far too cheerful, and he wondered what was going on. Davy began explaining that he had been breathing this gas. And the physician was curious enough to get his nose into it, so to speak, and smell it. He found it interesting and wondered if it might not have medical utility.

THE NATURE OF DRUGS

This kind of thing got out into the surrounding public, and he lost his practice. They were really closed down because they were experimenting with things that are obviously lethal and had no medical utility. The same story that you heard with morphine. The same story has come time and time again with the introduction of new things. You'll find it with ether; you'll find it with chloroform. These are now considered more or less old-fashioned things that are chemical things. But they were very distinct intoxicants and were broadly used.

In fact, on the nitrous oxide, I did dig out one thing over here. It's an ad. I've got to read the ad to you. This is about 1850 or 1860 in this country. Dentists were the first ones to really promote nitrous oxide. "A grand exhibition of the effects produced by inhaling nitrous oxide, exhilarating, or laughing gas, will be given at Union Hall this Tuesday evening, December 10, 1844. Twelve young men have volunteered to inhale gas to commence the entertainment. Eight strong men are engaged to occupy the front seat to protect those under the influence of the gas from injuring themselves or others. The gas will be administered only to gentlemen of the first respectability. The object is to make the entertainment in every respect, a genteel affair."

It really became the party thing. Same thing with ether—etheromania. I'm bouncing around; I always bounce around. Ether was consumed as a party thing, up to 100 grams a day. Consumed, I mean that it went down the mouth as well as breathing it. There is a material called "Liquor Hoffmanni," with two fs and two n's, that was used because alcohol was not considered particularly cultured to drink in public. This is back at the end of the last century. Especially by ladies. Any lady of culture would never drink alcohol, so she would carry a little vial of this Liqueur Hoffmanni. It was three parts ethanol and one part ether. And that was not considered improper, because a way of getting off of alcohol dependence was to use this Liqueur Hoffmanni, which was a mixture of alcohol and ether.

And I am just vaguely reminded of the treatments for people who are on methadone. You have this idea of supplying the drug under another name—in this case heroin under the name of methadone—to avoid the stigma, avoid doing something wrong in the social environment, but nevertheless allow you to continue what you're doing. Chloroform was introduced again about the same time, before it was used as an anesthetic. Chloroform's main

LECTURE 15 ~ *Intoxicants*

use in the latter part of last century was an anesthetic, but it was broadly used socially. Here you had quantities of chloroform consumed upwards of a liter a day. I mean, this is a quart of chloroform. A lot of it was consumed by smelling, putting it on rags and breathing. A lot of it was just drunk and caused irritation, vomiting, great irritation, and looseness of stools, and very bad physical problems went along with it, but it was exciting. It was actually a turn-on. And it was generally used. What is it now that they're banning the use of? Is it airplane cement?

STUDENT: Amyl nitrite?

SASHA: Amyl nitrite, there's a good example. Amyl nitrite, I haven't even thought of that one. Superb idea. Amyl nitrite is the nitrous ester of amyl alcohol. Amyl alcohol on nitrous acid is what's called amyl nitrite It's not even amyl—it's isoamyl in the medical use. There are little vials, so-called poppers in the trade, and they contain a little fluid, and then you breathe them. It's a very intense feeling of pressure in the chest. Your blood pressure drops way down, and your eyes can become somewhat reddened.

It causes just an immense drop of blood pressure and a drop of heart function. The medical use is straightforward in the case of angina, where you have pain because of inadequate circulation. You take a little bit of nitro anything under the tongue—nitroglycerin has been used, nitrotoluene has been used, trinitrotoluene, or amyl nitrite. It's the isoamyl ester of nitrous acid. All of these things act equivalently to drop blood pressure and by reflex dilate vessels, and if there's a small obstruction somewhere in the vessel, it is allowed to pass, and hence the relief of angina. And it's the standard. "I've taken my nitro pills." You've probably heard this said by people. "I want to go back and get my nitro pills in case I have an attack." It's standard fare.

The nitrite is standard fare. It is under FDA control, because it's a prescription drug. So, what has occurred in the last 10 years is some people turned to butyl instead of amyl; they'll make the butyl ester of nitrous acid. It has the same effect. It's sold very heavily in the gay community under weird names like "Oil of Locker Room" and "Oil of Jock" or something. I don't know the terms. I should get a collection of these bottles for the museum sometime. But it has a great reputation of being an aphrodisiac, which, in a sense, it might be, because it causes dilation, easy flow of blood, and hence sustained erection more easily. That's probably the mechanism of its reputation. But the government has sort of shied away from doing anything about butyl nitrite. Amyl nitrite, isoamyl nitrite, it's under medical control. Butyl nitrite, I don't hear about it.

I mentioned the idea of Favor Cigarettes, which are not cigarettes but are bits of nicotine on tubes that you can suck through to get nicotine, in which there's no government regulation. It lies outside of anyone's purview. They don't want to get into it.

STUDENT: There is no difference in the effects at all?

SASHA: No. Slightly more volatile, slightly more inflammable. Will last less long physically, but to a large measure, nitrite in any form is equivalent in the body in causing dilation. So equivalent effects.

But what are you going to do? As I think I mentioned another time, the narcotics agents are not the people who go into head shops, or especially in head shops where they sell things for the gay community. Not so much that they object to going in—they object to be seen coming out. [Laughter.] And that is an image that cannot be there. So what goes on in there is really quite colorful, from the point of view of drugs, because there's no regulations on these areas. There are a lot of drugs that have gotten on the market and have been sort of excused. I mentioned this with the phenylpropanolamine, which is not that innocuous of a material. It's heavy on the heart. It's a stimulant. It's been used as the so-called peashooters—the materials that are sold as look-alike pills through *High Times* and a thousand other sources.

They are stimulants, usually ephedrine, and are a little bit heavy on the body. But they do cause a heart rush and a stimulation rush. Overdoses are not that uncommon at the SF General Hospital with people who've just taken too many of these look-alikes. But the FDA says, "Really, under proper usage and with proper amount of usage, they're not hazardous." And so, they officially disclaimed any connection whatsoever with phenylpropanolamine.

There's a fair amount of industry pressure because the industry wanted a nonprescription appetite suppressant thing and phenylpropanolamine will serve that purpose. So these materials are being sold with absolutely no regulation whatsoever. I'm not a keen one on regulation, but on the

LECTURE 15 ~ *Intoxicants*

other hand, I'm not a keen one on the hypocrisy when they will regulate amphetamine or not regulate ephedrine.

The state of California regulates ephedrine. You have to sign to get it. But on the federal level, it's a mixed bag. It's an inconsistent pattern. I really believe that none of them should be—and this is my personal belief—that none of them should be brought under any sort of an advertising control, such as it is they're more widely known because of the restraints on their trade. But there are some that are and some that are not. Mixed thing.

By the way, if anyone is not familiar with ether or chloroform, I brought in samples of each of them. Smell, but if you smoke, don't smoke in the presence of the ether. Both will burn. Chloroform is not very inflammable. Ether is extremely inflammable. Chloroform has, by the way—yes?

STUDENT: Is that diethyl ether?

SASHA: Diethyl ether. Yeah, okay. That is a terminology that just became sloppy, but common. Ether is the euphemism for diethyl ether. There are millions of different kinds of ether. Ether is a general chemical classification of a type of compound. So the compound I'm talking about is diethyl ether.

ANN: Are those two bottles going to just evaporate?

SASHA: In time, yeah. They are quite volatile. If you put a little bit out on a surface, you'll see the volatility and you'll see that it will disappear quite readily. In about three minutes, it'll be totally gone. But it takes air circulation to evaporate. They both boil well above room temperature. Ether is about 35 degrees; chloroform is, gosh, I guess about 50 or 60 degrees. But in a closed container, they will stay indefinitely. In an open container, they'll evaporate according to the amount of air they can get to them to carry the vapor away.

Chloroform is a trickier thing, and they both have their hazards. I want to mention hazards. Chloroform is strictly one carbon with one hydrogen and three chlorines. Chloroform does burn, but burning chloroform or chloroform exposed to air carries with it a very strange and very unexpected hazard. There is a tendency for it to oxidize. This kind of compound is totally unstable and will split out HCl to form another substance that is excruciatingly poisonous. This is where I want to put a big caution. How many people have heard of phosgene? One, two—

STUDENT: A poison gas?

SASHA: Darn right it is.

STUDENT: World War I?

SASHA: World War I. Phosgene is murderously poisonous. Let me tell you a firsthand experience I had with a person with phosgene. This is about 20 years ago at University of California. I happened to be nearby at the time. It was on a Sunday. If I had known then what I know now, I might have been able to help more than I did. A person was in the top floor of Gilman Hall, which is one of the old chemistry buildings, on a Sunday and he happened to be carrying a liter bottle—around the size of a big grapefruit—about half full of liquid phosgene. Phosgene is a gas, but it was liquefied—it was cold. He knew it was murderously poisonous.

How it came to be—one will never know—he dropped it. He held his breath. He went to that end of the hallway. No one else was on that floor. He knew he was the only one on the top floor. He dropped the flask. He went and opened the window at the far end, went out of the doorway, locked the door behind him (the other doorway to the stairs; there's another set of doorways over there), locked the door behind him so it couldn't be opened. He went down the stairway down to the next floor. Went up to the other door, closed that door and locked it. Then went over to the other chemistry building. I happened to be at another building; he came past where I was. That's where I happened to see him. And he went over to Cowell, the hospital, and turned himself in as having been possibly exposed to phosgene.

He was unaware of having been exposed. It has a smell. I forget what smell. I always confuse newly mown hay with geraniums. One of these gases is a newly mown hay smell and the other is a geranium flower smell. It's one of those two. I've not sniffed it to verify.

He went over to Cowell Hospital. They put him on oxygen for safety, made him inactive, and put him on oxygen. He was dead the next day, 24 hours later.

LECTURE 15 ~ *Intoxicants*

STUDENT: God.

SASHA: He diffused. The liquids came into the lung. The whole lung tissue was destroyed. There was edema. Inside the lung they put tubes down and sucked it out as much as they could. He suffocated. That little bit of exposure was all it took.

If you are ever in a situation where a person gets exposed to phosgene, they should be inactive. Do not move. Carry them, get a stretcher. Carry the person to a source of oxygen. Bring oxygen to the person if you can. No activity, no motion. No activity whatsoever, and get oxygen. But that was to me a real startling thing. You see that minor, minor exposure of a very poisonous, extremely poisonous gas. A World War I poison gas. One of our contributions to society is to work this out. I came across a fascinating book. In fact, I have a whole lecture on war gases and war crimes and war things in general called "A Higher Art of Killing."

Yes?

STUDENT: Is it true that phosgene is like chlorine gas, those kinds of gases, if you hold your breath it can be in your lungs and cause poisoning? Are you safe if you hold your breath?

SASHA: My gut feeling is you're safe if you hold your breath. But when you hold your breath, the instinct is to take a lot of breath first. And I think that's probably where it happened. Try this some time. Experiment. It's an interesting little thing. Try experimenting on how long you can hold your breath having taken the big breath first, or just quietly exhale, counting that as part of your breath-holding, and hold your breath without much air in your lungs. There's not much difference.

That big, big pile of breath doesn't buy you much time. So if you ever get exposed to a gas, I would recommend you don't grab a breath to hold. Just stop where you are. Now, I don't know that. I don't feel how it could.

The material is very reactive. It's an acid chloride; it goes after tissue. But that's what happens when you get chloroform in a fire. And so yes, chloroform will burn, but in the course of burning you're generating a little bit of a very ugly material. And to a large measure, that's one of the reasons that chloroform has been dropped out of medical use. It has that potential of generating phosgene. There are other chlorinated solvents that do not.

Yes?

STUDENT: People do burn chloroform using a hood. They do burn it in a Bunsen burner.

SASHA: Yes. A good clue in a Bunsen burner in general is what's called the Beilstein test. A Bunsen burner that is burning a relatively neutral, fairly hot flame in the presence of chlorinated hydrocarbon will have a green flame. Sometime when you're at a Bunsen burner, take a little drop of dichlor, a little carbon tet or something on the bench, and put it in the flame.

There are other allies of chloroform. What was the material? I think the original material that was in another thing they tried to ban because everyone was filling their handkerchiefs and breathing it was this Wite-Out stuff, the correction fluid for typewriters. The type you would dab on the paper and go back and type, or get it all over your pencil.

The fluid in that was trichloroethylene for a long while. These are kind of neat things. Trichloroethylene was actually used as an anesthetic—a very effective, very good anesthetic. It causes a giddiness, a lightheadedness, an intoxication. It falls in the area of the excitants. And this was discovered by people who found that they could go through a few bottles of correcting fluid. Ah, glue sniffing! The whole concept of glue sniffing.

Benzene was a major problem in Europe at the turn of the century. They had discovered petroleum but had no cars to use it. So it was quite a novelty. Benzine, with an *i*, is a term for petroleum ether, for the very low boiling part of petroleum. And gasoline sniffing, glue sniffing. Toluene is a major thing in the stuff you mix with polystyrene to glue polystyrene together with other polystyrene—this big gluey, gooey liquid goop is polystyrene dissolved in toluene. So, you breathe the toluene; again, same sort of thing. Many things along these lines are intoxicants.

Yes?

STUDENT: Toluene?

SASHA: Toluene.

LECTURE 15 ~ *Intoxicants*

STUDENT: It's a carcinogen, isn't it?

SASHA: Well, yes and no.

STUDENT: It's on a lot of lists.

SASHA: Oh god, you better believe it's on a lot of lists, but is it a carcinogen? I don't know. Benzene is the first to be given that label. And everywhere they used benzene, they moved to toluene because it didn't have that risk. Then they're beginning to say that toluene has that risk; there's not many other places to move it. And so they're probably not going to push it too much. Benzene is a simple aromatic ring, six carbons. Toluene, it has a methyl group on it.

STUDENT: What about cyclopropane?

SASHA: Cyclopropane is an anesthetic gas, no smell. And it's sufficiently bizarre and rare; I don't know if it's being abused. I'm sure if you can get it, you'll find it being used, because almost all of the anesthetic gases at lower levels cause the lightheaded intoxication. And that's true of ether, chloroform; it's true of nitrous oxide. Cyclopropane is an anesthetic gas. But I don't know of it ever having been used. It would obviously have to be under a tent.

STUDENT: *Goodman & Gilman* described it being used as part of the training for anesthesiologists.

SASHA: Yes, it's an anesthetic gas. It's used medically. Why are these things dropped out? One of the negatives to cyclopropane or to some of these gases, like ether, is their inflammability. And there is no more damaging occurrence than to have a lung filled with a mixture of air and ether and have a spark nearby because then the whole ignition occurs right in the lung. It's very damaging. And so, the move was initially to form nonsparking instruments in surgery. Gosh, I forgot what the composition is. But it was a metal that did not have the ability to cause any sparking when striking other metal, when you handed two instruments to a surgeon with ether all around.

How many people are familiar with the smell of ether? Boy, I'll tell you, that's the first anesthetic I ever was given for an appendectomy at the age of 10 or so. And it took me a long while to disassociate the smell of ether with

THE NATURE OF DRUGS

that memory of being in a corridor at this hospital over in Berkeley, going down in a wheelchair toward surgery. I tied that into the smell of ether. It's very hard to shake. Ether was discontinued for a couple of reasons. One is the inflammability. And secondly, there is a small but real permanent brain damage from its use. It is about—who knows numbers—like one in 1,000 or one in 10,000 people will have neurological damage from it that doesn't seem to reverse. And when you see it, it's tragic and that kind of tragedy is not needed. There are other anesthetics that don't carry that risk, and the ether has largely been dropped for those two reasons.

Another point of ether, and this is purely from a chemical point of view, but it's a hazard, nonetheless. We're talking about hazardous chemicals which are hazardous drugs. There is a phenomenon of ether of getting old. And people who have worked in the laboratory will pick up this knack of being able to open a can of ether and smell it. "Aha, good." Or, "Yuck, it's old." It's a combination of getting peroxides in ether. It's extremely prone to form peroxides.

It somehow associates with oxygen, forms peroxides, which destroys the intactness of ether. But also, it gives an odd smell to it and chemically it's a dangerous thing to use. Physically, these things become extremely dangerous, especially with compounds that have branched chains. Isopropyl ether is one of the most vicious. This is a problem with ethers. Again, isopropyl ether is a good intoxicant, but something that is extraordinarily dangerous from the point of view of hazard, and explosion. Anything having this group on it will tend to form peroxides very rapidly. So if you ever see a bottle of isopropyl ether, and obviously it's been around for a while, stay away from it. Get someone who knows how to handle it. Treat it as an explosive bomb.

I saw a rather gruesome picture in a trade journal about 5 or 10 years ago, in which a person had tried to undo the stuck top of isopropyl ether bottle. Some of these ethers will develop white crystals growing in the bottom of them. You are looking at the actual peroxide that is explosive, and it will actually crystallize in the ether. If you ever see ether of any kind with crystals in it, get away from it. The isopropyl ether in this one case, the crystals were forming in the bottom. It was a one-liter ground glass bottle. The thing at the top was stuck. And there were crystals growing at the bottom but there also had been crystals around the ground glass at the top.

The person wanted to get at the ether. It would not open with a normal

LECTURE 15 ~ *Intoxicants*

effort. And what he did, he took it as you're prone to do with one hand on the glass top and the rest of the bottle he held tight into his gut. Because there you can get at it, that feeling of "I'm really going to open this son of a gun." He got in there and it detonated. It detonated and disemboweled and killed him at the spot. Just the crystals that were around the top of that thing was enough to kill him. If you see crystals in a bottle of ether, get away. There are people who make their livelihood, they get their excitement going to things that are extremely explosive and extremely hazardous; they are the people who love going in there and knowing which wire to cut to defuse the bomb. I mean, that's their business. Let them earn their fun and their jollies by dealing with ether that has crystals in it. Don't empty it out yourself.

What they will usually do with something like that, they will carry it inside of a meshwork of steel, flexible, that will keep the fragments from going anywhere. They get it out somewhere in the open and then open it up and burn it. They destroy it with a bullet, open up the container to ignite it and burn it. Almost anything that's explosive, under the right conditions, will burn relatively safely. The idea that you have an explosion of ether, but you can put a match to it. I could put this into a little beaker and put a match to it and have a nice flame going. It could be perfectly fine, but under certain conditions, it's explosive. It has to have that openness. You can even burn dynamite with great safety. Old dynamite is extremely hazardous because of the crystallinity that sets into it. But it can be burned more safely.

Okay, that has nothing to do with excitement. Yes, it does. Spices. How many people use spices in cooking? A cabinet full of spices. Marvelous things. Do you realize where they originated from? Way back in the Marco Polo days, they had caravans that went to China and brought spices back to Europe, sold for tremendous prices. Obviously you're not going to send a person through all that wild country with a bunch of camels to bring back something and not make it worth his while. A lot of them never made it back. What's so great about spices?

STUDENT: They mask the taste of spoiled meat.

SASHA: Yeah, I've heard that. I can see this bunch of people somewhere in the slums of London who have an old, somewhat tainted deer that they poached from some royal garden saying, "You know Maude, this meat is kind of high. Why don't we go and get some spices and hide the smell of this meat?" God, the spices sold for a king's ransom. I don't have a feeling they came back for meat. I have a feeling they came back because they had effects, as some spices still do today, and they were a novelty of course. The idea of a novel effect is worth something. I mean, that's part of the excitement.

"I've never tried that before. I think I'll try that." The first time I ate kiwi I said, "You're out of your mind. Green things with seeds in them?"

"Give it a try." You know, that novelty and excitement is in there. And the spice had that excitement but with a high cost. A lot of the chemicals I've worked with, and made psychoactive materials out of, have come out of the spice bottle: nutmeg, parsley, dill, apiole. All have been very rich starting materials. Once you add ammonia to them, they become psychoactive compounds. So I'm wondering if there might be a fermentation that's been lost? There may have been spice sprouts that have come in and diluted the content of the original spices. Nutmeg is still a standard thing in prisons. People find the nutmeg out of a kitchen that's using the nutmeg, and it puts them into a strange place. Your body temperature drops to about 94, 95 degrees. And you don't try it very soon again, but you are in a different place with nutmeg. It's a rough one. You can try it if you want. Take about 20 grams, oh maybe one whole can of nutmeg. And it's heavy on you. But it does produce some strange effects. And I'm wondering if spices as a group had had a history that we've lost in this way.

Yes?

STUDENT: How about smoking spices, like cinnamon?

SASHA: Probably. It would be consistent with this. I would expect smoking to be a fairly standard way. Except, remember, smoking didn't exist in Europe at that time. So they would not have done it by smoking. Smoking as we know it was a new-world invention. But most of the oils that I think are involved are volatile. I mentioned clove cigarettes, same idea. Cloves are a real flavor.

Yes?

ANN: Oh, what is the famous ointment that the witches were supposed to put on themselves to take a ride on their broomsticks?

LECTURE 15 ~ *Intoxicants*

SASHA: Oh, okay. This is datura. A plant broadly used. Again, I mentioned datura in the handout on parasympatholytic agents. But in the idea of riding the broomstick, the datura was made into an ointment and rubbed into the soft tissue. In the case of witches, the vagina and the broomstick riding, and the feeling of flying, the feeling of losing sensibility was an outgrowth of that and was ascribed to the witches' actions.

Datura is one of these delusional drugs I talked about. It falls into this PCP/ketamine world where you become quite dissociated from your real world and go totally into your mind. And a lot of it is not recalled. This was not really the spice category. But nutmeg, certainly mace, which is identical for all intents and purposes. It's the outer portion of the meg. Parsley. Parsley is a weird thing. Why is parsley so favored? Why did parsley get into the role of being the symbol of correct cooking? The idea of the dish coming from the kitchen, the head chef looks at it, he approves, he puts a piece of parsley on there as his signature of it having been approved. Why parsley?

STUDENT: Breath freshener?

SASHA: Pardon?

STUDENT: Breath freshener at the end of a meal?

SASHA: Why do you put it in the meal then? Why don't you get a mint as you walk out and pay the bill? I mean, it's in there. It's got a sharp taste. How many people like it? I don't particularly. How many people don't like it? Half and half. And yet it has some fantastic oils in it.

ANN: It's supposed to be very high in vitamin C.

SASHA: Oh god, everything is supposed to be very high in vitamin C. I mean, how many people go out and have rosehips for breakfast? They're very high in vitamin C too. I don't know, but a lot of these things have come along as tradition; the spice cabinet is filled with things that contain chemicals that with just very simple modifications become psychoactive. Maybe it's a change in our metabolism. Take—I guess you're not allowed to take it now—the safrole, isosafrole thing, of root beer.

Root beer they sell now of course is root beer flavoring and some caramel that's been added, put in a can of carbonated water, sold in a thing that says "grand old original-style root beer." It has nothing to do with root beer whatsoever. The root beer of 40, 50 years ago was a fermented-root beer. It actually came from a root. It was put in there, it fermented, and it had marvelous flavors known as safrole and isosafrole. Safroles were flavors that were brought in from centuries ago[11], through the Near Eastern trade.

And safrole—imagine a metabolic transformation in the body. If you can take an essential oil and add an amine to it, add ammonia to it—ammonia, ordinary ammonia, the stuff that you wash out your kitchen sinks with. It's a good old-fashioned chemical. Add ammonia to the active position in safrole, you make MDA, which is a well-known psychoactive drug. So, in essence, safrole is one molecule of ammonia away from MDA. Myristicin from nutmeg is one molecule of ammonia away from MMDA, which is a moderately well-studied psychedelic over the last 15 years.

Anethole, which is the acting component of anise, or what is finocchio, fennel, which is that real frothy green stuff that grows in parking lots, and you push it and it smells like anise, like licorice. Anethole, which is the active component of it (I call it active because I think it's one of the contributors to the action) is one molecule of ammonia away from PMA—para-methoxyamphetamine. That was very broadly used about 10, 15 years ago. It's synthetic and a moderately heavy chemical, a lot of pressor effects, but it does have mental effects too. So a lot of these essential oils—dill, the active component of dill that you put into pickles for flavoring, is one molecule of ammonia away from dimethoxymethylendioxyamphetamine, which is a big long pile of names without a trivial name, but a very active hallucinogenic that has been published and been explored a little bit.

So, there's a lot of stuff in spices that are very, very close to this. Eugenol is certainly known to be an active material in its own right. Eugenol and isoeugenol are essential oils in the same area.

Yes?

STUDENT: Aren't they in cloves?

SASHA: They're in clove; I think clove has a good known reputation of killing a toothache. You put a clove in and around a tooth that has pain and

---

[11] Safrole is most commonly thought of as being derived from New World sources such as *Sassafras* and some *Piper* species, but it is also present in some *Cinnamomum* species.

LECTURE 15 ~ *Intoxicants*

Essential Oil	Plant Source	Compound — derivable from essential oil by addition of NH₂
Chavicol (1) Anol (1)	Oil of Bay, of Betel Leaf:	Hydroxyamphetamine (stimulant)
Estragole (1) Anethol (1)	Oil of Basil, of Anise, of Fennel:	4-MA (stimulant)
Methyleugenol (2) Methylisoeugenol (2)	Oil of Snakeroot, of Ylang Ylang, of Nutmeg, of Pimenta:	(inactive)
Safrole (2) Isosafrole (2)	Oil of Sassafras, of Camphor:	MDA (psychedelic and stimulant)
Croweacin (3)	Oil of certain Citrus:	MMDA-3a (psychedelic)
Elemicin (3)	Oil of Nutmeg, of Elemi:	TMA (psychedelic)
Myristicin (3)	Oil of Nutmeg, of Parsley:	MMDA (psychedelic)
Asarone (3)	Oil of Calamus, of Rat Root:	TMA-2 (psychedelic)
Apiole (4)	Oil of Parsley:	DMMDA (psychedelic)
Dillapiole (4)	Oil of Dill, of Bamba:	DMMDA-2 (psychedelic)

it numbs it to some extent. It's a remedy that has an old reputation, but it also works.

ANN: Is that oil of cloves?

SASHA: Oil of cloves, as a topical anesthetic. And that's exactly what was used in the clove cigarettes, where you mix the clove into the cigarette. Real tobacco, but you numb the respiratory bronchi and so you can smoke a lot of it. Hence you get a lot of nicotine in because you have in essence anesthetized yourself by using the clove. Yes?

STUDENT: What's the one that produces MDMA?

SASHA: It's MMDA. MDMA is a modification of the isosafrole system. MMDA is from myristicin, which is the principal component in nutmeg.

Alrighty. What other things do I have on this big list to talk about? Ether, chloroform, benzene, nitrous oxide. I think in the very earliest part of the course I mentioned betel nut, and someone in the class raised their hand as having used betel nut. Is that person still here? Yes, you have tried betel nut? Why? What does it do to you? I've never tried it.

STUDENT: It's kind of like the first couple of times I used tobacco. It's a euphoric kind of high. Some friends had it. I wanted to try tobacco, I tried chewing tobacco, at the same time I tried betel nut. I kind of got like a reddish liquidy spit the same way that you do with tobacco. I smoke cigarettes very seldom. If I take a hit off a cigarette, I get very high, a kind of sick, euphoric feeling. The same way, that's how it made me feel. A little high, kind of nauseous.

SASHA: From your face it was not the terribly nicest high, but it was there.

STUDENT: Yeah.

SASHA: Okay.

STUDENT: It was very numbing, almost painful.

SASHA: Yeah. I think I mentioned then it was the fourth major psychoactive drug in the world. But I thought it was used over there: Asia, Korea, Southeast Asia, India. Every time I turn around, I find it's used in this country broadly. I got into a fascinating discussion between a Sikh and a Hindu, two people working over at the Lawrence Radiation Lab. Really, it's a pleasure to talk to the two of them when they're together, because they're both Indian, but they're not quite in the same caste. And it was unfortunate as the person who's a slightly higher caste, who's the Sikh, only has a bachelor's degree. The person who is a slightly lower caste, who's the Hindu, has a PhD. So it's a charming thing to watch them talking.

One will say, "I stopped using betel nut." This is the Hindu who is the PhD in botany. "I stopped using betel nut when I went to graduate school because my parents told me it would impair my memory and would cause me to think a little funny. And so, I didn't use betel nut at all through graduate school. I got my PhD and it didn't matter anymore, and now I use it. I have superb teeth. It's kept my health excellent." And this, that, and that. The praises of betel nut.

LECTURE 15 ~ *Intoxicants*

At which point the Sikh said, "Well, in the northern part, in Punjab, we have excellent teeth too, and we don't need to use drugs." What a marvelous—the snipe, snipe, snipe, all the time. It's marvelous to talk to the two of them. Anyway, this botanist brought me in a supply of the stuff he had squirreled away, betel nut. He said you can get the leaves down at the Ninth Street Trading Station. I did find the name of that place. I've lost it again. Someone asked me about the Bombay Trading Station.

ANN: On Valencia?

SASHA: I'm talking about the one in Berkeley. University and Ninth approximately, where they have the betel leaf, which is from a pepper. It has nothing to do with the palm tree. The palm tree is a source of the betel nut, which is the seed of the palm, which you take the husk off and inside is the nut that you break up or chew or what have you.

It's wrapped in a leaf, which comes from a pepper plant. An entirely different source. It's called the betel leaf as opposed to the betel nut. But the betel leaf, they've never gotten around to making illegal. So you can raise it domestically. In fact, it's raised in Modesto. I think I mentioned I found it comes in to Berkeley on Wednesdays, several hundred pounds every Wednesday. It's all signed and paid for ahead of time. By Thursday it's gone. And so there's very active use in Berkeley.

I got a sample of this betel from this Hindu who brought it in. I said, "How did you get it? The FDA prohibits it."

"Well, they kind of overlook it. About every six months when it comes in, we squirrel it away." Okay. I got a sample from him. I couldn't find arecoline in it. I looked for arecoline, which is the active alkaloid in betel nut, I couldn't find it. So, is this really betel nut? Or was it something they were bringing in every six months that looked like betel nut, and the placebo effect with a taste of the pepper leaf that was authentic would make the job work? I don't know. And then I began to inquire a little more deeply, and I find the Thai community. Everyone uses betel nut. "How do you get it? Do you get it from the Bombay importers in Berkeley?" "No, no, we get it in our Thai stores." I'm getting closer to the mark.

It is flown in frozen from Thailand routinely, daily. It comes in. The FDA has never heard of it. I'm sure they don't know that is coming in. Likely, they probably won't do anything about it. But it's coming in routinely. Now I have a package of frozen megs, these seeds of the *Areca*, the actual palm tree seeds, in a freezer over there in San Francisco, and next time I get the GCMS fired up, I'm going to extract and see if I can find arecoline in them. I have to use that as a source of getting at whether something I've seen is authentic or not because I don't know what it tastes like. I don't know what it does. In fact, one of the projects I'm working toward right now is to go after the ginseng. How many people use ginseng? I have not. Two, three. How many people not? Okay, about four or five have used it. What does it do? Who is a ginseng user? Who raised their hand? You use ginseng? What does it do?

STUDENT: I've never felt the effects. I really think that it's all in your head.

SASHA: Who over here raised their hand with ginseng?

STUDENT: It's a mild stimulant at most.

SASHA: Mild at most. You're double hedging. Probably marginal effects.

STUDENT: Mild stimulant.

SASHA: Who else has used ginseng? Yes?

STUDENT: I didn't feel any effect whatsoever.

SASHA: Okay, sounds reasonable. Someone else from here?

STUDENT: I heard that the Chinese say that you have to take it every day for it do really do anything.

SASHA: That sounds like a good way of avoiding . . .

STUDENT: You mentioned also that the type that you get really . . .

SASHA: Very much so. All I know is what I have heard. North Korea, South Korea, China.

STUDENT: The stuff that most people would ever use in this class, I imagine, would be the lower price stuff that you buy at the health food store in the little tea box or whatever.

SASHA: Yes, one more on ginseng.

LECTURE 15 ~ *Intoxicants*

STUDENT: I've seen liquors made of ginseng.

SASHA: Oh, I've got a box of tea bags that says, "Contains ginseng."

STUDENT: It's liquor. It's really potent. It's, like, over 120 proof.

SASHA: The alcohol may have a contribution into the effect of that.

STUDENT: It's made specifically . . .

SASHA: For the flavor.

STUDENT: Yeah.

SASHA: I would like to know. What I have been able to find in the literature on what it contains, and what those things it contains do, has been near zero. Here's a material that has been used for centuries in China. People swear by it. I don't know what it does. I really don't know, maybe a slight stimulation. It just doesn't give me a picture of what it does. I want to inquire about that.
Yes?

STUDENT: Another thing that I have come across is that the older people are the ones that benefit from it the most and that younger people like myself wouldn't feel any effects from it at all. I don't know, but that's just what I've heard.

STUDENT: It's like taking vitamins. It just gives you a—

SASHA: Actually, taking vitamins is kind of a good analogy. There is a concept known in medicine as "hypervitaminosis." It's when you take so many vitamins you get into trouble from the excess vitamins. You can overdo it. They can become a drug effect at that point. But if you eat a reasonably proper meal, you're getting all the vitamins you need, probably more than you need.

So the idea of taking vitamins has been harmless. If you take too much, you're going to get into possibly upsetting things. If you take, for example, quantities of ascorbic acid, vitamin C, as a sodium salt as it is commercially available on occasion, you can get too much sodium. And if you're having blood pressure problems, sodium is not nice in that direction. Plus

THE NATURE OF DRUGS

the fact that your urine goes very basic and you have other strange things that occur. But to a large measure, I think a lot of these things are, as you indicate, mental effects. But I'm kind of curious. I also have, on the other side of the coin, a great respect for something that's been used intensely by a civilization for many hundreds of years. There's something there, maybe. I want to look at it. And you're not going to get through this blockade of the scientific ethic saying, "We can't demonstrate it as having an effect, therefore it has no effect." Don't know why.

Yes?

STUDENT: Also, if you do GC it, you will probably run across a lot of organic molecules, and you don't know what their effects are. To find out what their effects are, you probably have to have years of studies.

SASHA: Let me tell you how Hofmann discovered the psilocybin in the *Psilocybe* mushroom from Mexico. Same idea, they brought back a big supply of mushrooms from Oaxaca. Heim in France wanted to get the credit for the whole business and really went deeply into the mushroom. And he injected this into rats, injected that into mice, and couldn't find anything that looked hallucinogenic. Of course, I don't know what you inject into a rat or a mouse that makes it look like it's hallucinogenic. That's one of the basic problems of research, though, using animals for something that is only seen in the human hoosie.

So Hofmann said, "Don't use them all up. Let me have what's left," and he did his extracting. And when he took an extract, he got it into two parts—that which is soluble here, and that which is not soluble there. And one day he ate this and the next day he ate that. "Oh, that's where it is!" Then he took that thing, and he took an extract of it and ran it up a TLC plate. He nibbled his way up the TLC plate until he found, "Ah! That's where the activity is!" And zeroed in on the activity just doing that very thing, until what was left of something turned a sort of a bluish color when you sprayed it with para-dimethylaminobenzaldehyde. It turned out to be an indole; it turned out to be psilocybin. He ran an NMR and got a good guess at the structure and synthesized it and there it was.

But when you're running down something that you believe to be active, you do it by separating it, fragmenting it into things, and testing for

LECTURE 15 ~ *Intoxicants*

whatever gave you the belief it was active. Find out what the activity is and pursue it. Don't worry if it doesn't meet your current belief system of what it should be. But before you can do this, you've got to demonstrate that it's active, otherwise you are pursuing the will-o'-the-wisp. There's nothing to know. So, the whole concept of the scientific approach to uncovering these mysteries of nature is not how you do it at this stage; it's to be sure you've got something that's real at this stage. Early. And that's what I'd like to find in ginseng. If I can't find in me an action of ginseng that is worth looking at, I'm not going to burn calories trying to find out what's in there that isn't active. But believe me—it's not in the literature.

Good. We have a couple of minutes, a couple of other directions I probably want to tuck into the lecture.

Yeah?

STUDENT: I want to backtrack about oxygen. I don't know if it's Priestley, but I read that some person who isolated oxygen said, "Only my test animals and I experienced this ecstasy of breathing pure oxygen," and I've always wondered why oxygen is not used as a recreational drug based on that account.

SASHA: It certainly will produce an elevation and excitement. I don't know why. Probably no one has written up something and made it illegal yet. [Laughter.]

Now we're at the end of the hour. Let's dissolve it.

If anyone wants to smell chloroform or ether, just don't smoke. At the top of one of the flasks is a little raw glass; don't cut yourself.

LECTURE 16

March 26, 1987

Deliriants

SASHA: Ah, it's 10 after. Good. Two, 4, 6, 8, 10, 12, 14. Okay, oh, 15.

I want to pretty much handle only a couple of drugs today and they're in the background [referring to chalkboard]. We're into the area where I want to spend more and more time on individual drugs, picking drugs that are more known, that are more notorious, more used. And today I want to talk about the depressants. I'd mentioned them back in lecture five and in the handout on the nerve system, and the cholinergic and adrenergic nervous systems. The autonomic nervous system has its balance between the two branches, the sympathetic and the parasympathetic. It's easy to think that the sympathetic is the adrenergic and the cholinergic is the parasympathetic.

Actually, the cholinergic applies to both. But the adrenergic only applies to the sympathetic, and it is the dictating agent. There are cholinergic synapses in the adrenergic nervous system. Today I want to talk about that cholinergic side, acetylcholine. It is a primary drug. In fact, let me start by trying to give a chemical picture without emphasis on the chemistry, but with more emphasis on the concept.

All of the neurotransmitters—this would be true for epinephrine, norepinephrine, for serotonin, for dopamine, acetylcholine, and probably the amino acids—are, in essence, something that has a charge, or can get a charge, separated from a certain distance to something that is vaguely basic. It's a dumbbell, a two-ended dumbbell. If you look at the phenethylamines—adrenaline, noradrenaline, dopamine—in all cases, you have something that is slightly basic, named an "aromatic ring," separated by two carbons from something that can take a charge, a nitrogen.

That physical separation, what is at each end of the dumbbell, the weak negativeness and the strong negativeness, are the fixed charges. And the degree of separation is common to all neurotransmitters. A lot of things that act upon the neurotransmitter as agonists, or as antagonists, have that same geometry. We mentioned that this is often something like NH_2 in phenethylamine or dopamine. It's NH-methyl in epinephrine. Always with that electron pair, so that, under the appropriate pH in the body, it achieves a positive charge.

Positive charges cannot easily—and in many cases do not at all—go through the blood-brain barrier. Things that have a fixed charge or have a

LECTURE 16 ~ *Deliriants*

positive charge are blocked because the blood-brain barrier is, in essence, a lipid barrier. So it's got to be uncharged to get through, unless there's something active to take it through.

Amphetamine, which would have that sort of thing, or methamphetamine, all the amino catecholamines in there and all the hallucinogenics, almost without exception, have this kind of system built into the molecule somewhere. But in the case of all these, under the right pH conditions, that charge can be lost. And so, if you can buffer the compound to lose that charge, it can go into the brain, and then it can pick up the charge again in the brain. This is the general concept of that separation.

Now, the same separation is very much true in acetylcholine. I've drawn out the structure so it's not necessary to copy it down. This is acetylcholine [referring to drawing on board]. You have a plus charge, just in the molecule, but notice it's a fixed plus. The plus charge cannot be lost, as it is a quaternary amine. This means that you cannot change the pH and change that charge. And you have a weak base down here in the form of the carbonyl. Acetylcholine has that structure. Almost all compounds that act like acetylcholine therefore are agonists, while those that get in the way of acetylcholine are antagonists. All these compounds have that kind of structure.

The basic compound that was the very first one studied in this area is the compound atropine. And I think I drew out the structure on a handout. Has everyone gotten it? There are two handouts for today. Atropine—I'll just draw the cogent part of the molecule: nitrogen, carbon, carbon, carbon. It happens to have a more complex structure, but that's the material. The basic structure of the compound is a nitrogen, which in this case can take a positive charge, and a base, a carbonyl, separated by about that much distance. How many angstroms, I don't know. But it's about that kind of distance of the molecule, which is very close to the distance found in acetylcholine.

If you take atropine or scopolamine—these very potent, very effective anticholinergics—and put a fixed charge on there by taking the atropine molecule and making a quaternary salt, the fixed charge can't get in the brain. It can get in the body, it can be absorbed, but it cannot get in the brain. So you'll find that many of these anticholinergics that are very disruptive mentally, such as atropine and scopolamine, are yet very valuable physically because they also are parasympatholytics; they will quiet the

THE NATURE OF DRUGS

tropine (scopine) **tropic acid** **mandelic acid**

scopolamine **oscine (scopoline)**

(dl) ATROPINE   ATROSCINE
    dl-HYOSCYAMINE   dl-HYOSCINE
                       dl-SCOPOLAMINE

(l) HYOSCYAMINE   SCOPOLAMINE
                       HYOSCINE

Hyoscymus niger   henbane

Mandragora officinarium mandrake

Atropa belladona   belladonna, deadly nightshade

Duboisea hopwoodii   pituri

Datura meteloides   toloache
D. innoxia   Jimson weed
D. stramonium

    Peyotl   (mescaline)
    Teonanacatl (psilocybin)
    Ololiuqui   (clavine alkaloids)
    Toloache   (scopolamine)

$$N-C-C \mid O-\overset{O}{\overset{\|}{C}} \mid C \begin{array}{c} R_3 \\ R_2 \\ R_1 \end{array}$$

LECTURE 16 ~ *Deliriants*

saliva flow, they will quiet the gastric motility, they will cause dilation of the pupils, they will cause a general syndrome that is desirable from the point of view of medical treatment.

You don't want the person to go mad as a hen upstairs and get the mental effects, so by putting this fixed charge on there, you allow it to be effective in the body. It can still be effective at the synapse. But it can't get into the brain to be effective at the brain synapse. So the idea of a charge that is fixed will keep a compound from getting into the brain. A charge that can come on and off as a function of pH will allow a compound to go into the brain.

In the handout I have given a collection of various plants. I need to back up a little bit, to the last lecture. I mentioned the idea of the tomato being called the "love apple." In reading up on the various atropine-containing plants, I have found reference to the *Atropa belladonna* as being the love apple. The belladonna plant has shiny leaves, a little blossom with five petals, in the center of which there is a small black berry. That is called the belladonna berry, and that is what is eaten. But that has also been called the love apple.

Has anyone ever heard the term "love apple," and could you put origin to it? I can't, and I had the impression it was an old name for the tomato. It may have been, but I found in reading it is also the name given to the belladonna plant. So I hold that in abeyance on the tomato being called the love apple. That may be something I extrapolated and assumed some time ago, but I can't find documentation for it. I looked into the evidence for the tomato again to find out where it became nonpoisonous. And I found a nutritional book of the 1870s that said tomatoes can be eaten but they must be cooked because they cause some odd effects when eaten raw. So there was a sneaking transition around 1870 somewhere, when the tomato was going from not edible to edible. I don't know exactly. I can't go to the library and look up tomato and go through 50, 100, 200 years of tomatoes. It's not an easy thing to run down. But that is the unknown I have right now. I'm on a tomato kick.

Also on the atropine, scopolamine—the thorn apple—I brought in a picture. Not a picture in this case, but a real thing. I picked them in a Safeway parking lot in Pittsburg, California. I happened to park the car there and snatched them, picked them off the plant. These are the thorn apples. It is

THE NATURE OF DRUGS

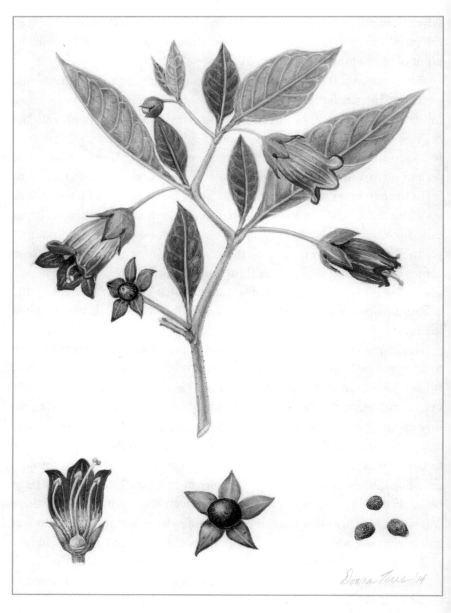

an atropine plant. It is a common plant growing in this area. If you look for it, you'll see it, I'm sure. Big, white, bell-like blossoms. In this country they come about three or four inches long. In South America they can be as long as a foot. Long, white, bell-like blossoms with these seed pods. How many people have seen them around here? Keep an eye open for them. They're really a dramatic thing. The leaf contains atropine. The seeds are exceptionally strong in atropine. The blossom contains some, the roots some; the entire plant is a toxic plant. Jimsonweed is one name; we got into that, I mentioned that before. Angel's trumpet. *Toloache.* Toloache is a California Indian name; I think it was largely the Pomo who used it heavily. The atropine is actually in the plant, the *Datura.* The one they use was not the *Datura stramonium*, it's *Datura inoxia*, but it's a variant that is probably not that different botanically. There are arguments about what the species should be called. They use it in the rights of puberty, in evolving from the child to the man.

THE NATURE OF DRUGS

In fact, the use of belladonna plants is quite broadly found throughout the entire Indian culture of the New World. I want to talk at some length in the next hour on peyote. But the Huichol Indians, who use peyote as part of their religious rite—and I want to get into quite a bit of detail on that—also have a companion dark angel that they use in balance, in opposition to it. And it is what they call *Kiéri*. The Kiéri and the peyote are opposite companions that, in essence, are the yin-yang of their entire religious structure.

There are shamans who are specialists in the Kiéri, and that is their world. They are the power side of the shaman structure. There are other Indians—I will talk a little bit about the Tarahumara Indians in Mexico—where the peyote is used both ways. The point is, these plants that are psychoactive do not carry with them a specific type of action that is embedded in the plant. The action can only be expressed by how the person uses the plant and what is done with the changes that are inspired by the plant. This is an outgrowth of what came up when I showed someone one of the

LECTURE 16 ~ *Deliriants*

show-and-tells—PCP. It is one of the compounds I want to talk about today. There is nothing intrinsically evil in PCP. It is a perfectly water-soluble crystalline solid. Here it is from Parke-Davis. Parke-Davis made a lot of money selling it.

Yet I talked to a person this last weekend and said, "What's your gut feeling about the problems associated was PCP?" And the answer was almost a shudder. They said, "You know, that's really a bad compound. It's really an evil compound." And I don't know quite how to address the question of a compound being evil. A compound is not evil. I think a lot of the information is self-explanatory.

I was off on *Datura*, now I'm back to PCP. PCP, Angel Dust, "KJ," as it's called in the San Jose area. KJ stands for "krystal joint." It is smoked primarily. It can be eaten. It is a very potent, quite volatile oil. It is a crystalline solid, but it's a fairly volatile crystalline solid, extremely lipophilic. From the point of view of the organic chemist, he will say, "Oh, you can

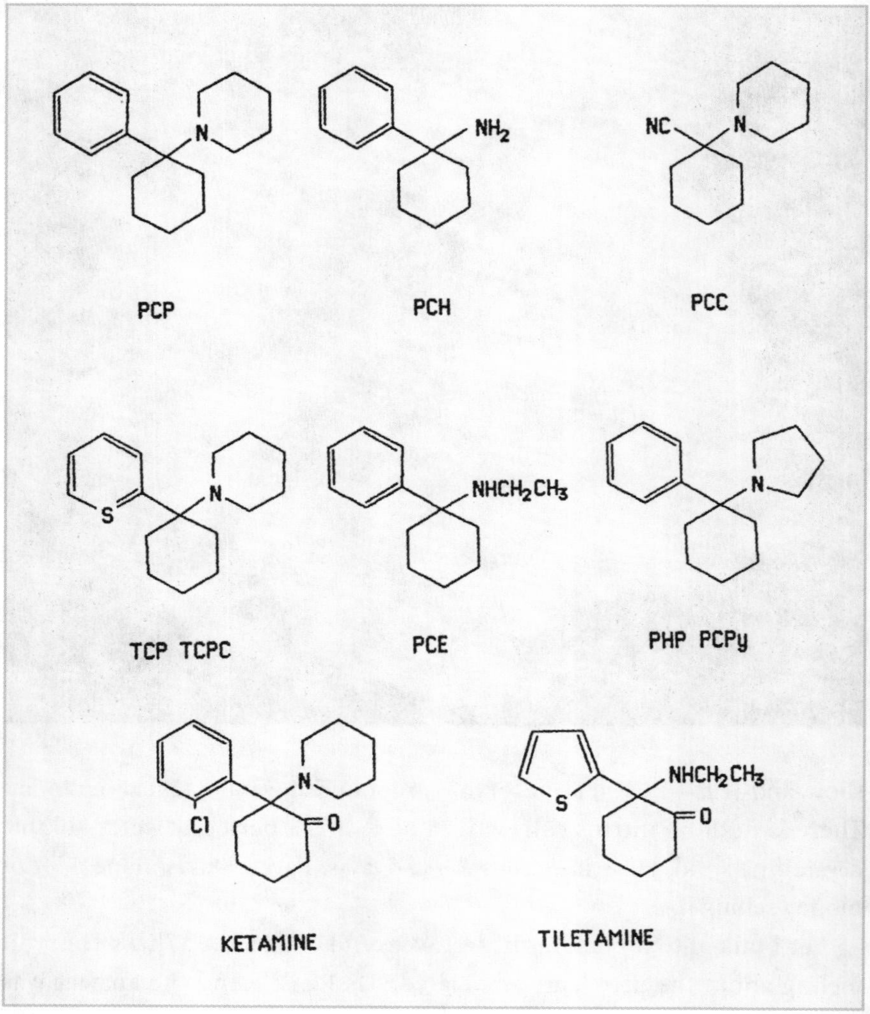

THE NATURE OF DRUGS

PCP　　　　　　PCH　　　　　　PCC

TCP TCPC　　　PCE　　　　　PHP PCPy

KETAMINE　　　　TILETAMINE

extract something from water by making something that's basic; by making the water basic and extracting it with an organic solvent. You can extract it back out of the organic solvent by extracting the organic solvent with an acid." This is Chemistry 1A—the basic textbook of chemistry: acids, bases, extract from aqueous into organic under basic conditions; extract from the organic into aqueous under acidic conditions. PCP under acidic conditions will not extract into water. If you were to take dilute HCl and PCP in two different solvents, the chloride would all stay in the organic solvent.

LECTURE 16 ~ *Deliriants*

PCP will not go into water if there is an organic around. If you take it into the body, it goes into the body tissue very thoroughly, very rapidly, very efficiently. It is there for a long time. Its actions may be fairly short lived, but it is in the body and being cleared from the body over a long period of time. It is so thoroughly invested in the body's repository that there are serious efforts being made now, and with some success, and with some problems, of assaying PCP usage not by going into the urine or into the blood or to the fat, but going into the hair. Because the hair, as it grows, carries traces of everything that the body has been exposed to. And I predict this is going to be the next big one down the line for violation of privacy. You know, drop your pants so we can take your pubies.

I mean, the hair is a method of not only telling what you are currently into, but your past record of what you've been in. How long does it take hair to grow from scratch out to that long? [Gesturing.] Several months? It contains an archaeological record of what your drug exposure has been for that period of time.

ANN: I suppose it might make baldness a new fashion?

SASHA: Yeah, but baldness—the pubic hair is always there. Pubic hair is almost always called for, for a couple of reasons. One, it is less often offensively dyed or destroyed or interfered with in its growth patterns. It is less often shaved, or it is always available, and it can be taken in a horizontal cut down to the base without being unsightly. Right? You take a chunk of hair and you cut it down to the scalp, you have a big hole going into the scalp. It's offensive. I don't know why it's any more or less offensive than peeing in a cup.

Anyway, the PCP is laid down in the hair. The problem of assaying it from hair is a matter of using a radioimmune assay system, such as with urine, such as with blood. And it has the same problems with the radioimmunoassay for being cross-reactive and being uncertain, but it can be done. If it's negative, there is no PCP there. If it's positive, you have reason to go on. I want to spend a whole portion of a lecture just on analysis later. So I'm not going to touch that now.

PCP was first introduced into public use—oh, I don't know which came first. Parke-Davis was the first to develop PCP. This was about the early

1960s. It was found to be an effective anesthetic, much more effective in primates than in lower animals. It was promoted as an anesthetic under the name of Sernyl. I don't know if I put these in the handout. Okay, I did. PCP, this is in the handout. The structure is usually drawn in this manner; it's as good as any. A very strong base, but a very lipophilic molecule. Sernyl or Sernylan is the trade name. It is currently still used for veterinary purposes. I brought in a bottle of it—oh, this is Ketalar. I didn't bring the bottle of Sernylan. It's used for veterinary purposes, but it's been used a lot in humans.

Its first human trials were in maybe '62 or '63. The military was the one of the first to really make extensive human trials of it, probably around 1965. The military has always had a very open mind toward getting at things that are psychoactive. They are continually doing research. I want to talk at some length about the whole chemical warfare approach, because there they use drugs in a very sinister way. But the idea of using drugs to disrupt a person's togetherness—all the drugs we're talking about today—we're talking about ketamine, PCP, some of the Datura-type compounds—have a single property in that it separates the mind and the body. That is why the dissociative drugs are used as anesthetics.

They're not anesthetics in that they make you not feel pain. They're anesthetics in that you have pain in the body and your mind is separated from it. When a person is under ketamine or under PCP in surgery, the person is not unconscious. The person is quite conscious—can be talked to and can respond. Not very well, but the consciousness is there. Not being out, eyes closed, asleep, like you see in many cases of anesthesia. The person will respond. The person, however, does not get signals from the body. The proprioceptive signals do not make it through. Hence, you can do surgery, you can cut, you can hurt, and there's no feeling of pain. The person often has a very fuzzy remembrance of what goes on. Because they're in their mind. They may be out traveling through the stars, the yonder planet, so to speak. But they are not in their body. There is no such thing as having a full bladder and wanting to go urinate while in a PCP or a ketamine anesthesia, because that's the body's problem. It's not your problem. The body doesn't exist. You're separated from the body and that's where the value of this comes in medically, because it is an anesthesia that can be used without

LECTURE 16 ~ *Deliriants*

the concerns of going unconscious.

I think I may have mentioned one of the basic problems of anesthesia. When a person is unconscious, they have no protection against the vomiting reflex. Hence, when a person is put into general surgery, you vacate the bowels, you empty the stomach, you get everything quieted down. What the anesthesiologist doesn't want is a mask full of vomit in the middle of things; you get the stomach empty. You don't want defecation everywhere; the bowel gets emptied, because all these controls get lost in anesthesia. And so, this is why you are cleansed thoroughly.

But if you take a person who's in on emergency surgery, he's been brought in from an automobile accident, you have none of this elegance of being able to take a person and make sure he hasn't eaten for 12 hours and all that sort of thing. It's an emergency and surgery may have to be done. Then these become superb anesthetics because they don't render a person unconscious. The person's reflexes are taking care of themselves, fighting back.

STUDENT: If they don't use PCP, then what do they use?

SASHA: Ketamine, it's very commonly used. PCP for a while was still used, but largely in children. Here's a little bit more about PCP. It got onto the street in 1967, I believe it was. The great heyday of the Haight-Ashbury when everything got on the street. And in fact, the military was exploring it at some length in '65 and then it became suddenly available on the street. I like to think there's a connection there. It's something you can never establish, but the military worked widely with LSD, and it worked widely with PCP, it worked widely with mescaline, with MDA, with phencyclidine—I mentioned phencyclidine, with quinuclidinyl benzilate, which I will mention later in this lecture—and all these things have made it onto the street.

I don't know. There's no way of finding a connection. But the timing has always been very much in that same sequence—about a year or a year and a half separation from extensive military trial to street use. It was called on the street in 1967 "hog." Well, PCP had a bunch of nicknames that I don't remember.[12]

---
[12] There are presently more than 200 street names that have been recorded or purported for PCP. Some names coming from law enforcement may or may not have actually been street names.

It swept through Europe. It was very widely used in Scandinavia. I don't know anything about the southern part of Europe; certainly the northern part of Europe, even in Scandinavia, it was widely used.

ANN: When did the term "angel dust" come in?

SASHA: Somewhere about the late 1960s. Angel dust is another name. Angel dust, PCP, hog are the names that I'm familiar with.

But at this time, it was still being used medically. It is a good anesthetic. It was used for emergency surgery, as I mentioned, where you don't have the prepping-before-time opportunity to get a person in good physical shape for surgery. When it was withdrawn from surgery, it was still used for children for a long period of time. Part of the difficulty in surgery, with it and with ketamine both—I'm talking about them more or less together, because there are very many similarities between them—part of the difficulty is in coming out of surgery, in what's called "recovery." In the recovery room, patients can be kind of hard to handle because there is a lot of reintegration of the body and the mind, and that can be in the form of things being believed to be seen—hallucination, attention-diverting imagery, fantasy in the mind's eye—and the nurses found it difficult to handle people. Surgery was usually not difficult. You use enough so that there's no disturbance, but in recovery that disturbance can be quite bothersome. Yes?

STUDENT: What about all this stuff, PCP making people strong? My brother works as a security guard at Bergman's, and some little guy came in on PCP, obviously really wasted, really skinny, and he was throwing four big guys around like it was nothing.

SASHA: Okay, why should a thing that is an anesthetic and a body dissociator make you strong? I don't know if it makes you strong, but it certainly doesn't give you feedback from pain.

STUDENT: You just don't feel it?

SASHA: You are anesthetized. You have a lot of strength in the body, you are able to carry a lot of things, but your feedback from the body says "Hey, that's gonna hurt me," or "Hey, I'm not that up to it. I don't want to strain myself."

LECTURE 16 ~ *Deliriants*

STUDENT: So the person just extending themself past what they would do normally?

SASHA: If you don't give incentive to a person, they're going to be in a lying back, almost in a sleep-type mood. They're not getting the inputs. And a person alone very often will lie on a couch for four hours with PCP or ketamine. The reputation that PCP has is probably about as vicious of a thing that I've seen. They tried to put it on to cocaine and it didn't quite stick. They sure did it with PCP.

Someone had mentioned that I had brought in "PCP Heart Wrenching Tragedy in San Jose: The Ravaged Children of Angel Dust." Well, yes, there are some grotesque situations with children. This stuff goes *in utero* into a child. The child can be born and show the effects of the drug from the mother's usage prior to childbirth. I think there was a big show on campus two days ago and yesterday on drug awareness, and they showed *Reefer Madness*. How many people have seen *Reefer Madness*? I have not. That's one big hole in my background. I've got to put that in there.

But the concept, the use of words "Ravaged Children of Angel Dust" has a little bit of a smell of that. You know, you're taking the paint and painting as dark and ugly and uncomfortable a thing as you can. Dammit! Yes, people can do strange things with PCP, but it's a white crystalline solid. It just really doesn't have that kind of intrinsic evil to it. Why would people use it medically for surgery and still use it with children for six years in surgery if it caused people to go berserk and crazy and real strong?

One policeman down in Los Angeles got called on the carpet because he shot and killed a person who was up a telephone pole without clothes on. It was a very weird situation. But the guy had been on PCP, and it was claimed he was a threat to the policeman. The guy escaped by going up a telephone pole, and, sure, it was not very normal. I mean, people don't normally go up telephone poles without clothes on. It's not what you would say is conventional behavior. But I can't see how that reads as a threat to the policeman. But it's a situation that could not be handled.

I could give a whole editorial on the Los Angeles police, but the response of people is one of immediate fear, rebellion. How many people have taken PCP? Okay, one, two—not very many. Three. How many people are scared to death, wouldn't go near it with a long, long...? Okay, three or four.

I don't like it. I don't happen to like being dissociated from my body. I don't like being out there in the outer galaxies, astral traveling, without awareness, without the awareness of what the body is and what it's doing and having that rapport. I like materials that make me aware of the body. And sometimes these are awkward, because they cause you to be aware of pains and aware of hurts and the uncomfortablenesses and incorrectnesses of the body. But I don't want that separation. I don't like the idea of anesthesia in surgery, because that's exactly what it does. It takes you away from the awareness. It puts you to sleep, so to speak, in a non-feeling way. I think if I were to have serious surgery, I would like to have an awareness situation. I don't care for the pain—that's not one of my turn-ons. But I don't think I want to be separated. I don't think I want that separation.

Some people do. They do at some length. I'm going to spin PCP and ketamine in together because ketamine, Ketalar—that is the injectable material I have over there—is still used surgically. It is one of the major emergency anesthetics that is used. It is used in animals, it is used in humans, it's used in children and adults. It is used quite widely, and it's a Schedule III drug, but it is one of the most highly abused drugs I know in the United States, and in the Bay Area it's very highly used. It's called "vitamin K." Ever hear the term "vitamin K?" Not the real vitamin, but the so-called vitamin K. That's a slang name for ketamine.

I don't know how to really paint the picture without letting my feeling of not being at peace with the drug myself, in my own use, color what I'm saying. That's not fair. I should be neutral on it. As everyone is biased, I'm biased. I do not like the experience of PCP and ketamine. You are biased. You do or you do not. Everyone is biased. One of the big arguments in giving any kind of an instruction, interacting with information, is to try to be unbiased. That's silly. You're biased. But try to know your bias, and speak around it to present things fairly. And that's what I'm going to try to do with ketamine.

Yes?

STUDENT: Isn't that what people who work at hospitals tend to use a lot?

SASHA: Very heavily.

LECTURE 16 ~ *Deliriants*

STUDENT: The doctors and nurses.

SASHA: Very broadly used in hospitals. It is available easily in this country from Mexico. It's made in quantity in this country where there are no restrictions in supplying international trade with drugs, and it's made by Parke-Davis. It's been licensed to about three other companies. The only one I know of specifically was Bristol. I happened to be at Bristol on a consulting task the day they bought the license from Parke-Davis to manufacture and sell ketamine. And I happened to be talking to the head of research and said, "Are you aware of the broad street paramedical use of ketamine in society?" He said, "No, I didn't know it was being used that way."

And I said, "Well, now you do, and forget about it," because obviously they're investing a lot of money in it. It's a very effective drug. A lot of research has been done on it. It's being made in quantity, bulk, being sold to pharmaceutical houses in Mexico. In Mexico, there are virtually no restrictions on the sale of poisons or drugs. People who ever had a toothache in Mexico go into a Mexican drugstore, and you're amazed what you can buy by walking in and saying, "I have a toothache," and you are supplied with drugs that would be triple prescription in this country. And you can go and buy liter bottles—they're brown bottles with syrup caps on them—that contain injectable ketamine with enough chlorobutanol as a preservative, but not enough to get in the way, and they'll bring it back into this country. It goes into Mexico legally; it comes back in by being smuggled. Ketamine is not a scheduled drug there.

The handout I got from the DEA about three weeks ago gives a current status of all drugs and where they are, what changes have been made to the federal drug law over the last 15 years, the name of the drug, the date, the authority for that change, when it was brought into fruition, and the current drug status with the schedule. I looked there, with a great deal of pleasure, to see where ketamine was. Ketamine was listed as Schedule II. It didn't have a date attached, nor an authority attached, but it said Schedule II. Ketamine is not Schedule II. It is a prescription drug in the sense it can be used by doctors only under the prescription laws, but it is Schedule III. I don't think it's a prescribed drug—it's used in the hospitals.

It is a very broadly used drug. How many people have heard of John

Lilly? Dolphins, teaching dolphins, learning, interacting with dolphins, a marvelous researcher, explorer for many years. The author of *Center of the Cyclone* and *Descent of the Cyclone*[13]. I can't remember the other books. Early in trying to interact the human mind with the computer system, at the time that was still a virgin territory and not much explored, he got into work with dolphins—communication with and interacting with dolphins, probably one of the seminal figures in that type of exploration.

Also into his own head. He was using his own head as a template for exploring the concept of the mental process and the concept of the intelligence process. And he was the first to develop and use the isolation tank—the tank where you are in water and magnesium sulfate, at body temperature and body density so that you float and are free from sensory input. He was the first to really begin exploring ketamine as a way of observing from within himself the degree of the mental extension that is possible. And there again lies the story that gets a little bit out of hand, because his use of ketamine became more and more and more; not just the exploring of the limits and the realms the human intellect, but a pattern that he chose to like and reinforce and continue.

He became, in the psychological sense, hooked on ketamine. He has used it routinely. I see him about once every year, year and a half. He has a little alarm clock in his pocket. When it goes off, he goes off to the bathroom and readministers ketamine. He does it several times a day, continuously, every day. He is continuously with ketamine in him. He explains how it has expanded this, how it has allowed that, and he's really learning a great deal, but all his effort is not in telling what he's learned about and what's being expanded, but in finding an opportunity to maintain that ketamine use. He is shifting his entire view toward that.

One book, Marcia Moore's *Journeys into the Bright World*. An unknown person who wrote a very beautiful book. She had written about six books. This one she coauthored with her husband who is an anesthesiologist—MD, I believe—in the state of Washington. It's a book about the virtues and the values of ketamine—the Mistress Ketamine, Goddess Ketamine, I think she referred to a couple times. A drug that really caught her attention. She used

---

[13] *The Center of the Cyclone* and *The Dyadic Cyclone*, the latter coauthored with Antonietta Lilly.

LECTURE 16 ~ *Deliriants*

it at different levels; her husband did too. They would use it before dinner in small amounts and let the dinner sparkle. They would use it in larger amounts and explore this out-of-body world. Almost like Leary and LSD, it was a promotional thing of a remarkable view.

You're talking about a 15-carbon organic compound. The compound is not doing it, but the person is allowing that compound to catalyze a sort of search. And it's interesting—talk about projection on PCP being ugly, here's a person who is projecting on ketamine—and it has been done on PCP—as being not only value but virtue, godlike, the remarkable savior of everything. I mean, it's a real promotional thing. Not that they're promoting as a selling argument—they're promoting it in the philosophic argument. That came to a very strange end just after the book was published. In the book you can see this commitment to the drug in the way that she was writing. She would say, "I find that I can do all this and find these areas of exploration without the drug. In fact, this morning, I only used 15 milligrams." And being without the drug is equated to using only 15 milligrams. So definitely the drug has spun its web into her behavior pattern, her husband's too.

Some few months after the writing of the book, the book was concluded in January. I believe this was in the middle of spring. While she and her husband were exploring this extraordinary outer place, she disappeared. She just plain disappeared. Her body dematerialized. That's what her husband said. And he panicked. She had gone. And when he was able to, he went to get help. Both police and medical help tried to find what had happened. She totally dematerialized. Her body was found in a swamp a few weeks later—murdered, killed, dead. I don't know the details. Don't know what went on there. Except I have a very uncomfortable feeling.

There had been a total devotion, a total giving, a total yielding of one's rights and choices and personality. Choice—back to the word of about three or four weeks ago. I have no objection to using drugs. I use drugs myself. I'm sure that you all use drugs. Know what you are using and what it can do.

If you get yourself into a drug use that has gotten to that extent, stop. Evaluate and say, "I'm going to look at it and see if I choose to do what I'm doing without choice." If you do that, bravo. That's all you can ask in the whole area of drug use. Choose, know what the situation is, choose to use,

THE NATURE OF DRUGS

choose not to use, have the choice. These physical and psychological—especially the psychological—commitments to the drugs tend to erode away that choice. And therein, I think, probably lies the true danger, the true negativeness of the use of drugs.

Yes?

STUDENT: Most people who are in that situation that you described, I find, see themselves continually, even with the idea of choice, they feel they have a choice still, but they still choose to use it. And so I still don't think that choice is essentially there.

SASHA: Okay, you make a very good point on the argument. I think Lilly chooses to continue using ketamine. I think Marcia Moore chose to use ketamine. What am I trying to say? Because choice is not the word then. Because a person who continually smokes and smokes without thinking and smokes continuously, chooses to pick up a pack of cigarettes, take out a cigarette, and light it. A person who injects ketamine 10 times a day chooses to do it. So it's not choice. What's being lost? What is there? Anybody? Any ideas? Yes?

ANN: We all have the right to refuse to take that drug, but in choosing to use that drug, if they don't have the ability to say, "I don't want to do it today," it's not a free choice.

STUDENT: It's more of an impulsive choice, I think. Because consider people who overeat. You know, it's like this real compulsiveness to do it. They just get this burning desire to do it. Look at a lot of people with sex. They get turned on and they've got to get it. And I think similarly, they have that choice, but it's not really a choice.

SASHA: It's a compulsive idea, something, you know—

STUDENT: I think there's a lot of self-deception going on.

SASHA: Really, I don't know the word to use. "Choice" is not the word. That was a good point brought up. People who choose to do something that they intrinsically know at some level is destructive and yet continue to do it, be it overeating or the using of ketamine routinely, they still choose to do it.

LECTURE 16 ~ *Deliriants*

So choice is not the word. "Compulsion" has a little bit of pathology—well, maybe pathology should apply.

Yes?

STUDENT: I don't know what the exact word is, but a person who is compulsive, let's say, for instance, someone who uses a narcotic or ketamine or a pack of cigarettes, you begin to move from an area of homeostasis, an area of non-drug use, a scenario of non-exposure, into an area of continual use. And so you forget where you came from, in a sense. You know one way only. You begin to learn how to use a drug daily. And that, that's normal. That makes you normal, when you light a cigarette in the morning, or you take your—

SASHA: Ketamine in the morning—

STUDENT: Ketamine in the morning, yes. It brings you to a normal state.

SASHA: I know people who cannot get down to breakfast without their first joint in the morning. I mean, any drug can fall in that pattern.

STUDENT: So the fear then lies in that if you don't have the drug to make you normal, then the place, being normal before drug use, is an off-place. For instance, for addiction to narcotics, withdrawal, starting about five or six hours later for cigarettes, the kind of cringy, tight anxiety feeling.

SASHA: Something is wrong, or—

STUDENT: The acidic stomach when you haven't eaten, when you're used to eating continual food.

SASHA: I'm glad you pointed out the argument that they are still choosing to abuse—they are. That's a very good thing for each person to kind of have a feeling for. I have a feeling inside of me of what I'm talking about, and I don't know quite the words to do it. But I think in some way you have that feeling too. Somewhere, you've gotten into some sort of habit. Maybe your idea of having forgotten what it was like before, or that there was a time when it wasn't there. It's become part of your pattern like chewing your fingernails. You've forgotten you once had long fingernails. How many people have you seen with fingernails down to the quicks? And they're continually doing this. They're not choosing to do it, and yet they are choosing to do it. It's a pattern. I don't know quite how to get out of it.

STUDENT: That's a good point that you mentioned about when you don't use drugs. When you move from that area of non-drug use where you say, "Okay, I want to smoke a joint or I'm not going to smoke a joint," you have that flexibility. But when you have to use it to maintain that homeostasis, I think that's a loss of choice.

SASHA: This is getting into the physical dependency.

STUDENT: But that's a choice also psychologically.

ANN: Yeah, as soon as you feel afraid that you might not have a certain thing or that you might have to go without it, as soon as there's fear involved, you are hooked.

STUDENT: I guess I could put out a personal example. I had never eaten a lot of sugar as a kid, but I moved from a time when I didn't eat sugar at all to a point a couple of years ago where I—just like a normal American—ate a lot of sugars because sugar is in everything. And I found myself, I mean, I chose to, I mean, I get this nice kind of—I would start smelling this nice chocolate cookie or something like that, and I would say, "Mmm . . ." and then I would get the desire to eat it. And I feel like I choose to eat, I want to satisfy the need. But I found myself in a position where I lost the ability to choose because I found that the desire to have it became so strong that I really couldn't choose anymore. Even though I was choosing to do it, eventually—because it was like a compulsion—I still didn't feel like I was choosing. I had the association of free choice with it. So my alternative to that was, I don't want to be in that position where I really feel that burning desire and feel that loss of control. So I just decided it get it under control. From my experience, I still had a choice, but I didn't have a choice. I don't know how to put that in words.

SASHA: I don't either. Any other thoughts on the matter? Because I really would like to find a way of getting what I intrinsically know in myself out. And I think in a sense, you can get the music of what is there, if not from your own experience, from knowing someone who has been in that experience. You have something that you lose when you get into a continual pattern of use, be it ketamine or be it tobacco, or be it alcohol or be it food, that you do it without thinking. Your voluntary motor motions say to the hand, "Go down there and take that chocolate out of the box and bring into your mouth and chew it up." So, it's a choice in that sense. And yet there's certainly a physical component, because, as you mentioned, you get a smell of a cookie or something. I think vanillin has been one of the largest contributors to overweight. Someone cooks a cake, the smell of vanillin goes through the house, saliva begins running, you know, you're heading back into the kitchen to get a piece of that cake. I mean, all these trigger responses are very real.

STUDENT: Getting back to the book about her saying that 15 milligrams, "I only took 15 milligrams of ketamine." When you get to the pattern of the people that use drugs and the ones that are recreational users—whether it be narcotics, stimulants, or hallucinogenics—the pattern where they think that when they take a certain amount just to maintain a normal feeling, whether it be 15 milligrams of this or 15 milligrams of that, then it comes into where you make a decision to indulge. "Friday is coming around. I smoke half a joint a day, or whatever. Now Friday comes, I'm going to smoke three joints." The indulgent pattern comes in, where during certain patterns of continual drug use you try and come down and try to slowly make your transition back to normal. But then there's certain things that start coming in, and you'll start deciding, "Well, am I going to shoot now? You know, this is Friday night. Can I?" Because people who do continual drug use kind of have a pattern, they kind of plan out when they can really experience what they enjoy about the drug, not necessarily whether it's certain levels of the drug that give certain characteristics and higher levels give physical—

SASHA: Well, every little gimmick that you use to try to get into a pattern of breaking patterns, the gimmick itself becomes a pattern, or it can be misinterpreted. I remember one of my ways of stopping smoking was I had a slip of paper in my pocket. I didn't actually have it there, but I pretended I had it there. It said, "Who's in charge?" So I'd reach for a cigarette and I would come out with a mental slip of paper that says "Who's in charge?" And that was supposed to indicate to me that I am calling the shots and I am in charge. That's why I don't need a cigarette. And yet, you can interpret that quite the opposite way. "Who's in charge? I'm in charge, which means I'll have a cigarette if I damn well want it." I mean, it can be used both ways. And so the idea of getting into a pattern of breaking habits, so to speak, in its own paradox can be a pattern. Yes?

STUDENT: That's very true. One point, though, is that in your argument of the idea of choice, the concept of choice, is going to vary so much just on that person's state of mind and their relationship to that drug, because I think you can see yourself down the block about it. And so that argument, I find, doesn't really hold. I understand what you mean, and I am really cued into the music of what you're saying. But as an argument I don't think that it really withstands much.

SASHA: Well, I think as an argument somehow it has to be made, to articulate, in order to break some patterns of drug abuse that are damaging. Yes?

STUDENT: It sounds like once you start using a drug, your free choice is conditioned.

SASHA: Oh yes. You get into patterns of doing it. Oh yes. That, of course, is the whole thesis of the edict of "Just say no." By not starting a pattern, you won't develop those patterns. Which is noble, but the thing is, the pressures to start those patterns are extraordinarily large, too. When you are in a group where everyone goes to the baseball game on Saturday, you will find a very strong pressure to go to the baseball game on Saturday. And that's for behavior patterns or bowling or the movies or up for dragging the main or whatever is going on. And the patterns are in the drug area just as much. "Just say no" is a very difficult thing.

The exact analogy, you brought the argument of sex behavior. The exact analogy to sex behavior, "Just say no," is a superb thing. And yet, by golly, there's an awful lot going on with hormones and peer pressure that's unbelievable. And I think the idea of instruction of facts is a very necessary component. And the "Just say no" is perhaps a laudable opinion, but the opinion can be listened to and respected or not. The facts should still be made available.

LECTURE 16 ~ *Deliriants*

I think facts such as there are a lot of people who have gotten into ketamine who have gotten into strange areas with it and have gotten into patterns of continuous usage and do not seem to be able to break those patterns or do not choose to break the patterns. That is a real fact. And yet there are a lot of people who are extremely devoted to ketamine. There are how many people in San Jose—they had a number in here somewhere—who are using PCP regularly? Why? If all it does is cause you to go real bonkers, climb telephone poles without clothes on, and end up breaking and throwing four people over a fence, why? It's because there's some good value there. It is a drug of choice to a lot of people, as is heroin, as is anything else that is considered negative.

There was an article that appeared in the *Journal of Psychoactive Drugs* about three years ago, "The Virtues of PCP." And it was a nice article. The sociology worker went into an area where PCP was widely used and interviewed people and said, "Why do you use it?"

"I use it for the following reasons." Very logical reasons. They weighed those reasons as being sufficient value to continue using the drug.

I consider the use of a drug in a person who is pregnant and nearly at the point of term, to where the child is hurt by it, falls into the another whole category, and that's giving a drug to a person without that person's choice, which I consider to be absolutely improper behavior and should be actionable behavior. There the child had no choice, but the mother did.

Yes?

STUDENT: What are the physiological harms of PCP?

SASHA: I can't think of any. I mean, if you take away the idea of jumping off of a telephone pole ... There is one I can think of. A very good friend of mine, Don McLear, who did a lot of work in Los Angeles with drugs. He had his own laboratory. In the course of doing some of the synthetic work for developing the methods of making PCP, to identify impurities, got a spill of PCP on him in his lab, and he went into a very strange mental place. He had to have psychiatric help. He happened to be, at one time, a part of the FBI. So he had access to mental health help of a sort, which was, for his case, very valuable. He could not go near that laboratory. In fact, he ended up sealing off that whole end of the laboratory. He became so extremely sensitive to it, that even getting near that laboratory would trigger a very bad psychological fear response. And I don't know what percent is physical and what percent is chemical. So there is the possibility that at least one person I know has a hypersensitivity, which means you're going to become increasingly sensitive. I don't think it's a general rule. I don't know other people who have shown that. This is one case I know. Physical harm from either drug, I don't know of any.

Yes?

STUDENT: Do you develop physical resistance to the drug?

SASHA: I don't know the answer to that, with PCP or ketamine either. Ketamine, I am not familiar with appreciable resistance being built up.

ANN: They tend to find themselves going lower.

SASHA: Yeah, but that means they're not losing responsiveness. If anything, they're becoming more sensitive. No, I don't know of long-term sensitivity or resistance. I don't know. I've not heard it reported. So in essence, here's a case where there are really no good physical negatives associated with the drug. But psychologically people get into very difficult dependencies, and their body is not really run particularly by their head. Hence, you get into bizarre behavior that the body is responding to, marching to a different drummer than the head is dictating.

PCP, I mentioned the precursors. One of the materials that I brought in is called the "nitrile precursor." Again, it's a white solid. Most of the PCP is made using it. In fact, some of the terms that are associated with it are worth noting because they've gotten a lot of legal publicity. A very simple compound, piperidine, hydrogenated pyridine.

I should have brought in some. A very fluid, ammoniacal amine, fishy-smelling base. It picks up carbonate quite rapidly. It's a very fluid, low-boiling base. Piperidine is probably the major precursor to PCP. Piperidine is one of the materials that has been brought under legal control. It is the one case I know of in which Congress passed an amendment to the law saying, "Piperidine shall be regulated in its distribution." As with many laws, they like to say, "So-and-so shall be regulated" or, "A drug test shall be given in sensitive employment positions." But nowhere do they say what "sensitive" means, and nowhere do they say how something should be regulated. These are left to regulations later on. And this is where the little martinets who are in charge of agencies that are in the business of enforcing law actually end up writing law.

A situation specifically on this is a person who is put in charge of making sure that the alcohol-assay laboratories—clinical laboratories that evaluate blood alcohol—should do a good job. The state legislator said, "Public Health shall be responsible for the quality of alcohol reporting, quality of alcohol analysis." Think of what an open book this is, "shall be responsible for the quality of the work."

This person says, "I've been told my job is to be sure that the work has a high quality. I think this has to be done and if you don't want to do it, I'm going to close you down." Suddenly that little person has got a lot of power. Same thing with regulations. The job shall be to collect income tax, not saying how, so the IRS in this situation will say, "This is how we're going to do it because we've been given the job of collecting income tax."

The PCP distribution shall be regulated. No one says how it is to be regulated, and suddenly you have a battle of big industry, because there are 55 tank cars going along the Santa Fe tracks (Southern Pacific now), and these big black tanker cars are filled with liquid chemicals. Every 20th one is piperidine. I mean, that stuff is immensely used in industry, everywhere, in quantity. And yet a finder's fee on a 100-gram bottle of piperidine is probably around $300 or $400 for the drug trade.

So the mischief that is involved in having one of the seven 55-gallon drums disappear, that is enough to supply quite a bit of PCP trade. So piperidine, bromobenzene, magnesium, cyanide, cyclohexanone. These are the things that are involved in the making of phencyclidine, PCP. The name PCP came from phenylcyclohexylpiperidine. The last P stands for piperidine.

It's a major component. It is restricted. It can only be sold on the assurance it's not going to be used for making PCP. Well, this guy who just sent a tank car from Dow Chemical Company over to Matheson over on the West Coast cannot be assured that no one's going to turn the tap and take some and make some PCP out of it. So at the industrial level, the law is not observed. At the individual level, it is observed, but they don't say in regulation you've got to report that you are getting piperidine. To whom? To the state government? The state government has a file of the little people who buy the piperidine, but the large companies do not report it. The federal law says you must, but they don't say to whom you must report. That's left to regulations. Who enforces that? The DEA is in the position of enforcing that, but the DEA has many other things that are much more important in their idea, so that is sort of slipping into a "gotcha" classification of law. If we need to use it, we will. Piperidine is a controlled substance now, Schedule III, as a precursor. But it was placed there by congressional law, not by administrative regulatory decision.

I wish I could remember for sure. I think piperidine is Schedule III. I'll look that up and get it to you next hour. But you have many of these materials that are items of commerce. You have acetic anhydride; that is a mandatory thing in making heroin. You have acetone. In fact, ether has really become a very difficult thing to get in South America because ether in South America is a standard fare for making freebase and manufacturing cocaine. They're going to stop the cocaine by stopping the manufacture and the importation of ether. So ether has gone from, I think it's in the order of something like $200 or $300 per drum to something like $6,000 or $7,000 per drum, so people are turning around and they're using methyl-t-butyl ether, using other things that are not regulated.

Acetone was barred from importation into Thailand because acetone is a solvent for the processing of opium and manufacture of heroin. Other things are going to be used. But the legitimate groups who use acetone in Thailand or use ether in South America, or use piperidine in this country, are being hamstrung by laws that are made to interfere with the drug trade and yet the drug trade doesn't get interfered with, because I think you'll find PCP labs are as plentiful now as they were before that law was passed. Yes?

STUDENT: What is piperidine used for?

SASHA: A catalytic base. It is very strong. It is very frequently used as a base catalyst. It is used in the manufacturing of several insecticides. It's easily made; one thing is, it's a very cheap base. You take pyridine, which is easily obtained, and hydrogenate it.

Okay, where am I? In the handout I gave the listings of the PCP analogs

LECTURE 16 ~ *Deliriants*

that have been regulated. The one additional thing I would mention in this area, in the same category of analogs of acetylcholine, is a material that is known as quinuclidinyl benzilate or BZ. It has a ring system in it like piperidine, except it has a bridge; it's a cage type of ring system. The whole compound is a bridge that I want to get into, toward chemical warfare. This is a material that was investigated and manufactured at quite a bit of length in Russia, in chemical warfare work. It does the same thing as does Ditran, as does ketamine, as does PCP. It separates the mind and body. It tends to make a person not aware of what he's doing and make him ineffective in doing things in the sense of conscious control. It's ideal for chemical warfare. Spray it over the troops. The troops go in strange directions and can't respond very well. And of course, these are called "incapacitating agents" from the military point of view. Not lethal.

The idea is to find some way in which you don't necessarily have a carnage of lethality, but you cause enough disruption that the people cannot be effective soldiers in defense. So a great deal of military effort went into this. That's why LSD was explored in the military, why PCP was explored. Quinuclidinyl benzilate was found to be explored quite widely in Russia as an incapacitating agent. So this country got into quinuclidinyl benzilate chemistry quite a bit. This whole general area, the compounds have this same separation. In this case, I'll draw the nitrogen and the weak base.

The JB compounds are in the law. This stands for John Biel, who was the head of Lakeside Laboratory research for many years. He made up a number of compounds, but his laboratory used his name as a coding of compounds. The federal law has made 336 and 318 illegal. I gave the structure of them in the handout. Very, very popular in the east. They never got particularly popular on this coast. But on the East Coast, and in Kentucky, Illinois, through the Pennsylvania area into New York, the JB compounds were talked about and very easily, very quickly referred to. They were very rare out here. But they were abused at the time the FDA laid down the BDAC laws, and they came into the federal law because they were written down by the FDA. Many of these compounds are easily made. The quinuclidinyl benzilate was of interest in the military because of the extreme potency. It was effective in man at a fraction of a milligram. Hence you can make an aerosol. They were studying how to make aerosols with vapor droplets that carry a few micrograms per droplet. You get a fog, and by breathing the fog, you are affected by this.

It was never very effective. The idea of distributing aerosols and distributing material in this manner was dependent upon the wind, the direction of the wind, the moisture conditions, what have you. They got much more excited when they found they could make little microscopic hypodermic needles, little darts. So they could put off a cluster bomb throwing jillions of little miniature needles everywhere, each of which is coated with a drug, and you get a few of them stuck in you and you get the effect by direct physical injection. This was dropped because, apparently, they couldn't get them evenly distributed even then. There's a whole chapter on chemical warfare I want to get into, but that's another chapter on drugs and drug use. But this is the foundation of a lot of the military inquiry into drug use, because these things are indeed incapacitating agents.

They do cause that dissolving of conscious control over action, but they are not lethal. Does that answer your question? No. With large amounts, you can have acetylcholine problems of shutdown, of gut motility and dryness of mouth and difficulty of vision. But these things pass. They are not long-term damaging agents.

That's about the hour. Any questions in this general area on ketamine, on PCP? I just tried to give it without getting too technical, to give the general feel for the drugs. I'm glad we got into that little area of choice. I don't know exactly where it goes, but the way I was saying it was a bad way, and I want to think that over a little bit more.

LECTURE 17
March 31, 1987

Peyote

SASHA: Someone give me a check on time, we've got a bad clock.

STUDENT: Five after.

SASHA: Five after.

STUDENT: Three after.

SASHA: Let's average out. Four after, plus or minus one. What's the consensus on when we officially start?

STUDENT: Four after.

SASHA: Four after?

STUDENT: Right now.

SASHA: Right now. Okay. We have two, 4, 6, 8, 12, 14, 16, 17. The handouts are the ones I gave last Thursday for those who were not here. They're not new. But if you haven't looked at the ones you got last Thursday, they'll look new to you. But if you take them for the second time, there will be others who will not get any. That's not so. I've made extra ones. Help yourselves.

I want to make the lecture today on a single drug, as I did previously on tobacco, on caffeine, on drugs that either are extremely widely used or are extremely fascinating and important in the area of the history and the development of drugs. And today I want to talk about peyote. Peyote, or in the Nahuatl name, which is more of a southern Mexico name, *peyotl* is a cactus. I brought in a sample from my collection at home.

Okay, before we get into that, how many people have tried peyote? One, two, three, four... okay. How many people have tried mescaline? About four or five. Okay, it's a cactus. It grows in the southern part of the United States in Texas. It grows in northern Mexico, the Chihuahuan Desert, and in San Luis Potosí. It is an ancient plant. The plant sample I brought in is not the prettiest. It's a little bit on the ratty side, but it's the only one I had that's just beginning to bloom. In fact, it's going to have two blossoms coming out, which gives you a chance to see the exquisite pink that is the color of the blossom. The books will say it comes in white blossoms and black blossoms and brown blossoms. Nonsense. It comes in pink blossoms. And I have a living example to show it, and I have seen many, many examples. I've never seen any white blossoms. It's always been pink.

It's easily fertilized. You can get pollen onto a camel-hair brush and do the thing that insects normally do, and they fertilize like a shot, they self-fertilize. And you can get seeds very easily. There are probably half a dozen seeds per blossom—little, black, very fine, round seeds. And in a moist petri dish, a little bit of sand, a little bit of moisture, they will fertilize. They will actually germinate in about four, five, or six days. You can see the little seeds having popped up a tiny something above the surface of the sand. And there it sits for about eight months. It doesn't grow. It doesn't do anything. It has sort of exhausted its efforts in having popped out something from the seed. It sits. That eight months is the disaster for probably 99 percent of all peyote plants.

If it can get through the eight months, then it begins growing. In its typical pace over the course of a few years, it gets to a size like this. When it does divide, it can go via the seed route and a few seeds may make it around the plant. More commonly it divides in the cactus area by a device known as "pupping." The plant, when it gets to a certain size and a certain degree of nutrition, or lack of nutrition, it will actually break out into pups—little plants will grow up around the circumference, around the periphery of the plant. And very often the central plant will eventually deteriorate and die. And so you'll often find peyote in the desert looking like that, in the field, if you find it at all now. It's increasingly hard to find. You'll find it in circles and sometimes you'll find it in complex webs, often rather hollow in the center.

The entire origin of it and of many of the plants that are involved in cultural rights, social rights, religious rights in Mexico, had their origins back in the Mesolithic period, back, perhaps 10 thousand years. I don't know; it had its origins way back. But they got into human usage, in the many

LECTURE 17 ~ *Peyote*

thousands of years ago, and into the Aztec culture, probably about 2,000 years. The record goes back to about 2,000 years ago, what few records the Spanish Catholics left when they came and went. About 2,000 years ago in the Aztec world, they had plants that they used in their ceremonies. A few of them are still known. Many of them had become lost. And some of the names have been sustained, but the actual identity of the plants have been lost.

Okay, peyote, tobacco, datura, what are the others? Jimsonweed... I had mentioned datura. The cacti, the mushrooms—the *Psilocybe* mushrooms—you know them now with the genus name *Psilocybe*, but they're the sacred mushrooms that came from the Oaxaca area, and ololiuqui, which is a morning glory plant that I'll talk about when I'm talking about the ergots and LSD. Today I want to lean primarily toward the peyote.

I'll do another one of my famous maps if I can get away with it. [Drawing on the board.] There's kind of Mexico. The Yucatan is a bit on the large side. [Laughter.] So I'm not a cartographer. Mexico City is in here. Guadalajara is in here. Mazatlan is right across from the tip of Baja California. I'll give one more and that's Chihuahua, which is kind of in the upper central area just across from the actual state that borders Texas. This area has basically three tribes of Indians that live in here [making reference to his map]. The Cora, the Huichol, the Tepecano. Those are the three that are in those areas. And the Tarahumara[14], a kind of an independent, very isolated group, which butts up against, more or less, Chihuahua and the Sierra Madre Occidentales. They're often called the Tarahumara Sierras.

These are the basic tribes that survived and maintained the use of peyote with the Spanish. When the Spanish came—Cortez and that crowd—they brought with them a very firm, very well established religion, the Catholic religion. And in the Aztecs they saw the broad use of many of these plants in conjunction with sacrifices, in conjunction with their own religious usages. And ah! Devil weed, devil this, devil that, and they just swept through. They wanted to expunge all records of the Aztecs, and they succeeded to a large measure. They destroyed thousands of documents. Then in the middle of

---

[14] Tarahumar/Tarahumara is a name that has been increasingly abandoned as a colonial name. In addition, in their language, its use is restricted to men. Rarámuri is now preferred, but because of the age of this material, the original use was preserved.

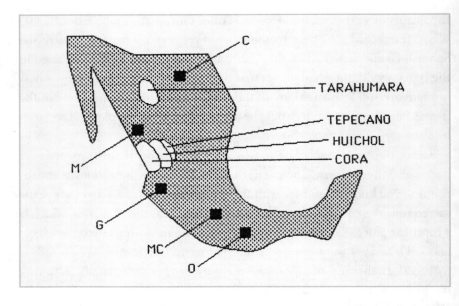

the 1500s, they brought in the Inquisition, and either you convert or you are punished as a devil and a witch or whatever the current ethic was.

And a test was, did you use ololiuqui? Or did you use the peyote? And if you did then you were obviously participating in murder and cannibalism, and everything else that is evil. They made their efforts with the Huichol[15]. The Huichol are probably the best studied of these Indian tribes. They are in the area just down from Mazatlan, and east. They established some missions with the Huichol, but to a large measure, they finally gave up. The Huichol went back to their classic story and tribal observations.

There is an area that the Huichol search out for the cactus; they have a pilgrimage to gather it, a ritual that has been unchanged for as long as the oral history has been maintained for them. And from this grew many of the other pilgrimages and the other searches for peyote that have grown into the Native American Church and its use by the Plains Indians in this country, and many of the practices of the use of peyote that have been instilled. Actually, it was not instilled in this country as a proper religion until about the turn of the century. Peyote was involved in the trade between the Northern

---

[15] Huichol is a name that has been largely abandoned as a colonial name. Wixárika (singular) or Wixáritika (plural) is now preferred, but because of the age of this material the original use was preserved.

LECTURE 17 ~ *Peyote*

Mexican Indians and the Southern High Desert area Indians of Arizona, New Mexico, Colorado, Utah. Primarily in that area, there was a trade that carried a great deal of peyote back and forth along with baskets of jewelry and other items of value.

The origin of peyote for the Huichol, as with many of the other tribes, was totally masked in myth. You must catch the flavor of the Huichol philosophy. The deer is to them the sacred animal, and the deer has given rise to everything they have. The deer is the source of the peyote. The peyote is said to come out of the forehead of the deer, the antlers were the forehead. From the forehead came the peyote, and from peyote came corn. And so they have corn, which is food, and peyote, which is medicine to treat the spiritual illness. The food is to supply the body's needs. And both have their origin in the deer. All the myth and all the verbal history of the Huichol tie these things together.

I think people may be familiar with the yarn paintings of the Huichol. They have become commercialized very much recently, but they are still obtainable and still seeable in collections from many years ago, before they fell into the commercial art direction. The Huichol still are maintaining a great deal of their original integrity, in spite of the fact that there are a great many people who are going there to guarantee they retain their original integrity. And of course, by being there, they are really interfering in many ways.

The Mexican government long ago said, not "Leave them alone, because it's a culture we want to preserve" but, "Leave them alone—they're not worth it. The land is not worth it." It doesn't have any commercial return, from the governmental point of view. If they struck anything like finding enough wood in the forest, that would be the end of the Huichol. But they have not. It is really very desolate country.

Their pilgrimage occurs in October, and as with many of the pilgrimages that are involved with their trek for peyote—to the peyote god country, the country to which you go when you die before you finally go west—it's deeply spun into their myth. The pilgrimage is about 40 days. It could take less now because you can do part of it by pickup truck, but still a great deal of it is done on foot, so it's a 40-day quest. There are always rules, such as no bathing, virtually a fast—very little eating for the entire process. No sexual interaction. Women may go. No salt. I don't know why, but in every culture that goes on peyote quests from the Mexican to the American Indians, salt is not allowed. Salt is just not a part of it. Tobacco is deeply invested on the other hand. They always take the tobacco with them, and in both the religious ceremonies and in the search for peyote, it plays a very integrating role in the social unity.

The quest goes to the east, to the high San Luis desert. And the original myth of how the peyote came to be was, again, with the deer. Many years before any records were kept, the deer was seen to take about four or five steps and disappear. And in searching out the four or five steps, they found the peyote at the footprint. The footprints were the peyote, and that spun the deer and the peyote into the awareness.

Every small group has its own mythos to its origins, and all of them deal usually with some sound, some voice that came from the peyote. The peyote was then gathered and was found to have its own vision, its own self-understanding that was maintained as a very tight cultural requirement in many different Indian groups. In many groups, not just Indian.

When you go into the high desert, the eastern desert just about 300 miles east of Mazatlán, the rituals are maintained. You stop at sacred areas for prayer, for what have you, largely fasting, but water is carried. The first peyote that is discovered by the leader of the pilgrimage is outlined by arrows as if you were shooting a deer; arrows are placed on all sides of the original cluster. And it is then gathered in about three days.

This is really the first serious eating. They eat peyote for three days. It is eaten continuously; it is celebrated continuously. Then they return. They paint their faces yellow with a plant I've never identified but apparently it is known[16]. They paint their faces yellow and return back. On the return trip they stop—this is now usually in November—they stop and spend about three or four days hunting deer, and bring the animals back with them. Then the actual ceremony—the festival—occurs in January, in which peyote and the deer and food are celebrated with a great deal of peyote eating.

The ceremonial use of peyote is really done in a basically nonreligious context. But it is in a context of great spirituality and great devotion. With-

---

[16] This is the ground-up root of agarita, aka *Mahonia trifoliolata*.

LECTURE 17 ~ *Peyote*

out the success of the peyote trek, there would be no water, there would be no crops, there would be no corn, and it would be the end of their life cycle. So, the assurance of the return of the rainy season, for plant growth, is made through the peyote.

I mentioned in a very early lecture about these very fundamental needs of rain, water, procreation, of fertility, of ridding of evil spirits, the curing of illness, of the possession of a spirit, that are tied in intimately with the use of plant drugs, where the drugs have played a role and had been invested themselves. All plants themselves are invested with spirits exactly the same way people are invested with spirits. So there is an intimate relationship. What better way to get at a spirit that has been messing with you than to use a plant that has a spirit which has the power and the strength to come back and relieve the negative, the undesired spiritual body?

The Tarahumara I might mention—which exist in between the Huichol area and the United States in the very desolate areas of the Sierras—are a bit more insular, a bit more isolated. They go to the outside to search out people and contacts. Their pilgrimages are a little bit looser, in that they will often tie them in with trading trips, tie them in with commerce, and they will go largely into the Chihuahuan Desert, which lies between the sea and the Huichol mountains, for their peyote.

They are long-distance runners—that's a good example of drugs in athletics. Apparently, that's one of the big flaps right now about people in the various athletic groups using drugs and obviously giving themselves an unfair advantage. I'm not quite sure why they have to pee in cups, as do airplane pilots and train operators and people in the government—but not the high officials, that's ridiculous, only intermediate officials pee in cups. Most drugs literally interfere with performance, and this is generally known, and although a lot of athletes over the years have given testimonials to cigarettes, not that many smoke. There is a lot of tobacco eating, but not that much tobacco smoking in the athletic area. And I think it is largely that the image is of giving the right message.

Athletes are people you admire, I guess. They're out there, they run this fast, or they have a high batting average, or they made three touchdowns or something. They are the image-makers, the image-people for those who are apt to misuse drugs. And if they don't misuse drugs, clearly the message is going to be given correctly. But I don't think the 99 percent of the people in the world who have at one time or another over the past history used drugs, such as the Indian tribes, and such as the people in India, people in Europe, people in Russia, people in Africa, have ever looked at athletes and said, "Oh my, drugs might be bad for me, too." They have been up to their ears in drugs because drugs have been an intimate structure of their cultural and religious development. And they are still up to their ears in it.

We have a society that's up to its ears in drugs, for whatever reason. Our compulsion to change or alter our states, change our point of view, change our consciousness. And I really wonder about the deprivations—this is an essay I've gotten on to before, I'll probably do it again—but I wonder if the deprivation of our freedom and independence, by such assays and such invasions of privacy, really carries a sufficient impediment to our drug usage and justifies the losses that come along with it. I'll be back to it later. In fact, I'm going to really go into the subtleties of urine testing in a much later lecture. I want to get back to peyote here.

Anyway, the Tarahumara runners are famous for being long-distance runners. They always run with almost no clothes on. There is a loincloth, but they go days virtually without stopping. They're famous for it. They always carry a bag of peyote alongside. They are stoking the fires with peyote all the time. They believe it lessens the pain and increases the endurance. And so, here is—really almost from prehistoric times—an example of drug use in athletics.

Anyway, it was traded into the United States—certainly hundreds of years ago—but it hadn't worked into the realm of a religion or into religious observations in the United States. There were many cultural and religious aspects that had been evolving in the Plains Indians—the Ghost Dance ceremonies, the various ceremonies that are conducted to the various gods that relate to fertility or to weather conditions for crop raising or to conditions for curing illness. But not with peyote. It's about the time of the Civil War that really the disenfranchisement of the Indians became quite apparent. They were often relocated to areas that were quite distant from their original areas, quite alien to them. Often, they were relocated tribe to tribe, in conjunction with where they had a history of being enemies, which led to a difficult integration in their new territory.

LECTURE 17 ~ *Peyote*

They were considered heathens, and they were considered pagans, and they were being affronted with the need of Christianity or whatever. Christianity was the in religion at the time. It still is. You had an insistence upon them speaking English. You had this "second classness" that was very offensive and very destructive to their self-image, and you'll see many of these influences on self-image are still alive today. And peyote was a vehicle that was found to be valid in the vision quest, valid for the searching out things from within yourself, that is really what religion is, to a large measure, all about. Finding some way of coming to peace with your negatives, coming to peace with where you're going to go, giving you some answer of what is ahead, giving some explanation of what is past.

This is really the heart of faith and religion. And the peyote served that role elegantly, because it gave a distinctly different view of things. It gave visual emphasis and sometimes very dramatic interpretations and viewing of complex objects from a point of view that is totally novel and often very informative, often very frightening. But it has this property. And this has spun into what has now come into being as the Native American Church.

Let me give a picture of this type of ceremony of the Native American Church. This was the church, by the way, that served as a focal point of one of the early contests between the United States law system against drugs and the constitutional freedom of religion. It became an interesting test where the Native American Church was formed to formalize the use of peyote in religious ceremonies by certain Indians who were of that faith, and yet the use of peyote was in conflict with the promoted thought that peyote caused people to become murderous, caused people to go crazy, caused people to die. I do not know of any death that is ascribable to peyote. It's been used over history by many thousands of people, hundreds of thousands of people, and it's been used by many on a weekly basis. And I do think there is more death that is ascribable to alcohol use in the Indians. And alcohol often ties in with peyote usage, especially with the Huichol.

When they come back to their festival of celebration in January, there is gorging on peyote, gorging on food, and some gorging on alcohol. And there is a lot of destructive use of it. In a sense, all these people, including the Tarahumara, can be very self-destructive in their interrelationship with drugs and with non-food chemicals—alcohol as well.

But the quest from the Plains Indians also requires no sexual interaction, no salt, no bathing. Tobacco plays a very large role in this quest. Very often tobacco is taken as a major community-together synthesizer. I don't know what the correct word for it is sociologically. But for example, if you're a visitor in one of the peyote ceremonies in the southwest, and you're asked to be a guest, you'll find an automatic, certain alien feeling toward you. What's your role? Are you there to be entertained? Are you there to spy? Are you there to get a free high or something? What is your role? Why are you a guest?

But if you are in such an interaction and participate in the ceremonial circle, and participate with the tobacco, which is the first of the integrating things, and then the peyote, by the end of the session the next morning, you are part of the group. You are accepted. You are a brother with the group. So all these things have a way of integrating behavior patterns and allowing trust and allowing fellowship.

This is in keeping with what we mentioned about why people are so enamored with PCP or with heroin. In the group where everyone uses it, it becomes a bonding. It becomes an extended family, which is something that everyone at some level wants. The fear of being alone, the fear of not being with someone when you are ill, the fear of dying without anyone being at hand is deeply embedded in the personality, so you want a family. And if the family happens to be knitted together by going to church, or knitted together by using drugs, or knitted together by the Oakland A's opening game or whatever, that is your extended family and that is with whom you share pizza and you share intimacies. This is knit together in the southwest Indians initially by tobacco and then by peyote.

So when the peyote is brought back, the ceremonies are held. There are really four people who are basic to the ceremony. There is what's called the Road Chief, the Drummer, the Cedar Man, and the Fire Man.

Let me talk a little bit about the ceremony. It's a rather interesting picture. There are variations in every tribe group—but the basic variations you'll find are between the Plains Indians, which are quite consistent in many threads, and the Navajo, who have what they call "the V Way," another name to their particular ceremony.

By the way, this thing is absolutely worthless [in reference to the clock]. What's my time?

LECTURE 17 ~ *Peyote*

ANN: Half past.

STUDENT: 11:30.

SASHA: So, I've got another—more than half an hour to go. Does anyone know what door you pound on to let the person who corrects this thing be aware of it? How long has it been like this? Three weeks?

STUDENT: Three years.

SASHA: Three years, marvelous. That shows my recent coming. Whatever, I'll ask periodically. I do not wear watches or wear jewelry.

In most Indian tribes, the ceremony is usually in a tipi, often constructed and certainly dedicated to the purpose of the meeting. In the Navajo it is usually in a hexagonal hogan that is committed to it, and it's used for that one purpose. The Road Chief is, in essence, not a minister, not a preacher, not a leader in that sense—he is the one who sort of puts it together and runs the show, but not in a superior sense. Usually in an old, familiar knowledgeable sense, and it's a matter of honoring tradition.

Let's get rid of Mexico; we're now in the southwest [referring to chalkboard]. The altar is here. Here's the rest of the circle, where to the east is the opening—obviously eastern to the rising sun. The altar is constructed in the center. It is a crescent-shaped thing, usually with the crescent arms going toward the east, and is mounded. I can't begin to describe it. It is like a curved banana with the points coming down to points at both ends. And the points come down in the east and it is mounded in the center. And it always has a track through the center of it, which is called "the road." And to the Indians, this is known as the "peyote road." It is a road that you must walk to become knowledgeable, to become a complete person, and it is peyote that will help you walk that road and begin teaching you about yourself and learning about yourself. And hence the term "Road Chief" stems from the fact that this is a person who sets up this pattern for the use of peyote.

He provides the satchel. Loads of stuff in that satchel. You have the peyote. You have a rattle made out of a gourd. You have a whistle made out of an eagle's bone. You have sage. You have a buckskin from which a drum will be made. You bring a metal pot, which will be the base of the drum. You have cord. You have stones that will make the drum. You have a staff. You have a bunch of feathers. You have your tobacco and about 18 more things I'm not remembering. These are all in the satchel. He brings this. He dictates the construction of the altar.

The fire is located in front of the altar. By the way, there are differences. This has nothing to do with drugs, just a fun bit of culture. In the Navajo, instead of the altar, they usually have about a three-foot-by-three-foot raised area of sand. It is not a construction. It is made out of sand, often a slightly lighter color, different color than the area itself. Instead of a fire, the Fire Man, who sits just to the right-hand side of the door to the east, will bring in hot coals. And so with the Navajo, there is no open fire, but there are coals.

The Fire Man will lay on this altar of sand a "V" of coals, which is the origin of the term "V way." The dress is quite casual, although some people will come with peyote jewelry. Often it is quite casual. In some areas there are men and women, some areas only men and no women or children. Different cultures call different dictums in this area. The people enter. They go, as you're looking down on it, clockwise. The Road Chief is up in this point, and on his left is the Cedar Man, and the Drummer to the right.

And then everyone else sits around in a kind of a pattern. You may have as many as 30 people sitting around the area of the fire. The fire is started. The function generally of the Cedar Man in all these cultures is to have ground up cedar—by the way, which also the Road Chief brings—and it goes on the fire, producing cedar smoke, incense smoke, which is the smoke that things are blessed with. So, in essence, he is the producer of the blessing environment. The Drummer is just exactly that. He is a source of drumming that goes along with the singing.

The first thing, a large peyote button called "Father Peyote" is brought out and usually placed on the center of the altar. And in the case of the Navajo, it is placed at the head of the V structure of coals after having been blessed. And the staff is blessed, and the drum is assembled. The drum is assembled by filling the metal pot partially with water and then stretching a skin over the top of it and tying it with the cord and stones. Stones are also in the satchel of the Road Chief.

And tobacco initiates the ceremony in almost every tribal ceremony of peyote. The tobacco is produced. In the marvelous container that the Road Chief brings are corn husks along with tobacco and also a fire stick. I'll keep

LECTURE 17 ~ *Peyote*

thinking of things as they come up. And the tobacco is started. It's rolled in corn husks, then it's smoked. And the things go around the group to the left, so on around goes the tobacco. And everyone rolls and makes his own tobacco. They smoke. Then they put the butts at one side of the altar. The smoking residues play a great role in the peyote ceremonies, not only in the ceremonial tipi, but also in the quest. In ceremony the butts are arranged in ways that are symbolic. I have no idea if studies have been made of this, but it is still a factual observation.

The tobacco goes around and the peyote comes out. The peyote goes around the circle, they take four peyote. Everyone takes four. Everything goes in fours. The Road Man takes four peyote, and the peyote goes around the entire group. In the Navajo it starts with two cups of peyote tea made ahead of time by boiling a great amount of peyote in water and letting it stand for a while until it's down totally in temperature. And then the peyote tea starts going around the circle.

They also have cut slices of the peyote button. A peyote button, I mentioned, is the top of the peyote plant that is cut off at ground level with a knife, and in many cultures where peyote is gathered, it must never be touched with a knife again. In other cultures it can be cut into various portions. But the peyote button comes from that and dries up to a withered little thing. These are fairly small ones, but they're just withered things with the tufts. The name of the plant is *Lophophora*, meaning carrying plumes, carrying tufts.

I'll get more into the botany as we get out of this little chapter of the ceremony.

ANN: Are you going to mention the taste?

SASHA: Oh, I might have mentioned that. Those who have eaten peyote know. Those who have not cannot imagine the taste of peyote. It is unspeakable. Literally. I won't even talk about it. It is the sort of thing that once you try it, you immediately spit it out because something in your head rings a bell that says, "Poison—do not eat." And yet, basically, you know that this has not a poisonous property to it.

If you can keep it down for 40 to 50 minutes, you probably deserve some sort of a prize. And when it does come up, usually it doesn't interfere with the experience. In fact, it's often the starting point of the experience, because

THE NATURE OF DRUGS

LECTURE 17 ~ *Peyote*

you're sitting there having voided your tummy in a very inelegant way, and you are looking at what is medically called "the vomitus," which is the peas, carrots, and what all that has come up, with bits of peyote chunks and what have you. It's fascinating. Really it is. This is the change of viewing things that comes with peyote usage. You're seeing things in a different light. And it's like a baby who has defecated in the toilet for the first time. They're fascinated by it. "Look what I produced!"

The vomit is something you produced, and for a lot of people they have not produced much in their life. And here's something, and you're looking at it, realizing that is part of you, an extension of you. It's a strange point of view. But it's a very real one. There's no regret associated with vomiting, because it is a natural function that is playing its natural role in a form of cleansing yourself, and it's looked upon that way. And at that point the active components have long since been absorbed and there's no loss of the experience from it. It often initiates the experience in a rather interesting way for a new person.

Vomiting, by the way, in the first use of the peyote is not that uncommon amongst the Indians and it's not sacrilegious to go outside the tent and do your thing and come back in. Peyote use—once it has been initiated and the singing is underway, especially into the heart of ceremony, which usually starts at nightfall and extends with a strange break, a change of pace at midnight—extends to dawn. Peyote is used as much as you wish. It's there to be used and a person knows their own background, where they want to go, and will choose their own amount of peyote.

Singing starts. The Road Chief will start the singing—again, four songs—with the drummer doing his thing, and the Chief will always carry in his right hand the staff, feathers, all of which have been blessed over the cedar smoke. And one more thing. A little bit of sage. The peyote button, Father Peyote, is always rested on sage. It does not touch the altar directly. It rests on sage.

And the left hand will take the gourd rattle. I mentioned there is a rattle in that marvelous traveling satchel as well. And with the rattle and the drummer doing his drumming thing, he sings four songs—strange songs. Very much in tradition, they have been there for many years. Very often not intelligible or a great deal of it is not intelligible. A little touch of the glossolalia that is found in some of the religious ceremonies in Christianity. Syllables are added and phrases are added that make no sense, but they make a characteristic fast-paced roll to the song. One of the major features in helping spread the peyote concept, the peyote culture, has been the songs that come along with it. It is a very ancient practice.

The Huichol have much singing in their festivals that comes in the same character. Very often the words throughout the Indian cultures that are associated with peyote and Peyotism are the same words, but they're all spun together in a pattern that is nonsensical, in the sense of it being translated. But that pattern is a consistent pattern.

Then the Drummer takes over the paraphernalia that has been blessed, and the gourd, and the Road Chief does the drumming and the Drummer sings four songs. And then the Cedar Man sings four songs, and then it goes right around the whole group, and everyone sings four songs. And anyone at any point can move around to get more peyote. This is more or less the open festivity portion of it.

At midnight the character changes. At midnight it is a sort of a break. You have midnight songs, an entirely new group of songs. Fresh water is brought in. The Fire Man, his role is sitting near the door all the time to go

LECTURE 17 ~ *Peyote*

out and keep the fire going. He leaves quite regularly. In the Navajo group, at midnight with the midnight songs, the coals are regenerated and changed. They're moved from the form of a V. (The V by the way, the point is toward the Road Chief and toward the altar. The V points toward where the Father Peyote is placed.) But the image of the coals is changed to the shape of a bird. And fresh coals are added to generate that new shape. The fire is maintained in the common tradition in the other ceremonies.
Yes?

STUDENT: How long does it go on?

SASHA: All night until dawn and even after dawn a little bit. It starts at nightfall, so it's about an 8 or 10 hour trip, depending on season. Winter is a much longer one, but it's still through the night. And there are the midnight songs. Then the actual praying. The whole function of the thing is a combination of peyote eating, of meditation, of social gathering, singing, praying; these are the roles that are played in the meeting. After this they are addressing people who are sick or addressing people who are away or addressing personal problems. And it's largely self-addressing. The Road Chief is not a shaman. He's not a guru. He's not a person who knows how to solve anything. It's put back on the person and upon the group's attention, and reinforcement of the person and his needs. That's the whole spirit and philosophy of the peyote ritual. At midnight the Road Chief will go out with his little eagle whistle and blow the four directions, four times, four whistlings.

During the night food is not allowed. Then food is brought and the food consists of water that has been there, of corn, of meat, of fruit, and of water that has been brought in. They are always arranged from east to west, water followed by corn followed by fruit followed by meat, symbolically. And with the closing of the ceremony, usually the beginning of daylight, or past the dawn into daylight, the drum is taken apart, everything is put back in the satchel of the Road Chief, everyone consumes some of the water that was in the bottom of the pot of the drum. A general celebration. People provide an immense amount of food. Whoever is hosting it provides a great deal of food, so often the hat is passed to help defray some of the costs that are involved. But the host group provides a large meal and provides the area in which this all takes place. The Road Chief will actually travel from place to place and is asked to come. He's not paid for it. It's not a commercial venture in that sense. It is a part of the cultural venture. It's a very interesting, developed thing that goes on today.

You tend to think all the religions in the country are based on the Judeo-Christian ethic. There's a lot of Christian aspect in this. In the Navajo, for example, the V concept is thought to have maybe come in from World War II, because it's quite recent, perhaps from something like "V for Victory." That's one of these little rumors; you don't know what truth it has. But they will often put pictures of the Last Supper or pictures of religious significance or little crosses carved on the staff. I mentioned the staff. The staff is one of the major things that the Road Chief brings with him. There will be Christian religious aspects that have been built into it, and it's partly to identify with another religion that serves as a persecuting body, and it's partly because it's out of respect to other religions that are searching in many ways for the same thing but using different vehicles and different symbols. And this has developed out of the struggle of the Native American Church to sustain this. In the 1920s and 1930s—as you had with every drug that has gone down the pike to be disallowed, to be made illegal—there were statements made that you would find in the newspapers quite regularly, "Peyote led to a series of murders from a bunch of Indians who used it, and it has led to this damage and that illness."

It was very heavy propaganda that came down against peyote. It has no fact, but it was just that it was heathen cultural usage and came out of Mexico, so it's anti-American. It's not what you want your culture to be involved in. It has all the negatives: it will kill, it will destroy, it will make you mad, it will be whatever-you-have. So peyote was put onto the drug list as an illegal thing. The church that had used it in ritual for a long time appealed, "Please allow us our ritual." The argument was made again that there wouldn't be interference with religious processes—just as the Constitution says, we will not interfere with religion—but anything that is a religious device which is an immense public hazard cannot be allowed. So, they replied, "Where's the public hazard?"

Well, everyone knew that it caused no murder or cannibalism or abnormal behavior. The route is quite the contrary. It usually is one of great

LECTURE 17 ~ *Peyote*

inward-going, and inward respect, inward search, meditation, and it is not associated with crime. But it was believed to be so, and it was not until just a few years ago that a successful appeal was made to exempt members of the Native American Church from the regulation that peyote should be a Schedule I drug. Peyote is a Schedule I drug in this country, and it was put officially in that classification of the current drug law when it was first written in 1970, and the thing goes on to say, "Peyote . . . the plants . . ." I get into a little bit about the plant.

The plant *Anhalonium lewinii*. I mentioned Louis Lewin as being the source of the book *Phantastica*. He was the one who brought peyote out of Mexico into the United States at a scientific level. He brought it to Parke-Davis, who were not interested in the fact that it was an intoxicant and had no interest in it whatsoever. He took it over to Germany where Heffter, a famous German chemist of roughly the 1880, 1890 era, isolated mescaline and four other major alkaloids from it. He was the first to experiment with mescaline himself and find that it was the agent that more or less duplicated the acts of the peyote. The *Anhalonium lewinii*, named in respect for Louis Lewin, and *Lophophora williamsii*—the two are interchangeable. There is a different genus and different species name for what is the same plant. I don't know which is *en vogue*. They go from one to the other. It's like spelling "analog" with a "ue" at the end or not with a "ue" at the end. Everyone has his way of doing it, but it's the same word.

Other plants; the *Trichocereus*. I didn't bring a sample. The smallest one I have is about two feet high and I was not going to carry that. But *Trichocereus pachanoi* is probably the major cactus outside of the peyote cactus that contains mescaline. It is a columnar cactus; I think people would recognize it if they saw it. It can grow 6, 8, 10—actually, in Peru where it grows quite wildly, it's a Peruvian cactus. It grows as high as 15 or 20 feet. It's used for fencing because you put them next to each other and they become impenetrable when they grow up tight.

It is a large, cylindrical cactus. It usually has six nodes, sometimes five, and little spines that come out in between. There are no spines on the peyote plant. Look at it closely. There are no thorns, no thistles, no spines. It has tufts—tufts of hair. I once collected a whole bunch of tufts because I had heard a rumor that the tufts contain strychnine, and I said, "That's absurd.

There's never been strychnine ever reported in the cactus plant." It comes from *nux vomica*, an entirely different plant.

But that's what they say: there is strychnine in the cacti. I got a bunch of tufts and analyzed them—there is no strychnine in the tufts. But the tufts are nonetheless often removed for aesthetic purposes. They have this fuzziness that is not considered appealing. I should have mentioned that in the peyote ritual, the four buttons are chewed and are very often—when they're thoroughly softened, as much as possible—then rolled in the hands during the actual ceremony to get them into a softer internal state and then swallowed. They can be chewed straight down. If you ever have an opportunity—it's not legal now—but if you ever have an opportunity to taste one little fragment of a peyote when someone has it, try it. The taste will stay with you for quite a while.

A very interesting observation. In the middle of a peyote experiment, the taste is very friendly. It's a very interesting, strange, allied taste. Here it's an ally. And there's no difficulty in consuming those additional buttons during the peyote ceremony and eating more will not lead to nausea. It is a perfectly friendly taste. Afterwards, the next day, the taste is just as bad as it ever has been. There's some distortion of the taste response in the middle of the session.

STUDENT: Is the *Trichocereus* thing the San Pedro cactus?

SASHA: San Pedro cactus, yes. There are about 15 San Pedros. There are about 15 *Trichocereus* species known. The *pachanoi* is the richest; *peruvianus* is another one. I shouldn't get much into botany, but I will. The *peruvianus* has lobes—six of them that have much larger spines in them. Both come from Peru. They're sold in the Peruvian markets. Whereas a half a dozen peyote buttons is an intoxicating dose, it takes about a foot or a foot and a half of the San Pedro to contain the same amount of mescaline. Mescaline is the active component in both. Mescaline is found in very small amounts in many other of the *Trichocereus* species, and they are of no use because the amount is vanishingly small. So, the amount of cactus you would need to eat is mountainously high. It is not doable.

But the *Trichocereus*, the San Pedro, is sold in the Peruvian markets as a medicine, as a healing item. It's sold in roughly foot-long sections. It's sold in

LECTURE 17 ~ Peyote

nodes, per node. When you propagate the cactus, you cut off a node and put the bottom end of it into moist sand, and it will root and grow. Sometimes it grows as pups around the bottom. Sometimes it grows as a pup off the top, but it almost always reproduces by pupping. I've never seen a *Trichocereus pachanoi* blossom. I do not know what it looks like. The ones I have never blossomed. They cannot stand frost. Even though they're raised in Peru, they're raised in areas that are protected, or they've grown wild in areas that are protected from frost.

To draw dirty pictures, the actual peyote component is a compound known as mescaline. For those who love these.

I'm not sure on the origin of the name mescaline, but it might have its origin as a corruption of "mezcal."[17] There has been a great deal of confusion between mescaline and mescal. They are different plants. The mescal bean is a legume. I brought in some mescal seeds, if you want to look at them. They're red, very colorful. And they were used as an intoxicant primarily by Southwest American Indians. It contains cytisine.

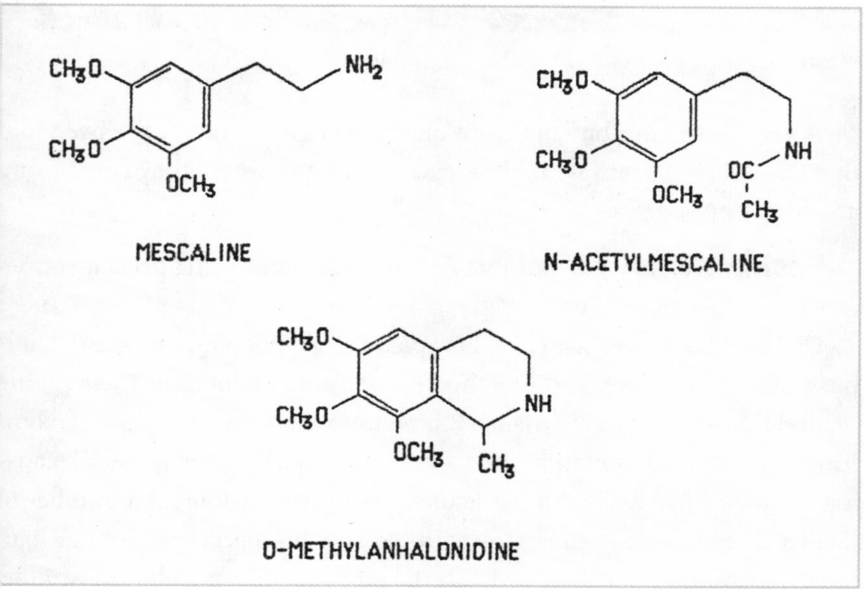

MESCALINE

N-ACETYLMESCALINE

O-METHYLANHALONIDINE

---

[17] Heffter named the compound mezcalin based on its origin being from what were then called mescal buttons or mescal beans (*s* goes to *z* in German). He also emphasized that it was not related in any way to the plant producing the beverage mescal, ie, the *Agave* spp.

THE NATURE OF DRUGS

Mescal beans.

ANN: Excuse me, but an urgent question comes to mind. I've been told that the little red beans with black ends, no matter where they come from, are deadly poisonous.

SASHA: Yes, they are. But that doesn't keep them being used as intoxicants, and they have been.

Cytisine, for people who like dirty pictures, I'll put it on the board. Double bonds are in there. Sort of a three-dimensional structure. This is going behind the board. This is cytisine. It is an intoxicant. It causes nausea, vomiting; it causes hallucination. It is an extraordinarily toxic material, causing unconsciousness, and it has actually been responsible for a number of deaths. The cytisine itself is mostly in the anticholinergic side rather than the sympathomimetic side, which is the area in which peyote—mescaline primarily—acts.

It was used quite broadly in the United States in the Indian culture between the Civil War and about the turn of the century. But it was not

LECTURE 17 ~ Peyote

Cytisine

liked. It was held to be effective, but dangerous. And it was used more heavily for jewelry. I think people have seen these necklaces and bracelets that are made by stringing these beads together? But it's a dangerous thing, because there is not much freedom of dosage. Peyote can be overconsumed and you may not be making a great deal of sense, but you're not in a life-threatening situation. Overconsumption of the cytisine or the actual plant—its genus name always slips my mind, Socorro, not Socorro, *Sophora*—is life threatening. And it was not liked, and that is one of the reasons that the Indians moved to peyote, because it achieved a much more easy, a much more friendly vision structure and interaction with each other, as opposed to the *Sophora*. I remember structure but can't remember the name.

As I mentioned, mescaline was first introduced as a plant in the scientific areas by Louis Lewin about 1880. The compound was isolated by Heffter a few years later. I think I had mentioned the word in passing, it's kind of a nice word in the chemical sense: tetrahydroisoquinolines. They are compounds of this general structure, nitrogen there, with various substituents on this ring, and the various substituents on those two positions. These tetrahydroisoquinolines are virtually unknown. And mescaline is totally unknown outside of cactus[18]. There are no other plants that carry it. If you find it in a plant extract and you are told this is an extract of a plant that contains mescaline, it was put there, as they say, administratively. It was put there by hand. It is not known outside of cactus.

The tetrahydroisoquinolines, there are five of them that were isolated by Heffter back in 1880, 1890s. There was anhalamine, there was pellotine, there was lophophorine and anhalidine and anhalonidine. Five of them. He nibbled them all. He tested them on frogs. All of these are without the benign aspect of mescaline. The *Lophophora*—the peyote plant—has these

---

[18] There are at least four analytical accounts in the peer-review literature purporting mescaline to be in other plants such as two Acacias (now *Senegalia berlandieri* and *Vachelia rigidula*), *Gardenia angkorensis*, and *Passiflora quadrangularis*. All of these are unproven, completely lack replication, and appear to be erroneous.

THE NATURE OF DRUGS

PEYOTE ALKALOIDS (1)

HO	NH_2	TYRAMINE
HO	$NHCH_3$	N-METHYL TYRAMINE
HO	$N(CH_3)_2$	HORDENINE

HO, HO	NH_2	DOPAMINE
	$NHCH_3$	EPININE
CH_3O, HO	NH_2	
	$NHCH_3$	
	$N(CH_3)_2$	
CH_3O, CH_3O	NH_2	DMPEA

PEYOTE ALKALOIDS (2)

HO, HO, CH_3O	NH_2	
HO, CH_3O, CH_3O	NH_2	
	$NHCH_3$	
	$N(CH_3)_2$	
	NHCHO	
	$NHCOCH_3$	
CH_3O, CH_3O, CH_3O	NH_2	MESCALINE
	NHCHO	
	$NHCOCH_3$	

LECTURE 17 ~ *Peyote*

PEYOTE ALKALOIDS (3)

	R_1	R_2	
CH_3O / CH_3O / HO	H	H	ANHALAMINE
			N-FORMYL
			N-ACETYL
	CH_3	H	ANHALIDINE
			QUAT. ME ANHALOTINE
	H	CH_3	ANHALONIDINE
			N-FORMYL
	CH_3	CH_3	PELLOTINE
			QUAT. ME PEYOTINE

	R_1	R_2	
CH_3O / CH_3O / CH_3O	H	H	ANHALININE
			N-FORMYL
	H	CH_3	O-METHYL-ANHALONIDINE
CH_3O / $O\text{-}O$	H	CH_3	ANHALONINE
			N-FORMYL
			N-ACETYL
	CH_3	CH_3	LOPHOPHORINE
			QUAT. ME LOPHOTINE
	CH_2CH_3	CH_3	PEYOPHORINE

	R_1	R_2	
HO / CH_3O / CH_3O	H	H	ISOANHALAMINE
	CH_3	H	ISOANHALIDINE
	H	CH_3	ISOANHALONIDINE
	CH_3	CH_3	ISOPELLOTINE

other alkaloids, and in sizable amounts, not as much as mescaline, but a matter of a fraction of a percent. Mescaline is present in the cactus to the extent of about 2 percent in different origins. It is present in the *Trichocereus* to about a 10th of a percent, hence it takes about 20 to 30 times the weight of the material to achieve the same effect.

These tetrahydroisoquinolines, there are now a total of about 30 of them known to be present. And there are a total of about 50 or 55 alkaloids known to be in peyote. These generally cause motor paralysis and toxicity related to muscular interference. They do not produce a high. They produce a physical inability, almost a curarizing effect, almost an effect such as curare, which is a muscular paralytic. But they are too small of an amount in peyote to show this effect. In fact, the question often has come up: "Can peyote be distinguished from mescaline itself, the active component?"

I believe it can. People say, "No, no. Mescaline is the only active component in peyote, and hence, they're interchangeable." There has been no, to my knowledge, no blind study made in humans of peyote versus mescaline. The people who have explored mescaline do not touch peyote because it is a plant source, so you don't know the strength of it, and these other alkaloids are present that are toxic. The people who use peyote, the Indian culture to large measure, say, "We'd love to be participating in such an experiment. Let us take our peyote, which we're familiar with, and give us an equivalent amount of mescaline to see if there are differences and see if it plays this role."

The experiment has never been done. I believe there are some differences; to a large measure, they should be interchangeable. Soon I'm going to get into the magic mushrooms of Oaxaca. I have lost the map now. Down in the bottom slump of the map, halfway between Mexico City going down through the drain and coming up in the Yucatan, at the bottom slump there is an area known as Oaxaca, which is a city and a state in Mexico, which is the source of the *Psilocybe* mushrooms. There, in native usage, they have indeed compared mushrooms with the synthetic psilocybin, which is the action component from it, and in the cultural usage, they're considered indistinguishable. So probably if such an experiment were made, it may well be that peyote and mescaline would show that same comparable action.

Mescaline has been explored as a psychiatric tool, as many of the hallucinogenics have been. Efforts have been made to see if it can duplicate the symptoms that are seen in spontaneous schizophrenia or various forms of psychosis, but not satisfactorily. This is the origin of the word "psychotomimetic," which I think I mentioned quite a long while ago. "Psychoto-," meaning psychosis, "-mimetic" meaning imitating: things that imitate the psychosis in the body. And the attempt to generate these changes with mescaline have been relatively unsuccessful, because mescaline's primary role is changing the visual field. And spontaneous psychosis primarily is expressed by changes in the auditory field, if there were a sensory distinction between the two. You give mescaline to a person who is schizophrenic, you get a very anxious and very tenuous, tight, uncertain, tense schizophrenic. You do not make their own personal experience much worse, nor do you cure. You merely lay on top of it a sort of distressing addition. Therefore it has not been useful in the treatment of—but it has been used quite a bit in the attempt to imitate—mental illness.

LECTURE 17 ~ Peyote

From mescaline has come a great amount of synthetic variation. And that's an area I want to talk about, as much of my research has fallen in that area. Efforts of modifying the molecule, changing the molecule to find out what aspect of sensory change, of interpretive change, what aspect of reality interaction, is due to what in the molecule? Try to take the molecule and modify it around to exacerbate something, to eliminate something, to see what can be found in changes of potency and changes of qualitative action. So both quantitative and qualitative changes have stemmed from the exploration of that molecule.

Any questions in the general area of peyote? This is sort of a bird's-eye view of the entire area. Thoughts or questions? I don't want to get into the chapters of the chemical variants of it, which has been more or less the exploration of the last 20 years, because that would be a part of another lecture. Yes?

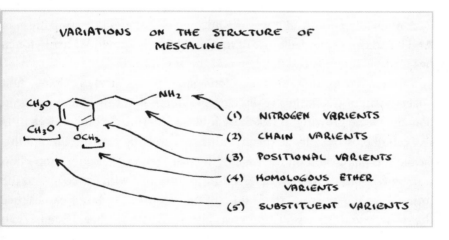

STUDENT: In Huxley's book *The Doors of Perception*, I think I remember he said that maybe peyote could be used as a cure for some mental illness.

SASHA: Yes, he held very much on the philosophical side. The question dealt with Aldous Huxley, a very well-known writer. I think the names of his books related to this. His experience was not with peyote—it was mescaline. *The Doors of Perception* and *Heaven and Hell*, I think this is probably familiar to many. Certainly, they were the vade mecum of the drug-oriented flower children in the '60s in the Haight-Ashbury. Everyone had read those.

In fact, you'll find copies in used bookstores with marginal footnotes comparing one's experience against Huxley's experience.

I guess one of the reasons I have an especially fond spot in my mind for mescaline is it was my first psychedelic drug many years ago, 30 years ago. And it was the one that really convinced me that there's no such thing as taking a compound of 6, 9, 10, 11 carbons; 3, 6, 9, 10, 11, 13, 15, 17 hydrogens; a nitrogen; and 3 oxygens, which is really a simple little pile of atoms multiplied and xeroxed a few billion times to make a white solid, could produce what it did. There's no way that could happen. There's no way in heaven's name you can take a simple molecule and see green things brilliantly green and colors in extraordinary color and with incredible interpretations.

My first wife, in her first experience with mescaline, could not get out of the car on Grizzly Peak Boulevard. She was a sober driver—that was not our problem. She opened the door and below her was the back of an alligator, a living alligator. She couldn't get out of the car. "I'm about to step on the back of an alligator." And it was the fine gravel that was reflecting the sun. And it had this bejeweled sense of that, and I finally convinced her to touch her toe to it, and it supported her, and she got out.

I remember in the same marvelous path up on Grizzly Peak. My first observation of *Ceanothus* really close. *Ceanothus* is this marvelous wild lilac they have growing. It has these intense blue-purple blossoms, a little like cattail blossoms. It's all through this area growing natively. I found this *Ceanothus* plant right on the edge of the place where that little choo-choo train goes around Grizzly Peak. And I was intrigued with the color and the intimacy of the blossom, the insides of the blossom. And here came a bee that went into the blossom, and I was intrigued with the bee. I had bees all around me. There's no way you're going to get stung, because you're not in argument against nature. You become actually a part of nature.

And I saw this bee, and I saw this great thing. I never knew bees carry their pollen goop on little hoozies coming out of their legs, but they had little pouches down there filled with all kinds of marvelous things. I saw a bee go in and selfishly get some more and stick it in; I was intrigued.

I never realized that eyesight could be so fine, that from a safe distance you could see a bee and what it's doing with all that marvelous stuff it's getting out of the plant. And then another bee would come in and it would

LECTURE 17 ~ *Peyote*

load its stuff. And I saw where the pollen plant kept producing all the things. I found subsequently, when I looked at the anatomy of a bee, I was seeing things quite correctly. I had been unaware of it. It's just that you are in a different relationship with nature and with your reality, but it is a friendly, constructive relationship. I'm not going to tout mescaline as a drug of use or abuse. But I'm just telling you, from my experiences, I found it to be a captivating attention grabber, because I was seeing things, interpreting things, getting into a relationship with what was around me, both in the visual sense and in the intellectual philosophical sense, that was impossible to rationalize as being the result of using a few hundred milligrams of a 25-atom molecule.

So where does all this come from? Well, obviously, it's all stored away up in there [pointing to his head]. The skill is in there. The ability is in there. The interpretation is in there. And this was catalyzing its release. And so, I said, "My golly, if that simple molecule can do it, what other simple molecules could catalyze my access to other aspects of myself?" And that is what launched my research work 30 years ago, and it has not been answered satisfactorily yet, so I'm still in the middle of it. [Laughter] But that is exactly where I'm going and what I'm doing.

Enough. I'm going to get on to the next two lectures.

LECTURE 18

April 2, 1987

Psychedelics I

SASHA: Why would moss be growing on only one side of a tree? Someone said moss was growing on the back of the tree. What about lichens? Do lichens only grow on the backs of trees? What's a lichen?

STUDENT: An algae and a fungus, isn't it?

SASHA: An algae and a fungus. The algae is absolutely dull; it makes chlorophyll. I mean, if it didn't make chlorophyll, it wouldn't be dull; but basically, the algae doesn't give anything of interest to the lichen. Algae can live without the fungus. The fungus can't live without the algae. So it's a halfway cooperative thing. But the fungus is absolutely loaded with fascinating chemicals.

Lichens carry derivatives of olivetol, and can be hydrolyzed with base to form olivetol. Orange peels contain citral. If you combine the lichen and the orange peel with a strong combining agent—phosphorus oxychloride will do the job—you make a derivative that has the action of THC. And marijuana, in essence, is a natural plant material that can be imitated very nicely from certain lichens and orange peels, which sort of gives me the feel that, even if we do get into a situation where synthetic chemicals are going to be more and more restricted, there are always a lot of things growing out in nature.

In fact, that's what I want to talk about today, things that grow in nature and have given rise to a number of drugs. It's a weird introduction—I wasn't planning it that way.

I want to talk about the phenethylamines. I will stop right there for a moment. I had originally put the "s" on the word, but I rubbed the "s" out, more or less. [Referring to chalkboard.] Phenethylamine. There is an easy way of speaking of things that is a trap. And that is to take something

THE NATURE OF DRUGS

that is unique and make a plural out of it to imply all those things that are related to it.

What you relate to something, and what someone else relates to something, might be different things, so be aware of that trap. I'm guilty of it myself because I am lazy. And I like to get into easy ways of expressing things. You get this term, for example, "amphetamines." "He was charged with possession of amphetamines." I'm not sure what they mean. Also, "These are psychedelic drugs," or, these are consciousness-changing drugs which are "phenethylamines"? Well, you can have chocolate ice cream, and you know what it is. It's a thing that comes in a little container and it's flavored with maybe chocolate. But what is meant by chocolate ice creams? You can't take a unique thing and generalize it to a collection. You can, but I just don't know what the collection means.

Take, for example, the amphetamine thing. A policeman comes up to the person and says, "I think I'm going to charge you with the possession of amphetamines." What the policeman means are those things that are classified in the law under the term *amphetamines* and related stimulants. Then you go to the pharmacologist and ask, "Could you tell me, what is the mechanism of action of the amphetamines?" And the pharmacologist will say, "Oh, it's very simple. This, this, and it goes into the synapse and releases such and such and activates so-and-so, and basically does something or other," while thinking of all the compounds that are pharmacologically like amphetamine, those that are stimulants—causes heart acceleration, kills the appetite, makes it difficult to sleep, whatever properties they commingle together. *Commingle* with two *m*'s—commingles together to form this family, that's what this person classifies them as.

Then let's say you go to a chemist and ask, "Can you tell me, is there a general chemical approach to the synthesis of amphetamines?" The chemist says, "Oh, sure." And then looks at this little organic structure that's a benzene ring with three carbons and an amino on it, and thinks of all those that are chemically related to it.

Each of these people has an entirely different collection of chemicals that they are embracing under this term "amphetamines." So when you get a lawyer talking to a chemist, they're talking about entirely different things. Though there may be a little overlap in what they're talking about.

LECTURE 18 ~ *Psychedelics I*

So be aware of these "collective, unique things." The term "unique." One of a kind, there is only one—hence the term "unique." Beware of such terms as "more unique than something else." You can't have "more unique." If it's unique, it's unique. "It was almost unique." That doesn't make a bit of sense. "Substantially unique." "Comparatively unique." Unique means one of a kind.

Amphetamine is a unique compound. Amphetamines is a collective that I would say I'm guilty of falling into. The term "amphetamines"—you'll find it used, but you don't know what it means until you know who the author is and what he has in mind. So be careful about using it. I tried to expunge myself from using it and I'm not entirely successful. I fall into traps, such as writing the *s* on "phenethylamines." What are phenethylamines? I'm going to back away. I'm going to say phenethylamine is the compound I drew. *Phen-*, a benzene ring; *ethyl*, the two-carbon chain; *amine*, the nitrogen at the end.

Phen-   ethyl   amine

There is a large number of centrally active drugs that are related to this compound. I want to start with mescaline. I drew it out—and I'll do very little drawing, I just want you to get the flavor of this. I will put all the alphabet soup that I'm talking about today into a formalized form. So if I get going a little bouncy and fast, ignore it. I will have all of this written up.

If you take apart the structure of mescaline, the main alkaloid we were talking about in the last lecture, you see very clearly that you have a slightly shrubby and modified phenethylamine. Phenethylamine: the benzene ring is still there, the two-carbon chain, the ethyl of the name is there, and the amine is there. So you can say that mescaline is a substituted phenethylamine. It is a derivative. Awkward term. I haven't talked about the term "derivative." Let me talk just briefly about it. It is used legally in one sense;

THE NATURE OF DRUGS

it is used chemically in another; it's used socially in another.

"Something derived from something" implies that you could make it from that thing by some little manipulation. You can make a bonfire out of kindling. The bonfire is a derivative of the kindling. Mescaline will be called, legally, a derivative of phenethylamine, but it is only a derivative in a sense that it has a certain amount of structure that resembles it. I don't know of any way, and I would challenge anyone to give me a way, in which I can take phenethylamine, which is a mobile liquid—and as the hydrochloride, it is a nice white solid—and modify it by putting these groups on it. It just is not chemically doable. There is no chemical reagent that will do that job. So I cannot consider mescaline a derivative of phenethylamine chemically. And yet legally it is considered that way because it is related to it. So the terms derivative, derivation, derived, are tricky terms depending on who you're talking to.

About five lectures ago I talked at some length about another material which does what phenethylamine should do in the body. This is amphetamine. [Referring to drawing on chalkboard.]

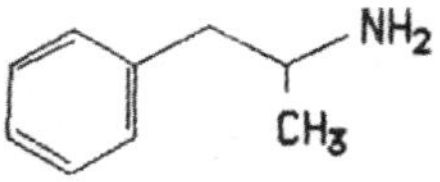

Amphetamine

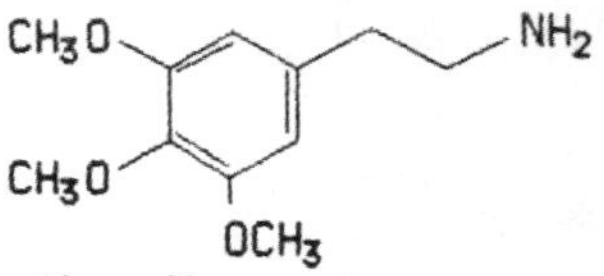

Mescaline

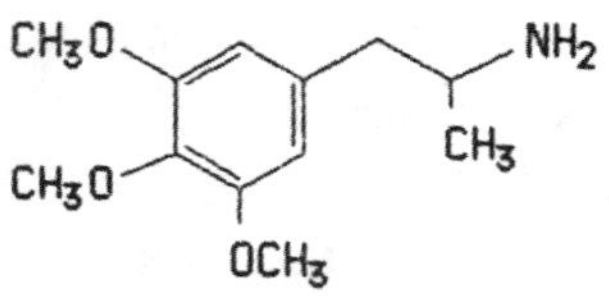

TMA

It is a stimulant. Phenethylamine should be a stimulant. It has all the right directions for being a stimulant. If you knew nothing about biochemistry, pharmacology, biology, or metabolism, and you were shown that structure and asked, "Would that be a stimulant?" and you referred to all the known stimulants, you would say, "Yes, that would be a stimulant." Yet, the body just chops this part off quickly; in the body, it is metabolized very

LECTURE 18 ~ Psychedelics I

quickly, and it has no chance to be active. So in the absence of some sort of a poison that will keep the body from chewing up the molecule, this does not have any central activity. If you put a simple group next to that amine which is the point of metabolism, this molecule is not metabolized as readily, and this material is active as a stimulant.

So this is a phenethylamine derivative, a phenethylamine analog. You have the benzene ring, here are the two carbons and the ethyl, there is the amine, but there's been a pimple stuck on—an added methyl group. CH_3 is known as a methyl. So you have a different compound. You can say, "Well, how could you predict the activity of either of these from the other?" Well, this is not active at all, but it should be a stimulant. But how do you know that it should be a stimulant if it were active? Because the material that is metabolically blocked is a stimulant. Okay, that's kind of an indirect way, but it's not an easily extrapolatable way.

Let me take the analogy that comes from this. What if I were to put this group (CH_3) onto this compound (mescaline), which keeps the amine from being removed metabolically? Now you're getting into the art of changing structures around. Let's consider the basic material. In essence, you have hybridized the two molecules. You have brought structural aspects of each of them into the same molecule. And you have a material that is quite rightly similar to both of them. You cannot make it from amphetamine, because you can't put the three methoxy groups on, but you can make it in the same way you make amphetamine and then you have a trimethoxyamphetamine. And really that was the first analog of mescaline that was made synthetically and found to be active. Trimethoxyamphetamine was among the first of the so-called phenethylamine psychotomimetics—or psychedelics—that was made and studied. This was back in about 1955 or so[19].

The compound was first synthesized in Canada. It was explored by a group of researchers associated with the University of Saskatchewan. They found that the compound TMA caused a tendency to increase the flicker fusion of a stroboscope. So they published the paper under the title of "Trimethoxyamphetamine, TMA: A Centrally Active Compound That

---

[19] It was 1947: P. Hey, "The synthesis of a new homologue of mescaline," *Q. J. Pharm Pharmacol*, 20(2), 129–134 (1947) PMID: 20260568.

Sasha: "What if I were to put this group onto this compound, which keeps the amine from being removed metabolically?" → CH₃

Mescaline

3,4,5-trimethoxyamphetamine (TMA)

Sasha: "...you can make it in the same way you make amphetamine and then you have a trimethoxyamphetamine. And really that was the first analogue of mescaline that was made synthetically and found to be active. Trimethoxyamphetamine was among the first of the so-called phenethylamine psychotomimetics, or psychedelics, that was made and studied. This was back in about 1955 or so." (Actually it was 1947: P. Hey. The synthesis of a new homologue of mescaline. Q. J. Pharm. Pharmacol., 20(2), 129-134 (1947) PMID: 20260568)

Influences the Stroboscopic Phenomena."

It was a psychedelic, but the term "psychedelic" didn't exist at the time. The feeling coming out of the University of Saskatchewan was something that caused this type of effect was not something they were totally at peace with, so they talked about the fact that it influenced the visual process as seen by a stroboscopic technique. The truth is it has the same type of action as mescaline. It's a little less benign than mescaline. It has a potency about twice that of mescaline, and it doesn't particularly have the stimulation of amphetamine.

So this is a good starting spot for the misuse of these words. This is known as a psychedelic amphetamine in the medical reviews. Psychedelic it is. Amphetamine it is, only if you look at the structure from the chemical point of view. If you look at it from the point of view of the stimulation—the eye dilation, the appetite suppression, the cardiovascular stimulation of amphetamine—this does not have it. It has a little bit but not very much. It is not very satisfactory pharmacologically as an amphetamine. But structurally, chemically, it is certainly right on top of it. This, however, was classified legally as a stimulant; this is a hallucinogenic. So the definition of

LECTURE 18 ~ *Psychedelics I*

amphetamine in the law slipped a little bit here. I'll say with certainty this is the first of the mescaline analogs that was studied and found to be active in humans. I had published a second paper on this about 1963 or 1964 in which I ran a clinical study and actually talked about its actions in humans.

This compound has given rise to others. This is going to be sort of the unveiling of the alphabet soup, bear with it for a bit. From a purely Teutonic point of view, from the disciplined point of view, you can say, "Well, if TMA, trimethoxyamphetamine, is a mescaline-like psychedelic, what can one do to the molecule to find out what part in the molecule is responsible for it being psychedelic?" The question is a general one. It is the Teutonic approach to the question.

The general question is, why is this a psychotomimetic? If you take off the three methoxy groups, you have it back to amphetamine. You've got a rolling stimulant, but you do not have a psychotomimetic action. With amphetamine you do get into a psychotic situation if you get too much. I mentioned this before, but it's good to emphasize it. A little bit of amphetamine, 5 milligrams, 10 milligrams, 20 milligrams: cardiovascular stimulation, some mydriasis, feeling of tingling, paraesthesia, hair standing on end, lack of appetite, lack of sleeping ability, but you become tolerant. The second time you use it, the third time you use it, less of a response. So one tends to up the amount, 30, 50 milligrams. And when you're conditioned with 50, you go up to 75, to 100. I mentioned you can easily get up in the area of a gram or grams of amphetamine and methamphetamine, the type of levels that would be lethal to a naive person—naive being someone who has not gotten conditioned to it. But at about 100 or 200 milligrams of amphetamine, you invoke a psychotic response. The person really gets quite psychotic. He is easily diagnosed as schizophrenic. And it is not a stimulant property. It is truly a mental disturbance property. And so, you do have lying doggo[20] in the amphetamine molecule, the capability of having some mental disruption in there. But you're usually blocked from getting there because of a very disruptive property, that of the stimulation. It keeps you from exploring those higher levels.

There are many cases where there are drugs that have a level that is known, but if you become conditioned to that level and go up to a higher

---

[20] "Lying doggo" is an expression that means "present but hidden or silent."

level, you begin invoking new responses. The body, for example, may be a contributor in its own metabolic process. I've done experimentations with mescaline, the one without the methyl group, in which I have made an experiment of taking mescaline at an effective level with a certain amount of radioactivity, then taking one one-hundredth of that amount, which is totally without any action on the body, but with the same amount of radioactivity, then in a third experiment, taking one ten-thousandths of that amount, which is obviously a trace, a trivial amount of chemical to the body, but with the same amount of radioactivity.

So you're taking three different levels of a drug, and you're getting no response at the lower levels at all. But you have the same seeability as to the fate of the drug. Is this drug doing the same thing independent of the level of action? You find the answer is very definitely no. At extremely small levels of the drug—but where you can see what's going on because you put in so many microcuries of carbon 14—you'll find the drug is totally metabolized. It goes to the phenylacetic acid, which is known to be inactive. At the intermediate level, it went largely to phenylacetic acid, but a little bit here, a little bit there, a little bit came out unchanged. At the higher levels, where it is distinctly centrally active, the amount of this metabolite was relatively small, and much more of it came out unchanged, and you got metabolites you wouldn't have seen otherwise because they were just not being made.

So you have the impression that the body has a number of ways of handling compounds, and each can be saturated. If you get a little bit of a compound, it's chewed up, fine. You get more of a compound, it's chewed up, and the overflow invokes some new process. Then if you take more of the compound, you begin invoking more new processes, and maybe some of its unhandled, because you run out of the capability of metabolizing it.

What you may be seeing is an explanation of why some compounds are very low potency. Because the body has good machinery for disposing of them. You may be seeing why with some compounds it's hard to work out their mechanism of action, because when you get to active levels, you're invoking new metabolic processes, and you may be generating the thing that is really doing the job, and there is no intrinsic activity in the compound you started with at all. So these are all different factors that can play into the question of potency and metabolic disposition.

LECTURE 18 ~ *Psychedelics I*

TMA got me into the curiosity of how you can jockey around the structure of a molecule and determine what makes it effective, or what of its effects are due to what in the molecule. But how to go about it? If you hold a hexagon by a point, and figure seeing the hexagon can spin, you'll find you have six totally different ways you can put three different substituents on that hexagon.

They can be opposite like this. They can be evenly spaced, two, four, six. A different round of positions. Let me do that ring system for a moment

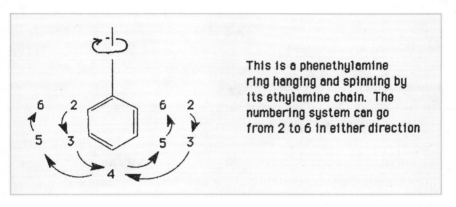

here—this is the point you're holding the hexagon, and let's call this two, three, four, five, six. [Referring to drawing.]

How many ways can you make what are called "positional isomers" of this type of structure? We have the three, four, five; the two, four, five; the two, four, six; the two, three, four; the two, three, five; and the two three six. There are six different ways. All these compounds are makeable. All of them have been made.

An additional tie to natural products comes with the second of these that has ever been synthesized. I'm going to cut down a little bit of writing the CH_2. A moment of pause for the people not in chemistry. A bend in a line is equivalent to a CH_2. It's a convention to keep it simpler. It's actually a strategy to corner the market on chalk. You don't have to write as much. [Laughter.]

The second of the compounds we studied was a material that had the methoxy group that was in the three-position relocated to the two-position. Again, from the chemical point of view, you don't go into the molecules and systematically pull a methoxy group off of the three, take your magic glue and stick it on the two. It's not done that way. You do not convert that compound to this. It is not done. You start with something from Eastman or Aldrich, and you make this sort of thing from scratch. But it can be made quite easily. And here, I'll give you the process as we go along.

We saw what happened. As you went from amphetamine to TMA, you lost some of the stimulant properties. You got a compound that was twice as potent. You got a compound that was a psychedelic, more like mescaline, which is the first material we talked about. About twice the potency, duration is about the same. This is the whole idea of what's called in the scientific trade "structure-activity relationships." The structure changes, and the activity changes with the structure. You tally the structure, you tally the activity, and you find a relationship in this graph that says, "Now I understand how these things are active. And knowing that, I can predict any further compound you wish. No point doing any experimental work, any chemistry, we just predict it all."

Here's another example. We know the amphetamine. We know TMA. The question is, what's the activity of this compound? [Referring to a picture of TMA-2 on the chalkboard.] It has the amphetamine chain. The chemist will say, "Ah, that's obviously amphetamine." The pharmacologist will say, "Oh gee. We went from a stimulant to a hallucinogenic." Does this look more like amphetamine or more like TMA? Both. How many people say it looks more like amphetamine than TMA? I don't think so. I won't vote yes. How many people think it looks more like TMA than amphetamine? I think so. Oh good! Good pharmacologists. And that's indeed what it does.

Okay. I'll give you another argument. Potency of amphetamine is 5 to 10 milligrams. Good oral dose. Potency of TMA is 200 milligrams in man, single dose, orally. What is the potency of the TMA-2?

STUDENT: About the same.

SASHA: You say, "About the same." That's ridiculous. Which one? How many vote for 20? I don't. How many vote for 200? So did I when I first created the compound. That brings up an interesting point.

An interesting side tangent which has nothing to do with this whole thing. I'll tell you how I discovered the activity of TMA-2. It is the second of

LECTURE 18 ~ *Psychedelics I*

Amphetamine
(high potency stimulant)

TMA
(low potency psychedelic)

TMA-2

High or low potency?
Stimulant or Psychedelic?

the TMAs. What better name do you have then to pull out the alphabet soup and call it "TMA-2"? The second of the TMAs. I agreed with you. I believed it was going to be hallucinogenic, and I thought the activity was going to be around 200 milligrams. It wasn't. This is a point in the idea of experimentation with compounds that are absolutely new, that have never been synthesized before, that have no specific record of anyone having found out their activity. This applies to the hallucinogenics, it applies to anticonvulsants, it applies to antibiotics, it applies to lead in the water.

Anything that has not been assayed before in man you cannot accurately guess. Guessing accurately—that's an oxymoron. You cannot accurately state or validly guess what the activity is going to be. So I was certain of where I was, I finally understood all that was to be understood about structure-activity relationships. So with great confidence I said, "Eh, 200 milligrams, but let's be conservative. I'll start with 100." Where are you going to start with a new compound? How do you glean 10 milligrams? Whooh! Oh, 5. All new compounds are safe at 5. [Laughter.] How about 1 milligram?

STUDENT: How about an animal?

SASHA: Ah, but what animal is going to have your response?

STUDENT: See if it dies first.

SASHA: Okay, all compounds will kill animals. Good argument. This compound kills rats at 10 to 15 milligrams per kilogram. A typical rat weighs maybe a quarter of a kilo, so 2 or 3 milligrams is fatal. This compound specifically kills mice at 150 to 200 milligrams per kilo. Clearly mice and rats do not share the same pharmacology. I'll give you the information. Where do you start?

STUDENT: Go to higher primates?

SASHA: Ah, gosh, monkeys are expensive.

STUDENT: How about dogs?

SASHA: I have not run this in dogs. The closest animal to humans amongst the animals you can reasonably afford is, strangely, the pig. The pig is a very close friend in pharmacology and in metabolic studies. But let's make believe. Let's say I run this in a dog. Has this been run in dogs subsequently? I don't recall. Very close compounds had and almost all the compounds in the dogs are in the 80 to 100 milligrams per kilo level for toxicity. It's certainly true with the compound TMA. It is in the 8 to 100 milligrams per kilo.

The question is rhetorical. You do not systematically, in any logical way, get to a level that is safe. There is no such thing. Because someone can say, "Well, what if this . . ." Okay, we'll go to half that level. And so you go down and down and down, pretty soon you have, you know, a homeopathy study and the stuff disappears, and you still maybe are not safe.

The truth is, I always take a factor of one-hundred-fold from where a reasonable guess would be. I start there. You can start way down the line and go up. You take what you think the dose is and divide in half and half and half and half and half—you double the dose each time. So you're not consuming quantity, you're only consuming time. There's clearly no action at one milligram. I started at about a milligram, two milligrams, four milligrams. All in great confidence. Maybe I'll go up to six. No action. Took 12 milligrams, with complete arrogant confidence that this thing was not going to be active until way up there at the bucket level. Twelve milligrams. One hour into it, no action. I know what I'm doing.

I take another 12 milligrams. Never, never double up. It's bad business. You don't know what's going on, and you don't know if you're doubling into

LECTURE 18 ~ *Psychedelics I*

activity. Well, I had no sooner taken the second 12 milligrams when the first 12 milligrams began to show some effect. And so I may have been a little bit hasty on that, you know, but there I was. It turned out to be a very interesting day. [Laughter.] And the active dose is somewhere between 12 and 24 milligrams. But the effects were most extraordinary. It was a hallucinogenic, but the active dose was, let's call it about 20 milligrams, that's about the level I would give this active dose. You can tolerate 40 or 50. And you can milk out something at about 10. But the dose is around 20 milligrams. Very, very strange effect.

Again, if I take two lectures on something, I take two lectures on something. I wrote the list of what lectures I'm going to give, and I'll give them another week. Okay, so I'm going to continue going with this particular one.

What I got into was a strange uncertainty of where I was and what I was doing. This, I found, is very common—when you're into something that you're not at all familiar with, you're not seeing a typical picture of what it does. Ann's son's child saw a movie for the first time at the age of eight. The whole response to that movie was a little strange. Once they've gone to 82 movies and 120 hours in front of the TV set, that world will become a more familiar world. But they will remember that was a very strange thing the first time. Or the first time you remember seeing a funny paper and seeing all these pictures of people in colors, moving around doing strange things and looking funny, and animals talking and all that, it has a strange impact.

Similarly, with drugs, the first time you're into a drug. When you try a new prescription drug there is a bit of something comes with the territory. The physician will say, "Well, occasionally, you'll have this effect, you'll have that side effect. Be attentive. Don't drive any tractors or something in the middle of this." Those sorts of precautions. You have access to other people's experiences, therefore you are a little bit prepared. You will, however, still always have a little bit of anxiety. This is really the first time you've put your toe into this particular pond. They say it's not particularly cold, but you go in very cautiously. And the shock of the temperature of that water on your toe is something you weren't expecting. It wasn't warm or cold, or neither warm nor cold. It was the first toe in the water.

If the drug, on the other hand, has never been looked at by anyone before at all, then you can build all kinds of projections into the effects that you get, and get some effects that never validly should be applied to that drug. In this case, I got an alternation of numbness outside of my little finger on both hands and a tendency for my visual field to close down looking at the clock.

Well, neither of these was particularly appreciated, and both of them are a little bit disturbing. I know from the neural netting involved that nothing can make the little finger numb without having the outside of the fourth finger numb because they are fed by a single nerve. So intellectually I knew this was absolute neurological horseshit. I'm not losing my thing, but there it was. Then pretty soon the visual field would close down. I would lose peripheral vision. You get this graying-out when you don't get continuous stimulation. I was looking at the clock and I didn't realize this graying out was something that you could aggravate by focusing too much, and the field would close in on the clock. Obviously I was watching minute by minute. I didn't know where all these things were going.

Once I was watching the clock, my fingers were perfectly okay. And once I looked at my fingers to make sure that they weren't numb, they would start getting numb, but the visual field opened up. So I got this alternation of input. Pretty soon I got fascinated with it. How are my eyes going to look at the clock enough to turn off this hysteric nonsense going on with my fingers? I got intrigued with this interrelationship. Pretty soon the whole thing opened up and I had one hell of a good time!

The first couple of hours were rather a strange territory, but it told me a lot of things. One, it told me a lot about myself. And, by the way, neither of these effects has ever occurred again, on that drug or, for that matter, any other drug I've used. So that happened to be my hysteric approach to that particular drug. It taught me that with a new structure you cannot reason by what you know of other drugs what it's going to do. To reiterate what I said 20 minutes ago, all compounds are unique, all actions are unique, all chemical entities are unique. You can apply your wit and your knowledge to make an educated guess, but don't take it as a personal affront if you are wrong or not very accurate.

This ring system, I thought at the time it was virtually unknown. I turned out to be wrong. It is a known ring system, but it's a very rare one. This is the common ring system that's found everywhere. But suddenly, in making one little trivial modification, I increased the potency of the compound by an

LECTURE 18 ~ *Psychedelics I*

order of magnitude. What else lies down the road? Well, this is the second of six, let's make the other four, which I did, and they're not easily made.

Let me go back, a little diversion because I'm going to tie this into essential oils. This is a compound that's found in oil of parsley. Going to take the same family, same trimethoxy groups on the benzene ring, the same three carbon systems. So this is a direct imitation of what's called the "carbon skeleton." Every carbon has its counterpart carbon here, but no nitrogen, but a double bond in it. It is a compound. It is known as elemicin and it's a component of oil of parsley. I brought in a sample if anyone wants to smell oil of parsley. It's a strange, interesting smelling essential oil.

I think I mentioned the term "essential oil." The term "essential" has a meaning in biochemistry, meaning you've got to have it, and you don't make it. It is required. It is essential. In botany, the term "essential" means it has a smell, it has an essence. Two entirely different meanings of the word essential. This is an essential oil in the sense that it has a smell. And you can believe me, you can go very well throughout life without eating parsley.

So here is elemicin, and it has a carbon skeleton very much like this. And I found that this carbon skeleton is represented in another material—I'll just draw it out here—known as oil of calamus. All the carbons in the same way. It turns out the double bond is there—nature has its little games it plays. This compound is known as asarone.

ANN: What is oil of calamus?

SASHA: What is oil of calamus? It's mentioned in the Bible. Who is a Bible scholar? Oil of calamus, well, it's in Matthew. No, it's not in Matthew. It's in the Old Testament. I don't know. It's called "rat root" in Canada. It's a turnip-like plant that is used as a medicine by Canadian Indians. But it contains this material. So I began tasting this and got nowhere, other than a slight stomachache. The body does not orally convert these things into the amines. But look how close the structure is.

If you were somehow going to add ammonia, which is NH_3, to this double bond, you would make this compound. [Referring to a different dirty picture.] You can take an animal and take a perfused liver—this is where you take a liver out of an animal, good-bye animal—and put goop going through the structure of the liver, and then pass chemicals in this goop. The

THE NATURE OF DRUGS

liver is doing this chemistry as something that is almost a living thing. All the natural biochemical processes can be invoked. Add ammonia to the molecule. You make TMA from elemicin. You make TMA-2 from asarone. And yet it doesn't appear to occur in the body. At least no one has systematically searched for and found it to occur in the body. I have not looked at it.

But this is an interesting hypothesis of the reasons for the interest in spices, and herbs. Because these materials carry these essential oils that are

$$\text{ELEMICIN} \quad \text{TMA} \quad \text{M.U.} = 2.2$$
$$\text{MYRISTICIN} \quad \text{MMDA} \quad \text{M.U.} = 2.8$$
$$\text{SAFROL} \quad \text{MDA} \quad \text{M.U.} = 3$$

structurally closely related to known compounds. Let's take one more example. Let me follow this general process. Going then into the plant world, if I can learn that much from nature, maybe I can go into nature and find out where nature has put these groups on the molecules. But let me finish up the story. I completed the other four of the TMAs, just for the record. The other one that had any action at all was the sixth of them. It was the compound that had the two, four, six orientation.

So of the TMAs, TMA-2 and to some extent TMA-6 were the most active. This locked me into the two, four, five orientation, which has consistently proven to be as active as any orientation in any of the phenethylamine psychedelics. There is another compound here. I'm going to take advantage of having a piece of chalk and not having to worry about taking notes. I do this kind of easy chemistry, switch things around. This compound with a double bond in the out position is known as myristicin. I got quite interested in it, found I could get a quantity from nutmeg, from mace. I brought in a bottle of mace as a show-and-tell. It's oil of mace, but in essence, it's oil of nutmeg. Really they

LECTURE 18 ~ *Psychedelics I*

$(CH_3O)_3$-C$_6$H$_2$-CH$_2$CHNH$_2$-CH$_3$ TRIMETHOXYAMPHETAMINES

ORIENTATION	CODE	ACTIVITY
3,4,5	TMA	2.2
2,4,5	TMA-2	17
2,3,4	TMA-3	2
2,3,5	TMA-4	4
2,3,6	TMA-5	13
2,4,6	TMA-6	10

are the same plant; the meg is the center of the seed. The mace is the very fine filigree cover of the center of the seed. The oil composition is very similar in both cases. *Myristica fragrans* is the name of the plant that produces them.

This is myristicin. I said, well, if you can add ammonia to elemicin and parsley or ammonia to rat root or asarone, what happens if you mentally added ammonia to myristicin? After all, nutmeg has a fantastic history. I thought it was most unusual because I went deeply into nutmeg's history in the use of medicine; it cures anything, it causes anything, it's used for every illness under the sun. People committed suicide with it. They had long lives with it.

And I discovered this for not just myristicin; this is true with any spice. You go into the literature and you'll find this wealth of medical use, and anecdotal records of virtue and evil that apply to every plant. You cannot use this as a guide. But the idea of green medicine is a very good one; here are some plants that have been used, let's look at what's in them, they've been used by the natives in the Amazon for the prevention of pregnancy and for the treatment of broken bones. You'll find a lot of these will lead to very valid things, but a lot of them are invested with many properties that are not consistent with what you would find. This goes back to the very early lectures when we talked about what a plant does and what a compound does. It is, to a large measure, what you read into it, what you project into it, as to what you expect it to do.

So this is myristicin. I added ammonia to it. I'll draw the molecule in a more conventional manner. And the product of this was. . . . And here is the

— 299 —

THE NATURE OF DRUGS

first of the chaos that came with the alphabet soup TMA. I'll put it up here to make a little tally. The first of the families of what I call MMDA, because it is a methoxy, there's your first M. Methylenedioxy is the "MD," and that famous carbon skeleton known as amphetamine. This is a real trap in the current popular literature, because there is a compound I'll talk about in

LECTURE 18 ~ Psychedelics I

a few minutes, MDMA, which is actually just taking the letters out of the alphabet soup in a slightly different order. So it's very easy for people to want to talk about this and end up talking about a totally different molecule. This is MMDA, methoxymethylenedioxyamphetamine, and it is a reasonably inactive compound. It is a very mild, slightly sedating, slightly visual, relatively inactive, easily tolerated material that takes literally hundreds of milligrams for an active effect. Much more like mescaline.

Yes?

STUDENT: What is the name of the one that you circled?

MDA   MMDA

SASHA: Methylenedioxy.

I mentioned this in a little chemical note. CH_3—which is carbon with three hydrogens on it—all carbon has to bind to four things, so whatever it binds to is the rest of the molecule. The name of this group is a methyl. So you name things out this way: methyl chloride, methyl cyanide, methyl bromide, methyl alcohol.

If you come back into it with a name from out here, the name of the parent system is methane. So you have chloromethane, cyanomethane, hydroxymethane. You can name molecules from either end. It's always allowed as one end can be unreasonable, and the other end is the only one you can use.

Okay, methoxymethylenedioxyamphetamine, we'll put this as one to six. It also has five possible. Let's leave it this way. There are six of them, but they can only go up to five because the two, four, six orientation cannot be substituted. You cannot put an adjacent ring on two positions that are not adjacent in the molecule. On the other hand, I'll just mention in passing, if you have the oxygens in this position, you can have methylenedioxymethoxy, or methylenedioxymethoxy. [Referring to two different molecules.]

You have two orientations within one substitution pattern. There's still a total of six. Of the six, the most potent one again is the two, four, five. I, and others, have maneuvered around this structure and have not found any higher potency and have not found dramatic central activity. It is rather interesting stepping-stone on the whole.

Another interesting essential oil is safrole. It has the character of root beer; it was the original flavor of root beer. In fact, root beer, as I mentioned, was a drink brewed of roots up until the time that it was decided that safrole and isosafrole were too hazardous. They are known to be carcinogens in certain animals; they will cause cancer. They're not allowed. Since they are not allowed, the plant system that makes them cannot be used in any food source. And root beer now comes from a synthetic flavoring in a carbonated water factory. Even though it may say, "Grand-Dad's Old Original Recipe Root Beer," there is no root beer in root beer. What you're getting is something of the flavor. But here you can smell what was the flavor of the original root beer. The character of it is still there, but the chemicals are now disallowed.

Safrole—on the same idea of adding that amine group, if adding ammonia to the double bond, which is still a theoretical step, although it does work in a perfused liver—gives rise to a material that's called "MDA." And this compound is not very potent, but does cause a very pronounced change of consciousness. It was one of the earliest studied, of the materials from the essential oils. Actually, when I said that TMA was the first published thing on human activity, I should back off. MDA was published earlier by a person named Gordon Alles, a pharmacologist at UCLA. He didn't actually publish, but he gave a report at a Macy Foundation meeting about 15 to 20 years ago on MDA, in which he ascribed unique properties to it.

Again, keeping in mind the idea that the first time you explore a material like this, you get your impression of what it does to you. And what it gave to him was acute hearing. People who have quoted this report refer to it as the "gray donut" report, because what he saw were gray donuts floating in the periphery of his vision. He got really taken in by the gray donuts. And they were only interfered with by the fact that he heard people walking outside the window. But he was on the fourth or fifth floor of a building and the outside of the window was way down where people were walking.

LECTURE 18 ~ *Psychedelics I*

But he heard their walking, and he heard their talking—it was acuity. And he had the gray donuts. I would put this character, gray donuts, and auditory acuity exactly in the character of numb little fingers and visual closedown. It's what got his attention and was noteworthy to him on his first experience. He never got either the acuity or the donuts again. But he did report in this one off-the-record report with the Macy Foundation about these effects. And MDA was looked at to some extent by Smith, Kline, and French. They didn't particularly need gray donuts or auditory acuity drugs, but they were interested in the possibility of it being used as a CNS stimulant. They explored it, and they found it could not be. It was not pursued any further.

MDA was the material known as "mellow drug of America." I think that's one of the names. These names have been given because of the chemical logic; MD for the methylenedioxy and A for the amphetamine chain. Once they get into proper usages, the names get distorted into some easy to remember mnemonic.

I'm getting a little bit on in time. In this general direction, I was going to talk more about parsley, but it's a rather uninteresting direction. I'm going to go back to that parent compound. I was talking about TMA-2. Here is our TMA-2. [Drawing on board.]

One of the very early approaches to TMA-2 and this whole area of structure-activity relationships, the SAR study, is what could be changed on this molecule that would make it more active, less active, have a different type of action? One of the early thoughts was that the four position is where many of the metabolic attacks on these compounds are made. Amphetamine is hydroxylated at this position. Phenylalanine, one of the principal amino acids, is hydroxylated in this position to form tyrosine.

There is an easy approach to this position. The thought was what if one were to take a compound without anything in that position as a stimulant and nothing more? Dimethoxyamphetamine does have a little bit of psychedelic effect, but as you're appreciating the effect, you find you have a very pounding heart and a very intense pulse increase. There have been deaths that have been validly reported to be a consequence of using this material. It was sold in the Kansas area and in the Nebraska area, under the name of "chicken power" about 15 years ago. I refuse to even guess how it got the name "chicken power." There were a series of deaths through the Midwest and into the Canadian Midwest, and the thing fell very quickly out of use. It was known chemically under the name of dimethoxyamphetamine. So add this onto the list.

But going back to the "tri." With the exception of MDA, the tri substitution is necessary for biological activity in this particular type of central action. The question came up, "What if one were to take the oxygen out and put a methyl group in there?" This is a material that is totally unknown. The ring pattern is totally unknown in nature—it's purely a man-made ring pattern. The methyl group is sitting there like a callous wart on a metabolic scene. It is not going to be gotten at. It's got to be metabolized, maybe oxidized, or attacked in some way, but invoking different metabolic processes. It's not a hydrolytic step. It's going to be a real chemical-doing step. So the question is if this compound could be really potent and an active hallucinogenic because the body can't get rid of it. But if it fits into that slot that the compounds must fit in to be active—the so-called receptor site—if it's not active, for some reason, because it doesn't fit quite right, but it gets in there, then it can be a potent antagonist, and it might be an agent that could block that action if given to a person before he took something that caused a central change. And if there's anything to be said about a person who is spontaneously schizophrenic or mentally ill because he and his chemistry produces something that activates that receptor, then it could be a therapeutic agent. It might be a prophylactic or it might be reversing agent.

Take a person who is schizophrenic for reasons that you have no knowledge of, because his own chemical factory is making something that is a "schizogenic," if you want to make up a word, then this might be a therapeutic agent in blocking that internal process. Unknown. But putting the methyl group on it might be a very potent thing in its own right, and it may be a blocking agent without action but blocking the activity of materials that

TMA-2

DOM (STP)

LECTURE 18 ~ *Psychedelics I*

2,4,5-TRISUBSTITUTED AMPHETAMIMES

R_1	R_2	R_3	CODE	ACTIVITY
OCH_3	OCH_3	OCH_3	TMA-2	17
OC_2H_5	OCH_3	OCH_3	EMM	<7
OCH_3	OC_2H_5	OCH_3	MEM	15
OCH_3	OCH_3	OC_2H_5	MME	<7
OCH_3	H	OCH_3	2,5-DMA	8
OCH_3	CH_3	OCH_3	DOM (STP)	80

would have the action.

Vote! How many people think it's going to be active in its own right? I frankly didn't know. How many people think it's going to be a blocking agent? Okay, about five for blocking and one for active. I just didn't know. I could argue it either way. I kind of wanted to know, so what do you do? You start with an emphatically small amount. We go through that whole harangue again. The compound turns out to be very active. Its activity is easily seen at three milligrams and it's active from about three to six milligrams. And the compound—getting back to the terms in the alphabet soup—is called "DOM."

STUDENT: Do all these compounds have similar properties?

SASHA: All water-soluble salts. All white. All approximately the same or relatively the same in spectral properties. The infrareds are different. The NMRs are different. The UVs are quite similar. They are all toxic. In rats you often get tremors with the compounds, and tremors precede death at the lethal levels. Dogs, you get salivation and motor ataxia.

STUDENT: Do the rats have seizures? Or do the mice have seizures?

THE NATURE OF DRUGS

SASHA: Mice have seizures. I don't know the cause of death of rats. You definitely get seizing with mice. They are very similar. You cannot use an animal as a way of evaluating.

This material was called DOM. It went onto the street, introduced in San Francisco—which is an embarrassment to me because I was in and around San Francisco at the time, although I was not the introducer—under the name of STP, which has been a rubric. The name apparently stood for "serenity, tranquility, and placidity," and no one could handle "placidity." I can't handle it either. And so, it became distorted to "serenity, tranquility, and peace." And it's not the peace pill; the peace pill came out at the same time. That was PCP. Hog, or Peace.

But the difficulty was that it was introduced on the street at levels of about 20 milligrams. Why? I have no idea. Apparently, the person who had evaluated it before he put it on the street took 20 milligrams to do whatever he did. So the pills came out at 20 milligrams, which is a whopping overdose. Another property is this material happens to be unusual amongst all the ones I've talked about in that the onset is fairly slow. It takes a while to build up to a full effect. The full effect lasts for a number of hours, and it takes quite a while for things to disappear and quiet down and recover to a psychological baseline.

The result is some people took 20 milligrams and in an hour said, "Hell, this is nothing here at all. I haven't been wiped out. I'll take the other two." And so, you had people with 40 and 60 milligrams on board, although the active level is 3 to 6. You had overdoses, you had panic. The Haight-Ashbury Clinic was in a very bad place, partly because they were getting people in there who were really stoned out on something. They were panicked and in a very bad affect. And they had no idea what it was. I was vaguely around there—I was over at the medical school at the time, and it was in the area—and I heard about this STP that was really laying people low. They said it is unusual because you can't use the usual tranquilizers to quiet. I don't know why that came out, because it turns out you can. But these rumors were spreading around. No idea what the material was. I was kind of embarrassed, because if I had known and gotten a sample, I might have been able to identify it and give some information about it. But it came out in the name of STP. What other terms came with that? Stop the police?

LECTURE 18 ~ *Psychedelics I*

[Laughter.] STP.

STUDENT: Is it the "businessman's quick"—?

SASHA: No, that's DMT. That's the next hour's lecture.

ANN: How long does it take to see the effects?

SASHA: Usually about an hour you're launching into it. At three to four hours you have peaked.

ANN: And how long would be the total effects coming down to baseline?

SASHA: At normal dosages you're at the peak for about 4 hours. It takes another 8 to 12 hours.

STUDENT: What about people who experiment with psychedelics? They come on very slowly, and about halfway through they change their mind. They don't want to go through this. Is there any way to shut this down?

SASHA: Yes, you can do it with the heavy tranquilizers, tricyclic thorazine.

ANN: Compazine.

SASHA: Compazine. What it does, it doesn't shut down the experience. It shuts you down.

STUDENT: You just can't do anything?

SASHA: You can't do anything, and you weather it. Psychologically it is of no value whatsoever, but from the point of view of your physical anxieties, it turns them off. Haldol is a disagreeable compound, but it does a good job. I would probably almost lean toward Haldol.

ANN: How about Valium?

SASHA: No, too light. It would not give you that much help. You want to get to an antipsychotic.

STUDENT: Miltown?

SASHA: No, I wouldn't. Miltown is a paralytic. It doesn't really do that much to the mind, except when your body is relaxed, your mind follows.

ANN: Wouldn't you think they would want Compazine?

SASHA: Yes, and they found it did not help. Hence the word got out that this was a new class of compounds, which it was not.

STUDENT: Maybe they were just using too high of a dose, and the Compazine they were using wasn't—

SASHA: It could very well be they could not cut it because they were fighting a great big bonfire.
Yes?

STUDENT: How did it get the name STP?

SASHA: I don't know. At that time there was a popular name, it sounds like an apple cider, who ran cars in the racetrack. A person's name, Martinelli? One is an apple cider and one is a person who sells oil for—

STUDENT: Martinelli's Apple Cider.

SASHA: Martinelli's Apple Cider. Who was the one—it's a name with an *M*, a person who sells oil and goes around Indianapolis 500.

STUDENT: Mario Andretti.

SASHA: Okay, that may be it. It's a weird connection and I can never remember the name. STP was very popular at the time as an additive and there were millions of these little stickers. STP with red things on a blue background. They were everywhere. And I'm sure it was an advertising gimmick on the part of whoever the drug dealer was who started the thing going.

STUDENT: Do these compounds have similar properties to lysergic acid?

SASHA: Yes. Not lysergic acid, but LSD. Yes, this one is quite similar, except LSD has a much faster onset, more intense, and drops off faster. It's a different time chronology, but there are similarities to these higher doses with LSD.

LECTURE 18 ~ *Psychedelics I*

STUDENT: With LSD you get the term "peaking."

SASHA: Using the LSD term, with LSD you will peak usually about an hour, hour and a half, somewhere in there. With this material you will peak about the third or fourth hour, using that terminology.

Other questions? Oh, another one on STP, "too stupid to puke." [Laughter.] That was the policeman's term for it. If anyone knows any others from the Haight-Ashbury days, you can add it. I love collecting things like this.

Okay, so you have a material such as this. If methyl does it, what about larger and longer groups? Ethyl, propyl, butyl, amyl. These are synthetically available things. And there's a whole family of materials in which M became "Et," "Pr," "Bu," and so forth. The most potent of the group was the ethyl compound. Ethyl and propyl have comparable potency and are of comparable duration. A little bit more potent, I would say. The dosage for the ethyl is probably two to five milligrams.

The ethyl has an interesting story attached to it. This material, DOEt, was originally DOE. But DOE is a name that has been sequestered away for desoxyephedrine, which is a synonym for methamphetamine. Hence the term DOEt. This was originally studied at Johns Hopkins by a psychopharmacologist there by the name of Solomon Snyder. He had originally gotten quite active in the DOM area, and then it hit the bad press. And believe me, the press in the late 1960s was lurid with this being a new thing, something made by the army and released—which it was not. They dumped everything on the army at the time. He was exploring DOEt and he decided, "You know, we've really kind of dirtied the nest for research in this area with the bad press that DOM got. Why not take DOEt and use it at smaller levels that just opens up the person's rapport with the investigator, and see what it might have at that level."

They found at small levels, it caused an interesting interaction. It didn't have any debilitating central effects. They called it a "sensory enhancer," which is a nice euphemism. The world is filled with euphemisms where people know that psychedelics are bad, so you want to find something that can be used where the terms don't produce a negative effect. So the cognitive enhancer, the sensory enhancer, was used by Snyder, and all of his early publications on that only use levels at two and three milligrams and

THE NATURE OF DRUGS

indeed got this effect. If you do go to five or six milligrams, you can't really tell it from DOM, but at low levels it has this. To the degree that that's valid, I don't know; it's certainly in the literature. I think low levels of DOM would produce the same thing. With many psychedelics, low levels have that type of loosening, ease of rapport with others and with yourself, as low levels of many intoxicants have.

I mentioned the first drink of alcohol is a marvelous disinhibitor and the opener of conversation and interaction. It's when you get to the larger levels that you invoke—in the case of alcohol—the depressant. In the case of the psychedelic, the less readily controlled type of mental change, with a naive person.

So this family—DOM, DOEt, DOPr—was explored as chemical extensions to this chain. Another approach to the systematic changes of the molecule was what I call the exploration of the "10 Ladies." Considering the molecule as it stands, there are 10 unique hydrogens on it. That is 10 hydrogen atoms or more than 10 hydrogen atoms, some 15 or so. But there are 10 hydrogens, each one of which can be held as a unique entity. For example, there's a hydrogen that's on the methyl; there's a hydrogen that's on the nitrogen; there's a hydrogen that's on that carbon. There are two hydrogens up here that are both different because of three erythro isomers. I won't get into that chemically. There's a hydrogen in this position. A hydrogen there, a hydrogen there, and a hydrogen there. A total of 10. One more. Here's the 10th. [Referring to drawing.]

There are 10 unique hydrogens, and they're called the "10 Ladies," namely you could have 10 classic ladies from A through whatever the 10th

9. Iris
10. Juno
4. Dorothy (the threo isomer)
5. Elvira (the erythro isomer)
3. Charmian
2. Beatrice
1. Ariadne
6. Florence
7. Ganesha
8. Hecate

THE TEN HOMOLOGOUS LADIES OF DOM (STP)

LECTURE 18 ~ *Psychedelics I*

letter is—*A* through *J*. And a systematic approach was made to synthesize the 10 compounds and see if they were the same or different. If they were different, how they were different, and if you know how they're different, then why they are different.

The very first of those produced a startling effect. This hydrogen was replaced with a methyl. The first of the methyl homologs. The term *homolog* in chemistry means a molecule that differs by one carbon atom from another molecule. "This is a homolog of that" means it has one more or one less carbon atom. *Analog* means it resembles. *Homolog* is specifically based on the term *homo* for *one*— one carbon atom. And not nitrogen in place of an oxygen or sulphur in place of oxygen; that is an analog but not a homolog. So this is the first of the homologs. The first of the Ladies was Ariadne. And it turned out to be quite a startling thing. Not a hallucinogenic.

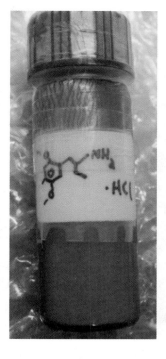

Oh, I should have asked. How many people want to bet it's hallucinogenic? [Laughter.] Not hallucinogenic at all. It is not a CNS-altering thing in that sense at all. It has nothing of the potency of the parent, STP, but it turns out to be an antidepressant, which is not that different in some ways. I mean, if you are down, a stimulant will tend to push you up. And if you call that down a depression, then the stimulant becomes an antidepressant in the popular use of the term.

Medically they pulled back. "No, no. We've got to understand the mechanism of depression. We've got to get a chemical that goes in there and modifies that mechanism, then you'll get an antidepressant that is pharmacologically kosher." But since a pep pill will sometimes put you up if you're down, and you can look at it as an antidepressant, it went into clinical trials. The company was Bristol, in Syracuse, New York, and it went through the IND process with the FDA; it went to Phase Two and was dropped by them. But it's a chemical that did actually have a great deal of chemical investigation. It's being reexplored now. But it was dropped for about 10 years.

ANN: Why was it dropped?

SASHA: Economics. Bristol was basically into penicillin and cephalosporin—beer brews and bug brews business—and not much in the synthetic chemical area, so they dropped it.

Yes?

STUDENT: Is it—in the medical profession nowadays—all right to prescribe amphetamine for depression? Are they looked down upon?

SASHA: Looked down on. Too much of use of amphetamines.

STUDENT: What is an allowed use of amphetamine?

SASHA: Narcolepsy, I think, is one that is medically allowed. Narcolepsy is when people tend to fall asleep spontaneously, despite the fact that they've had adequate sleep.

ANN: Also appetite suppressants.

SASHA: Not amphetamine.

ANN: It's not?

SASHA: Amphetamine is really frowned on. It's Schedule II. They would love to expunge it entirely, but there are companies that are holding on. Methamphetamine I think is still Schedule II, but methamphetamine has practically no valid medical use.

STUDENT: There are some that are Schedule III. I work in a pharmacy and I see that being written all the time.

SASHA: How recently? Quite recently?

STUDENT: Yeah. I'm just wondering for what reason—

SASHA: It has bad press. And what reason they could use it? It's a good appetite suppressant. It's amongst the best, but it can be abused and has been abused quite extensively. People will get multiple prescriptions and

LECTURE 18 ~ *Psychedelics I*

just keep loading up on it.

ANN: Narcolepsy would be another.

SASHA: Narcolepsy is a valid use, but that's trivial amounts.

STUDENT: Is that Dexedrine?

SASHA: Dexedrine is amphetamine. Dexedrine is the *d*-isomer of amphetamine.

STUDENT: Less of the heart-pounding, stimulating . . .

SASHA: Right. A little bit easier on the body, and it takes less of a dose. Okay, I may touch on more of this when I get to the next lecture.

The Ariadne with the methyl group here. The methyl group in the second of 10 positions led to some interesting developments. That methyl group B, the Lady whose name is called Beatrice, was a loss in activity. It dropped in activity quite distinctly over STP. It had some psychedelic effect but was way less potent and it had a lot of physical uncomfortableness with it. It was never explored very widely. But the concept of putting a methyl group on a nitrogen is something that rings bells from the discussion we're having right here about amphetamine and methamphetamine. Amphetamine has an NH_2. Methamphetamine is exactly that—meth, a methyl group on the NH_2. So, quite a study was made of all the psychedelic materials and maybe their N-methyls, such as you go from amphetamine to methamphetamine. Maybe you can go from DOB to N-methyl-DOB; STP to N-methyl-STP; to all of these different items. And they were done. TMA to methyl-TMA. Mescaline to methyl-mescaline. TMA-2 to methyl-TMA-2.

In all cases with one exception, the activity was way down, it was substantially lost. And when active levels were reached, if they were reached, there was a lot of physical difficulty with it, indicating a general global toxicity in the body. The only exception was the one material that I mentioned a little bit ago. It was a methylenedioxy, and when the methylenedioxy was methylated to an N-methyl, you had an entirely different type of compound. The psychedelic effect may well be there, but it's way up in the overdosage range. You have a new property that came in and hid the access to the psychedelic thing. It's a property of a general stimulation and a general openness. This material was called methylenedioxymethamphetamine from that ally up here. [Referring to drawing of MDA on board.]

And it has recently been very much in the news under the name of "Ecstasy" or "Adam"—there are other names I keep forgetting—or MDM. But in this case, the methylenedioxy is the methylenedioxy of amphetamine. The MD, methylenedioxy of MDA. But it is a methylated material. This is a methamphetamine. Hence the letters are put into reverse order.

Probably quite a few questions will come in on the MDMA thing because it's been a very popular polemic and palaver the last few years. Actually, MDMA first got into clinical usage about 10 or 12 years ago, and its primary usage was in the psychiatric area. What it did was cause a distinct stimulation, which is of no great value in the patient-doctor relationship, but what it also does, rather strangely, it has the property of dropping paranoia, dropping the fear of saying the wrong thing or being told the wrong thing. And it turned out to be an ideal thing from the point of view of the patient-physician relationship because much of the interaction between the patient and therapist is getting over the fear of that relationship and getting a rapport started, getting a communication started. And this material, in more people than not, achieves that very quickly. It's short-lived, which is ideal for therapeutic use. The effects are noted in about a half an hour. They usually peak, to use the psychedelic term, in about an hour, and are on the wane at 2.5 hours, 2.25 hours. So it really had immensely broad usage up to 10 or more years ago; but with popularity, it also got into street usage, and that is what spelled its doom from the point of view of legislation. It got very widely talked about as "Ecstasy." If I had a chance of giving the term a name, I would have named it "Empathy" rather than "Ecstasy," because it has much more of that property. But I don't think people recognize the word *empathy* as broadly as they recognize *ecstasy*. And how "Adam" got started, I'm really at a loss to say.

Yes?

STUDENT: What kind of dosages were they using in that?

SASHA: Most dosages in the therapeutic area range. There are two general philosophies of its use. One is a philosophy of patient and therapist in which the patient, having taken the drug, will interact with the therapist

LECTURE 18 ~ *Psychedelics I*

in a verbal one-on-one arrangement. Sometimes the therapist will use the material because the loss of the paranoia, the fear of open exchange, can be as beneficial to him in the relationship as it is to the patient. Typical dosage, 80, 100, 120 milligrams. The second general approach is quite different. It's used by some therapists, in which he sits over there, and more than one person may take it simultaneously. And they go into themselves—the eyes closed, earphones, motor-sensory isolation—and let them use this freedom to talk to themselves with a bit more of a group therapy thing later. Not so much in this country as it is in Europe in that usage. Here the dosage is going to be more like 120. Because one does not have to mind one's behavior quite as much, because you're not interacting.

It is a stimulant in the sense that there is a cardiovascular effect, typical systolic rise in blood pressure of 40 or 50 millimeters. This will occur usually in the first half hour. Usually, it is beginning to drop already when the effect of the drug is achieving its maximum. There's a refractory hypotension at about the 24-hour point of a loss of about 10 millimeters below normal. So probably the overall cumulative strain on the heart is not that serious, but 15 millimeters is pretty big. And some 10 to 20 millimeters of diastolic increase is probably more of a hazard. I would say people who have high blood pressure or cardiovascular difficulty would be very well advised not to use this material.

STUDENT: It is a prescription drug, though?

SASHA: It is not a prescription drug. It is a Schedule I drug with no recognized medical use.

STUDENT: But it used to be a prescription?

SASHA: No, at no point was it a prescription. It has never been medically recognized.

Okay, it's the end of the hour, if my little portable Swiss watch is valid. If there are more questions, I'll start at this point and go on with this area. I may screw up these three lectures entirely to get to the information that I want to get to.

LECTURE 19
April 7, 1987

Psychedelics II

[Ann and Sasha talk about MDMA and more ahead of class, partially with the instructor from the previous class. A good portion of that conversation was duplicated in the class later.]

SASHA: Okay. On the triptych I handed out, the article from *Playboy*, on page three, I must have leaned my elbow on the "enlarge" thing inadvertently and it came out enlarged, which means it is only partially readable. But the whole thing is only partially readable because it was black against dark blue or something in the original and it's not legible. I made a new page three that is also illegible, but at least it's the right size. And then someone came up with a reduced copy of something in the current *Playboy*, the same information that *is* legible. There's that legal-size thing. So you can complete your inadequate thing with a third handout if you want and wait until I get this copied if I can remember next week, and have this as a free thing and throw the other two out, or keep them all or discard them all.

STUDENT: It's a historic piece. That one is 14 years older.

SASHA: Okay, I've not read this through. All right, fine. You can be a research person in your own rights and see how ethics have changed in the course of 14 years. I have not read this one. I'd like to read it through. But if it follows the basic quality of the other one, it's accurate.

I'd like to continue where we kind of left off on Thursday last week, with the phenethylamine hallucinogenics. This is the middle portion of this three-part course, a triptych in its own right. The first was the kind of introduction: wiring, plumbing, structure of the body. The middle portion will be a trip through the stepping-stones of the actual drugs and talking about individual drugs and their histories, their backgrounds. Then the third portion is where I'm really going to have fun, in which I get into areas

THE NATURE OF DRUGS

that are more broad than just the drug issues, such as everything from nerve gas to radioactivity to nuclear warfare to other aspects of human and social behavior such as drugs, or social things that are equivalent to drugs in their action, and in the social control and the establishment rebellion against. That's a bad sentence, but I'm not a grammarian.

I just finished with MDMA in the last lecture, and two or three people asked me to go into more information and more detail on it this time. This is the material. [Sounds of chalk on blackboard.] If you remember the generalized dirty picture of the phenethylamine, it was this parent compound. This is one of the two basic classes of all centrally active drugs—phenethylamine, the one that's based on phenylalanine, tyrosine. Again, if within this living sphere I get around to writing out this particular lecture—which will be coming down the pike, maybe in three or four weeks or maybe to your address when you have finished the course—I will give a lot of the technical detail on it and its relatives, so again, don't worry too much about taking notes here.

The amphetamine was the one-carbon extension of that chain. This is the amphetamine molecule. The compound that preceded this was MDA. Again, the *MD* is a general terminology in the alphabet soup of drugs for methylenedioxy, *A* is the abbreviation of amphetamine: MDA. Its relative that we had ended up on was an N-methyl compound, which of course would be methamphetamine, and often called MA. Hence the combination of the two is MDMA.

1-(benzo[*d*][1,3]dioxol-5-yl)-*N*-methylpropan-2-amine (MDMA)

1-(benzo[*d*][1,3]dioxol-5-yl)propan-2-amine (MDA)

This is a material that received quite a bit of notoriety. It probably really hit the press about two years ago, along with a lot of legal and social difficulties. It's a material that is known as "Adam" or "Ecstasy." Oh lordy, if I turned

LECTURE 19 ~ *Psychedelics II*

around, there are probably half a dozen more code names for it. But its first initial name was MDM or MDMA. I believe "Adam" was the first name it had been given in clinical and medical use, and "Ecstasy" is the name that had been given in street use. The first place I ever saw that carried the name "Ecstasy" was an issue of a magazine called *Wet*. I don't know if any people here would run into that one. There's a whole series of publications; *Wet* was a superb name. Oh, I guess within the last decade there is *Wet* and *Flash* and a couple of others that are almost unspeakable in public.

They are counterculture magazines, and they carry this type of information. One that's currently sold is something called *High Frontiers*, which is now in issue three. Each issue is a totally different philosophy from the previous one. *High Times* is an ongoing monthly thing that's now in about its seventh or eighth year and has had its own philosophy swings from extremely pro-marijuana to extremely pro-cocaine, then anti-establishment to pro-advertising. It has its various swings of philosophy. But it's a valuable source of information. I first read about Ecstasy in *Wet*. I didn't know what the material was. I was a little startled because I found my name in the article. [Laughter.]

It always brings you up short when you realize that things are being put in your mouth that are not necessarily supposed to have come out, and didn't. Anyway, the term "Adam" was the term MDMA had initially, in its clinical usage. The drug received a lot of press when the government initially proposed that it be placed into Schedule I. This is approximately two to three years ago. It was supposed to be put in Schedule I because it had been seen over the course of five or six years periodically in illicit drug seizures. I did make a handout that gave the actual arguments of what justifies a drug going into Schedule I.

The two main justifications are that a drug has a high potential for abuse and the drug has no recognized medical utility. These are, in essence, the mainstays of Schedule I. "No recognized or accepted medical utility" is a little bit fuzzy because it could have medical utility in Germany. But what they really mean is no recognized utility in the United States. It doesn't say so in the law, but that's understood. You have the argument of "What is recognized medical use?" or "Recognized medical utility?" That is extremely fuzzy, because it has not ever been defined legally. What it means is that, by and large, you went out and talked to a bunch of physicians. They say, "No, I don't think I've ever heard of that" or "I don't believe it has medical use" or "I've never used it."

The FDA would like to say that recognized medical utility is something that they have approved, or have exempted, which is nice and dandy, but the FDA has never been empowered to preach the practice of medicine. So physicians say, "I don't care what you approved or what you exempted. If I want to take earwax out of my right ear and shove it up my patient's nostril to treat his sinus problem, and it's my ear and my earwax and I do the pushing, I am the creator of the drug, and I can do as I wish in the practice of medicine. And if you think I'm not practicing good medicine, then you go and look at my collection of peers. I've got eight good buddies out there who are all members of the AMA and all licensed in the state of wherever the state is, and they are all willing to tell you that they approve of my practice of medicine. Namely, I have peer approval. So if I am the maker of the drug, and I have peer approval for how I do it, I'm practicing good medicine, and the FDA has no voice on what drug I use. They can't inspect my ear."

"Well, you can't sell it. You can't make it available."

"I don't have to make it available. I'm using it in my practice of my medicine."

So this is the physician's stand. Well, the FDA sitting over here will say, "Look, we have been empowered by Congress to verify this, to check that, make sure it's not going to make babies come out with short arms, to make sure that it's not going to be toxic, that it's safe, it's efficacious, correctly labeled, and the kindergarten kids can't get into it. That's what we have been told, to make drugs into that kind of classification."

"Fine, do your job. Don't get in my way of practicing medicine."

It's a nice conflict. And it's really gone on back and forth without ever getting to an actual sticking point. It came close, and it may yet come close again in the argument of making THC—nothing to do with MDMA, I'm off on a tangent; you have to live with it.

THC, which is the active component of marijuana, is available for the treatment of nausea or illness associated with chemotherapy in terminal cancer cases. Just recently a big polemic, a big palaver, has been issued on this point. THC: no medical use, high abuse potential, Schedule I. "I don't

LECTURE 19 ~ *Psychedelics II*

care if there are 80 million pot smokers in the US, it is a dangerous drug, and it's going to stay Schedule I. It's a crime to use." That's the stand that's written into the law. That's what's hewn into granite.

It's been known for a long time that THC in certain instances can relieve nausea, in certain instances can relieve aspects of glaucoma (internal eye pressure). It can relieve nausea that is a result of using chemicals for chemotherapy, which is of major value in the treatment of cancer. When you have cancer, you want to destroy the cancer. Cancer is a rapidly growing cell. You try to destroy the cancer, but you tend to destroy the body of the person who has the cancer. You want to destroy cells that are growing, but all cells grow, all cells divide. You want to get in the way of it without hurting the host. As a result, most cancer treatment agents are really pushing a person to the brink of death. I mean, they are really doing all they can to destroy one aspect of the body without getting at the other aspect. And the result is nip and tuck.

The treatment of cancer has had fantastic developments in the last decade. Lymphocytic leukemia was lethal to children 10 years ago. There's now a 90 percent recovery rate. That kind of change has come from the development of drugs that, in essence, almost destroy everything, but happen to get to the cancer more than the person. As the result of this, many of these agents produce nausea. Just the idea of the illness: upset stomach, your liver has gone to pieces, you've lost weight, you're in bad physical shape, your hair is falling out, you vomit. Relieving that nausea has been a difficult problem because some of the standard antinausea agents, antivomiting agents, have not been terribly effective.

Marijuana has shown in some cases to be effective. It's been known. Marijuana is Schedule I. This is a medical utility. It's a conflict. So what was done—which incited NORML—was to take the THC that is in marijuana out of Schedule I and put it into Schedule II, which is an area of high abuse potential but has medical utility. There was no desire to acknowledge that any part of this drug that is so widely used had any medical utility, otherwise you open the floodgates. You know, it gives a bad message. It gives a message of approving drug use, all the negatives you could imagine.

But it was effective in this. So they gave it a name. I forget the name—Drabinol? I can dig it up if you want it. They said, "We'll put it in Schedule

II. However, it has to be in sesame oil." Okay. "It has to be in a gelatin capsule." Why? I'm not sure. "And it can only be used for the treatment of nausea that is refractory to normal treatment in chemotherapy of such and such and such." So that is how it got into Schedule II. NORML says, "You want to put it into Schedule II? Put all of marijuana into Schedule II." They said no.

Well, the thing is, suddenly the DEA is in the position of dictating the practice of medicine. The physician will say, "Hey, you mean I can only use this drug if I use it for this specific purpose? Who is telling me how to practice medicine? The DEA, a bunch of lawyers? Nonsense!" So the conflict is there. It has not been resolved. But it's an issue that's going to come up.

You have the same sort of thing with MDMA. Let's go back in its history. It was first reported in the literature in 1912. It was patented in 1914 in Germany, Christmas Day, just at the time of World War I. You'll find in all of the press releases that have come—*Newsweek*, *Time*—all the flap that came out of MDMA, everyone copies from whatever else is published. So any information that gets instilled that is a little bit strange gets copied down the line. You'll find every single report says it was patented as an appetite suppressant. Everyone knows that. You go back to the original German patent, there is no such claim there at all. You go back in the Merck files, there's no such claim there at all. But somehow someone said, "Gee, this is like an amphetamine. Amphetamines are used to suppress appetite. I bet this was used for an appetite suppressant. Why else would they patent it?"

And it's in there. No, it was not patented for that purpose at all. It was not patented for any purpose. There's no utility in the patent. It was patented because they could make it. It could be made, it's a new material, we'll put a patent on it. All patents in the chemical area are basically in two categories. There is a patent that's called a "composition of matter," which is a new compound, a new structure, it's never been made before. New properties, you've described it for the first time. There's a patent that's called a "utility patent." If you invented a new compound and you find it kills gophers, you can patent the compound and its use. You have a much stronger patent because no one else can patent that compound. Other people can find other uses. It also kills moles, and so they can patent it as a mole killer, but they still have to go back and give credit to the person who owns the compound. So "composition of matter" is a very powerful patent. A "utility" is a weaker

LECTURE 19 ~ *Psychedelics II*

patent because it depends upon something that's known and that may be owned by someone else.

A lot of chemical structures have been patented as new compounds and maybe a few scattered utilities to cover a few bases. But the strength of the patent is in the owning of the compound. MDMA was made from safrole, isosafrole. Four or five other compounds were made in the same area. There's no evidence that they knew it had any action whatsoever, and there's no evidence that they had any intended utility for it. So if you go back to the original, it was not patented as an appetite suppressant. It was patented because it was a new compound.

And about 10 years later, someone found a way of making something with a methyl group on a primary amine using chloral hydrate or chloral something, and patented that as a method, gave five examples. One of the examples was MDA that could be methylated to MDMA. No utility. Again, in Germany. Many years later in the 1960s, 1950s, it was published in a Polish journal, its synthesis and two or three other materials. No utility, no suggested use, but merely in the literature as a Polish chemical paper. The first pharmacology on it that was ever published was about 1955. No, it was published in the 1970s. It was done in 1955 at the University of Michigan by a group of pharmacologists who studied seven or eight compounds, MDA, MDMA, MMDA—I mentioned in the literature the vocabulary problem with that material—mescaline, alpha ethyl-this, alpha methyl-that, methylenedioxy-that, and trimethoxy-something-else.

Eight compounds, one of which was MDMA. They studied these eight compounds in five animal species—the rat, the mouse, the guinea pig, the dog, and the monkey—for toxicity, for behavioral patterns, as potential hallucinogenics or stimulants or what all. Centrally active compounds. A large, massive study. It's fascinating because that large, massive study which was published was declassified, but that was done then. The fact that it was declassified means that when it was done, it was classified. And if you look in the very fine print at the bottom of the article it says, "We wish to thank the US Army for the funds that made this work possible."

So suddenly you realize this work was done by the army. We're into chemical warfare. And I'm going to talk about chemical warfare in one whole lecture, on nothing but chemical warfare. Chemical warfare was the source of this material to this group of pharmacologists under security classification to study eight compounds in five animal species. There has been a long study of mescaline, of TMA, certainly of LSD, quinuclidinyl benzilate, I've already mentioned a host of different hallucinogenics as chemical warfare agents, because they all have the potential of being disruptive. They are called "disruptive agents." The idea is to get something you can spray over the enemy, and get them to act silly or act irresponsibly, or to pass out, or somehow to keep from necessarily killing them, but definitely not leave something lingering that keeps you from going in. So you want short-term disruption. You don't want stuff to be there for the next decade. The idea of interfering, disrupting, is a very desirable process. It's very big in the chemical warfare needs.

Here are materials which are closely related to hallucinogenics. Hallucinogenics are generally accepted as being things that cause alteration to the integrity of the cognitive and the sensory process. So the work was done in Michigan. It was declassified and published. This is the first work on this compound that was ever printed that is toxicological. They found it was mediocre in the flow of the eight compounds, it was kind of in the middle. It was so-so in toxicity to the dog. It caused some behavioral disruption in the monkey. It did not stand out in any way. MDA—which, in this list—was more toxic and more effective in the animal screenings, primarily in the monkey, in which they determined behavioral patterns that could be defined as centrally disruptive.

MDMA got into clinical usage probably about 6 or 8 years ago, perhaps as long as 10 years ago. The use of it was for a particular property that it had. It has good properties, and it has bad properties. The particularly good property that found clinical use was what it did and what it didn't do.

What it did do is it would largely decrease the paranoia, the fear of interacting. It would allow easy communication. It would allow a searching-out and acceptance of trust. This is one reason why in its clinical usage it was often used not just by the patient, but by the patient and the physician. This is one of the major methods of using it. There would be an exchange of more or less unrestricted and restrained—uncensored, I guess is the term—uncensored speaking about oneself, and it would go both ways.

A second major use was where you'd have a number of people participating and let them go inward, let them do their own searching-out without the physician interacting. Here, a larger dose was used, and there was more of a process in isolation. Both schools of therapy had quite a few followers. I would estimate that by the time the first legal actions were taken, it was probably used by literally thousands of psychiatrists and psychologists, from what I can gather, in virtually all the Western countries that I have been able to get information from. In Europe and in this country.

What brought it into the official view was it getting onto the street. It got into a very broad usage. When I saw the term "Ecstasy" and the term "Adam" come out in *High Times* and in *Wet*, so to speak, I realized that this was going to be an increasing street problem. That is the evidence of abuse, by the way. High abuse potential is—from the government's point of view—justified by illicit and broad social usage. I mean, obviously, if it's going to be sold in the bars in Texas, and it's going to be passed around at parties, it has a high abuse potential. I cannot argue against that. It has one. It has an abuse reality, and when they saw this wide distribution in Texas, they realized this.

It had been seized a number of times before, usually in illicit laboratories, and had been commented on. "We seized this, and this, and that material was also seized, not illegal." The amount of seizure was very small, but it hit at a time when there was a great deal of disturbance about fentanyl, about heroin analogs that were being made that were not illegal but were being promoted as heroin. The so-called designer drug syndrome. This was around the same time that it achieved its notoriety in Texas. And it was proposed for Schedule I as being a drug that had high abuse potential and had no recognized medical use.

I've already commented on what "abuse potential" may mean, on what "medical use" may mean. Abuse potential is a difficult one to evaluate. In about the third lecture—which I handed out in about the seventh lecture—the written text talked about the classification of the different scheduled drugs and what is necessary in order to assign them to Schedule I, II, III, IV, or V. One is high abuse potential, then the next one, third is less than that, the fourth is less than the third, and the fifth is less than the fourth. There is no place in the entire scheduling for a drug that has no medical use and does not have a high abuse potential—something other than high, medium, low, so-so, 32 percent—that's not defined legally. It is an oversight in the writing of the law. Clearly, I think you could argue that the legislators that wrote the law didn't intend this gap to be in the structure. And so, from the hearings that were held, the judge says, "Well, clearly, if it has no medical use but it has no or little abuse potential, it should be maybe Schedule III or something." This was a suggestion that was ignored.

The government proposed that the material be placed into Schedule I, as a drug that fit all the requirements of Schedule I, gave the classic 60-day period for hearings, and to their amazement, they got letters in from a number of physicians saying, "Hey, this thing has a good medical utility. Sure, it's abusable. Most things are abusable, but it has medical use. Don't put it in a form that will take it out of medical use or take it away from human experimentation or research."

As was required by law, enough people gave this response that hearings were held. Hearings were scheduled; one in Los Angeles, the second to be in Kansas City, the third to be in Washington, DC. And they were scheduled about three months apart.

STUDENT: Did you testify at them?

SASHA: Pardon?

STUDENT: Did you testify?

SASHA: I did not testify, no. But I attended the first hearings in Los Angeles. The first hearing is to determine whether this drug has a high abuse potential or not, if it has medical utility or not. Should it be Schedule I? Remember, Schedule I, in essence, is like the reactor at Chernobyl underneath 55 million tons of concrete. It is in there, and it will substantially stay in there. Research cannot be done, because there's no justification for doing research on drugs that have no medical utility and have a high abuse potential. What's to be gained? You only want to do research on drugs that have the potential of medical utility. In essence, it's a catch-22. Once it's in there, it's hard to get it back out. If you can keep it from going in, then you can establish an argument that it may have abuse potential, but it may have medical utility. Let's find the correct category. So the hearings were held to keep it from going into Schedule I.

Just before the first hearings, the government invoked a new law that

LECTURE 19 ~ *Psychedelics II*

had been passed just about eight months before. It's called the "Emergency Scheduling Act" of 1984, which allows the government to put anything they feel is an imminent threat to public health, or is an imminent public hazard, into emergency scheduling without argument for a year, while the discussions take place, and then it can be extended for six months if it's wished, and then they will make a permanent scheduling or abandonment.

This emergency scheduling was invoked on MDMA, just prior to the first hearing to find out if MDMA should be Schedule III or not. They invoked a Schedule I emergency scheduling. There's a technicality that is an interesting one. I know one person who has gotten his entire charge dropped in Colorado because of what the government did when it passed the law. When Congress passed the law on emergency scheduling, they empowered the Attorney General to invoke this emergency scheduling act. The Attorney General normally transfers this power over to the DEA and did in the case of the original Controlled Substances Act of 1970. But in this case, the Attorney General had not yet transferred the power over to the DEA, and the DEA invoked this emergency scheduling act anyway without authority, so one person had his whole charge thrown out because of that. It's moot now because it's a permanent Schedule I.

So MDMA was under emergency scheduling. It was kind of a statement—I'm reading into this with my own bias, but I see it this way—"You may have your hearings to see where it should eventually go, but we initially said it was supposed to be Schedule I, and we're going to make it Schedule I. And you can go ahead and have the hearings, but it's going to be Schedule I as a de facto reality during the hearings, and it's going to have to come out of it one way or the other if the hearings produce a result that has to be adhered to."

So the hearings appeared under a cloud, in that the DEA had already invoked a conclusion. The judge was not particularly happy at this process, as you might glean. What the judge was sitting in hearings to decide had been preemptively decided for him by the administrators in the DEA. The hearings were held. The judge made a long decision that said, "On the basis of everything I'm seeing, there has been a medical utility." The DEA was unaware of any of this medical utility, so they were acting at least on a good faith on the basis of no information. So I will not remove that from their pattern of action.

And the judge, Judge Young, made the conclusion that this material has medical utility. To say it has none puts the total power in the FDA's hands for calling what it does have, and the FDA has already said, "We don't want that power." And so, he thought that Schedule III would be a very excellent compromise that allows research to continue and would still give the DEA full control over the scheduled drug. The DEA said, "Thank you very much for your opinions. We'll leave it in Schedule I and make it permanent," which they did. They had no requirement to follow the judge's opinions. It is now on appeal. I have very negative feeling that it will ever come out of Schedule I and that little research will be done on it.

The argument was made at the time it was put into emergency scheduling that all efforts will be made to expedite whatever research can be done in clinical studies to evaluate and to determine its potential medical usefulness. As of now, no permission has been given to any clinical group to study it. The DEA has found that an IND cannot be rewarded until much more animal work has been done. There is no argument of how much more animal work will be adequate, but none has been issued, and I have reason to believe that none will be issued. I believe that MDMA is legally and medically a dead issue and will stay so.

It has a great potential. I think the potential will not be realized with that drug. The concept has great potential of using a drug in therapy, and I think that will grow. But that has yet to be seen.

Yes?

STUDENT: How about the therapists that are using it? How are they getting it and how are they getting around the loopholes? Or how are they dealing with the law?

SASHA: I can't answer accurately because I don't know the answer. I'm not privy to therapists who are using it. I have been told that therapists are using it. If so, they're committing a felony, and I would consider it to be a very dangerous act on the part of a professional person.

STUDENT: Any idea where they get it?

SASHA: I sincerely hope it's not from the illicit street market because a lot of stuff is being sold as Adam and Ecstasy that does not resemble Adam and Ecstasy.

LECTURE 19 ~ *Psychedelics II*

STUDENT: Is there any test to determine the quality of it?

SASHA: Yes, there are instruments, there are tools that are quite satisfactory. They're not the sort of thing you have in the back of your VW. There are instruments that will do it. There are spot tests that, if they fail, the material is not it. But they are not easily done. There used to be a laboratory that did anonymous analysis. That was PharmChem down the peninsula, somewhere around Palo Alto. They have gone out of the business and turned it all over to a group in Florida. Drug something, "Drug Alert," "Drug Awareness," drug something in Florida, who I have heard now has gone out of business. To my knowledge, there is no laboratory that will take a drug anonymously and tell you what it is. If someone knows of it, I would like to learn of it. I don't know it. I think it is a public service that does not exist and should exist. To answer your question, how do you know what you have? I don't know. I don't know a way of finding out.

Okay, more questions.

STUDENT: Go to the chemistry lab and run an infrared.

SASHA: Oh, go to the chemistry lab and run infrared on it! Yeah, sure. You may find lactose and you may find sucrose and you may find underneath it MDMA. The material I have seen has either been very good or totally wrong, what I have looked at personally.

STUDENT: How come if the MDMA has methamphetamine in it is it more sedative or not as speedy as MDA?

SASHA: It's not as speedy as MDA, but it's pretty speedy.

STUDENT: Why is that? Why is it less?

SASHA: It's a different compound. All compounds are unique. And I think the fact that it has an amphetamine structure is only incidental to the fact that it causes cardiovascular stimulation. I know materials that have that amphetamine structure that are psychedelic which do not. Every compound is unique.

Yes?

STUDENT: Is it true that MDMA drains your spinal fluid?

SASHA: Okay. Rumors. I was just talking to a person here who gave the last lecture, who saw the *Psilocybe* mushrooms and said, "What's all that about?" We got into MDMA. And I used the analogy of the fact that the Ganges runs red because there are dyes in the Ganges. The Ganges is not normally red. There are more rumors. I'd like to know who starts these rumors. Everyone has someone who knows someone. I know a person who took MDMA and he only has one leg. [Laughter.] That's the truth. I mean, he has a thing truncated at the right hip. In fact, he's very fortunate. He came down with cancer of the leg and had to have the whole leg amputated, the bad leg that got cancer, so he at least has one intact leg, the other was taken off at the hip. That has nothing to do with MDMA. But he also is a user of MDMA. I can say, "I know a person who has used MDMA and has lost his leg," and I'm absolutely accurate. But you don't extrapolate from that one example to an epidemiological fact unless you really want to get in trouble with referees or don't want to publish.

Does MDMA deplete the spinal fluid? My god, I can't see how it could. There is no record of it. There's no reason it should. There is no record of any way it depletes the spinal fluid, except by removing it and injecting air or something. But nonetheless, it makes a honey of a rumor to indicate that MDMA is going to be bad for your spinal column. Does it give you back pains? Not to my knowledge. Does it goof up the immune system? Not to my knowledge. I don't want to contribute rumors. I will say that I don't know of it. There is certainly nothing in the published literature, and I know no firsthand experiences of that. It's got its negatives.

STUDENT: What are the short-term and long-term side effects?

SASHA: Short-term side effects. It pounds your blood pressure way up. You are getting a blood pressure rise, systolic—I think I mentioned this in the last lecture—maybe 15 millimeters, diastolic 10 or 15, or even 20 millimeters, which is very concerning if you happen to have a high blood pressure situation and you have a heart that is not as sound as it should be, you have veins that have big swollen sausage-type aneurisms in them and they're held together by a marginal fragility. Accidents can happen. There was a report in the JAMA—*Journal of the American Medical Association*—just about two weeks ago on five deaths associated with MDMA and MDE.

LECTURE 19 ~ *Psychedelics II*

They named it "Five Deaths Associated with Ecstasy and E." MDEA, they call Eve. I've heard it referred to as MDE, and I've heard MDMA referred to as MDM. These are presumably interchangeable.

"Five deaths associated with"—one death that was ascribed to MDMA, normal dosage, 150 milligrams or 120 milligrams was the reported dosage that was taken. In these anecdotal things, you very often don't know what was taken. The patient, if he lives to say so, says, "I took what was sold to me as, it was sold to me as being so much." And so, you can't go into the published literature and say that this much of that drug caused such and such, because you're getting it from anecdotal information. But what you can do—and was done in the case of the one girl, 18 years old I believe her age was, 150 milligrams is what she said she took in the form of MDMA and died of heart complications, and they ascribe it to the drug. They took body tissue samples—blood, urine, liver, kidney, heart, I believe was the extent of the tissues—and found MDMA in these tissues by what sounds like a reasonably good analysis.

I have no question that the material that had been taken was MDMA. The amount they found was within some levels I have seen of normal usage. I have no reason not to suspect that this girl had taken the normal amount under normal circumstances, had heart problems, and died, and the cause of death being MDMA. I have no issue with it.

ANN: Maybe she had heart problems before she took it.

SASHA: Before, right. And this was complicated with MDMA, and the drug legitimately could be called the cause of death. The other four were deaths associated with MDMA, or MDE. In each case the amount of drug was fairly small and was complicated by other drugs being present, including alcohol. You don't know what the cause of death was, so it was not assigned to the drug. This one was assigned to the drug. I know of a death in Seattle that was incorrectly assigned to MDMA. The material in the person was MDA. MDA has a fairly narrow margin of safety from the animal tests extrapolating to man. There probably have been other deaths.

On the other hand, there have been probably tens of thousands, if not hundreds of thousands, of usages of MDMA. So I don't think it's a large risk, but it's a real risk. And I think the report of the death ascribed in this case was valid. So I say this is a negative.

The physical symptoms of using it. The jaw clenching, the grinding of the teeth. This tendency is never disturbing to the person, but it looks and feels terrible when you have a sore jaw the next day. Clinically when this occurs, it's relieved by putting a wet washrag in the mouth and let the teeth clamp on the washrag. It usually eliminates the sore jaw. Eye nystagmus. Nystagmus is the tendency for the eye—when it's put into an extreme position or even when it's in a normal position—to experience a certain amount of tensor effects, and some of the ocular motor nerves to go twitch, twitch, twitch, twitch. Sometimes it can be aggravated if you're looking straight ahead and move your eyes off to the left to follow a finger, and your nose is that way but your eye is going that way, your eye will try to recover itself, twitching. Very real. Happens in maybe a third or half the people who use MDMA. It doesn't seem to bother the user, but it's the sort of thing that could make focusing on an object difficult and hence makes driving hazardous. These are effects. Appetite killing is a universal effect with MDMA. It just destroys the appetite very much like amphetamine or methamphetamine in this sense. Eye dilation, mydriasis, reflexive adrenergic stimulation is very much part of the experience.

It's fairly short-lived. It's effective in half an hour, full effect in about an hour. It begins dissipating in about two hours and is largely back to where it started in about three or four hours from the initial taking of the drug. Hence, from the therapeutic point of view, it is quite desirable because it occupies a fairly short period of time. You do not want a 42-hour psychotherapeutic agent to use in the practice room in therapy.

So, any more questions on MDMA? It is more than I expected to get into, but several people asked for more information. Other questions on it?

STUDENT: Is there a supplement when taking MDMA, not really to prevent any of the side effects, but to help with them?

SASHA: There have been a number reported on these things. Calcium and magnesium supplements beforehand, mineral supplements to try to ease some of the jaw clench. Phenylalanine has been reported to be good going in to lessen the adrenergic effects. Tryptophane has been argued afterwards to allow sleeping and reduction of some of the stimulation.

LECTURE 19 ~ *Psychedelics II*

Long-term effects: The animal studies that were used for justification of making it illegal were based on the depletion of serotonin and damage to the serotonin neurons. Serotonin tends to be depleted and interfered with in a way that's perhaps not permanent, but certainly persists on a long-term basis in rats. If it occurs in man, it may account for the sleeplessness and the alertness that the person experiences.

STUDENT: Insomnia?

SASHA: Insomnia.

STUDENT: So, it doesn't last very long?

SASHA: No. Other questions on it? How many people have used Ecstasy or Adam or MDMA? Good grief. I won't ask how many haven't, because I won't get a valid hand raise, but over half the class. It's interesting. I got a call from someone at Stanford in neurology who says it is everywhere on the Stanford campus now. I wonder what's really there, because if it's there, it's not coming from legitimate sources. Yet, in all honesty, what I have seen when it is the right material has been good, pure, and without adulterant. On the other hand, a certain small but real percentage has been totally unrelated. So I don't know how to assay what we're seeing.

STUDENT: Are you saying that it's not often cut with other things?

SASHA: The stuff I have seen has usually not been cut. If it's right, it's been right and good. But if it's wrong, it's been the wrong material. I've seen MDMA cut with borax for example, this kind of thing. Sugars are commonly used. Borax—I've recently come across a lot of borax. Why? I don't know.

ANN: What sort of other things have you seen that were the wrong material?

SASHA: There are quite a few things that I've seen. PMA—para-methoxyamphetamine. I've seen 2,5-dimethoxyamphetamine as MDMA. The para-methoxyamphetamine is quite a heart stimulant and a dangerous drug. I put it as a true dangerous drug in that it causes heart stimulation before actually getting into the mental effects. You try to push it to get the mental effects and you get a blood pressure that becomes dangerous. There's a question over here.

STUDENT: It was basically answered.

SASHA: Okay, let me bounce on. Is the minute hand roughly correct?

ANN: Yes.

SASHA: Okay, so I've got a feel for where we are at. I want to talk a little bit more about the phenethylamines and get into some of the indoles. By the way, one thing that did occur to me, does anyone here have the original handout that gave the schedule of lectures and their dates?

STUDENT: Yeah.

SASHA: Good. See if I put marijuana on there? I don't know if I did or not.

STUDENT: You did.

SASHA: I did. It's still ahead of me somewhere?

STUDENT: Yeah.

SASHA: Okay. Walking up here it suddenly occurred to me, my god, I don't know if I put that on there. That deserves a fair bit of time on its own. Okay. I'll find my own at home.

I want to go to the other phenethylamines, or ones of this general family. There are a lot of mescaline analogs that have been made. Mescaline, remember, was the active component of the very first cactus lecture. I'll do what is fun to do on a blackboard—it's miserable to do when you take notes—use an eraser and chalk.

This is the molecule mescaline. A large number of studies have been made on mescaline. I talked about changes on mescaline to search out where in the molecule the activity can be assigned, what is in the structure that correlates with the activity, the SAR—structure-activity relationship—study. The amine group is a total wipeout. Any substitution on the amine destroys the activity. The alpha group, only if it's an alpha-methyl to the amphetamine is the activity maintained. Other extensions wipe it out. The beta position, very slight changes can maintain activity. Hydroxylation is not

LECTURE 19 ~ *Psychedelics II*

active, but methoxylation is, and any alkylation destroys activity—any alkyl group. The ring has been the treasure for research work. And I've mentioned already the relocation of the groupings on the ring—the two, four, five; the two, four, six to give you numbers. Mescaline is three, four, five, which is a relatively inactive position. Yet, in the human body, mescaline, which should not be active at all, is active, and this orientation is the only orientation that produces that particular quality of effect.

It should not be active, as the body—remember when I was talking about the adrenergic synapse—has an enzyme that destroys these primary amines—monoamine oxidase. It's the one that takes the amine that's released into the synapse and gets rid of it so the synapse does not stay transmitting and go into a situation of neurological tetany. This amine oxidase is a desirable material to get rid of neurotransmitters that are amines. It is called monoamine oxidase—MAO. And many of the drugs that are used for treating certain neurological difficulties are inhibitors of this enzyme. Well, this is a monoamine and this material should be oxidized and destroyed. This material should not be active at all. In fact, phenethylamine and methoxyphenethylamine and the very simple molecules are destroyed. Tryptamine, tyramine, all these basic simple amines are destroyed by this enzyme system. This material with the three methoxy groups is not attacked by that particular enzyme system.

It's as if these three methoxy groups made this sufficiently basic that the monoamine system would not work, but a diamine system is called upon to be effective. The truth is that mescaline is not an effective substrate for the monoamine oxidase system. And in general, the basicity in this end of the molecule is the type of thing that brings the material into this particular type of central activity, the so-called psychedelic or hallucinogenic activity. The shorter, less substituted ones that are destroyed become stimulants, if you interfere with a destroyer and use a monoamine oxidase inhibitor. Those that are well substituted over here are in the psychedelic area. The magic is in this four position which I mentioned with the substituted amphetamines. This chain has been extended; it has been modified. As a rule, an aspect of the intoxication is lost, but the visual aspect of the intoxication is maintained. And in general, the compounds become immensely more potent.

Mescaline is a relatively inactive compound. Let's say 300 to 500 milligrams is a total dosage that is required in a human. Compounds with the longer chains on here drop by an order of magnitude. They're active in the 20 to 30 milligram level. A type of variation has been made in which sulfur atoms have replaced oxygens. These materials are called "thiomescaline." They become yet more potent; they become longer lived. They are not particularly with the same type of chronology of mescaline, which is not a terribly long-acting compound. Di-thio compounds have not been made. The thio analogs that have been made ... to change the molecule is a real mixing of the spice cabinet here. But these two directions are the major ones that have led to maintained activity.

I want to make one aside of another class of compounds that are available. I've heard quite a bit about their availability. I want to mention them so that you know what their structures are. If you remember the original argument of the two, four, five, number two, three, four, five, six. [Drawing on the board.] I'll just change the numbers and leave the groups alone. Benzene rings can number either way. The two, four, five substituent pattern is the most potent of all patterns. The two, four, five where the amphetamine is, is TMA-2. DOM, which I had mentioned, is the compound with a methyl at this position. Again, the four position can be modified in many ways. One compound that became—for a short period of time—a street abuse problem about the late 1960s, early 1970s, dropped out of availability. It pops up about every year or two with another little enclave somewhere that uses the material; it's a compound with a bromo at this position, called "DOB."

It is a very potent compound. The active dosage of DOB is, say, one to three milligrams. So it's a material that's probably not more than about one tenth the potency of LSD. It is very broadly used in Germany. It had been used on the east coast. It never got widely used on the west coast in this country. It was associated with deaths on the east coast, about five or six years ago. I found a little information on it. It was being sold as MDA. The person thought it was MDA and had taken a line of it and snorted it and got 30 or 40 or how many milligrams, I don't know, of DOB, went into a vascular spasm, and died. The companion—there were two, a boy and girl—the girl died and the boy survived. But apparently a gross overdose.

LECTURE 19 ~ *Psychedelics II*

One report that came out of the Napa area, a girl had gotten into vascular spasm on DOB. A vascular spasm is, in essence, cutting down circulation to the periphery. It is not an uncommon thing with some of these substitute amphetamines, and it's very much like ergotism in which you lose circulation, and as such, you eventually have a problem such as gangrene, which is due to the lack of oxygenation of tissue. She had taken what she claimed to be this, she had gotten gangrene, had gotten amputated, and brought a suit against the emergency physicians who were in the hospital she had been in. I got involved a little bit in the case. It can be distilled in one simple sentence. There's no knowledge that she had taken DOB. She had been told that, she believed it to be that, but no samples available, and no one knows what actually went in. No analysis, no urine was saved, no blood was saved. No analysis could be made. But it's in the literature that this compound in experimental animals will cause some vascular constriction. And so that was the basis of her suit. I don't know how the court case came out.

But the point—oops! This is today's lecture. Brazil, Colombia, Amazon. [Drawing on the board.] This is going to be one funny lecture. Here is Orinoco. [Laughter.] An analog with the 2-carbon chain is a material that has been called "2C-B." The B for the DOB and 2C for two carbons. And I have heard this material being talked about periodically. I know it's been seized in a number of illicit laboratories, and it's a material that I don't know what its common name is. The literature name is 2C-B. It's a material that is quite a psychedelic. How many people have heard of 2C-B in one form or another? About six or seven. It emerged first as a problem in New Orleans where it was added apparently to a punch bowl and a number of people got very ill and were sent to an emergency hospital. They found it there, isolated the material, got into the literature and found that I had published some material on it. I sent them a sample and they sent me a sample of theirs, and it was 2C-B indeed. So the material is occasionally appearing in the social street market. I don't know what the authenticity of the material is there.

STUDENT: What is the difference between 2C-B and the DOB?

SASHA: DOB and 2C-B differ by the methyl group in this position. The 2C-B is without the methyl group and DOB is with it.

THE NATURE OF DRUGS

1-(2,5-dimethoxy-4-methylphenyl)propan-2-amine (DOM)

2-(4-bromo-2,5-dimethoxyphenyl)ethan-1-amine (2C-B)

1-(4-bromo-2,5-dimethoxyphenyl)propan-2-amine (DOB)

ANN: What is the difference in the duration of them?

SASHA: DOB is a 24- to 30-hour experience; 2C-B is about a 4- or 5-hour experience. It's much shorter lived. Dosage of 2C-B—whereas 1 to 3 milligrams of DOB is, I'll say, 16 to 24 hours. Dosage of 2C-B is probably something like 12 to 20 milligrams, and the duration will be 4 to 5 hours. Shorter lived, less potent. Similar in general character of activity. Yes?

STUDENT: DOB, could that also be known as bromomescaline?

SASHA: It would be a tricky thing to do. Bromomescaline is a difficult thing to do. If you brominate mescaline, you get an inactive compound. So probably bromomescaline would be a misnomer to give. I don't know what is meant by bromomescaline. If you replace a methoxy group with bromo in mescaline, you get into chemistry that you're not going to find being done in the street. It's very difficult. Yes?

STUDENT: When these things appear, are they trying to make MDA and ... ?

LECTURE 19 ~ *Psychedelics II*

SASHA: No. I think they're going into the literature and finding things that are psychoactive and duplicating them. All these things have been published in the medical literature.

Okay, other questions on that?

STUDENT: What is the status of 2C-B? Is it Schedule I?

SASHA: It's not recognized legally. DOB is Schedule I. 2C-B is not recognized legally. There are a lot of things in that category. In fact, that's worth . . . Ha! I'm not going to get to indoles at all. Next hour.

I want to bring up a point on that. There is a law that's just been passed. Have I talked about the Designer Drug Law?

ANN: A little bit.

SASHA: A little bit. Let me mention it here. Is this legal or illegal? 2C-B is not named in the scheduled drugs. There are a lot of drugs like MDE that are not named in the scheduled drugs. MDMA is; MDE is not. The Designer Drug Law—the Analog Bill says that anything that is similar to a designer drug—and clearly 2C-B is similar to DOB, and MDE is similar to MDMA—anything similar, either in structure or has the action that is similar or is promoted as having the action that is similar, is an analog drug. And if you have possessed it with the intent to give it to a human, then it has a classification; one can be prosecuted as if it were a Schedule I or II drug. So no, it is not illegal. But if you have it and you say, "Hey, take this. You're going to find it's a real neat turn-on, very much like . . ." you name something that's Schedule I. For that purpose, you are committing a felony as if it were a scheduled drug. So there may be, over the next coming few years, no more drugs being put on the scheduled list, because they don't need to, because it's the similarity and the intent that has been passed into law, not the de facto explicit reality. It's a dangerous law to enforce for that reason, because it is not explicit. It's an open-ended thing. And the first court cases have not yet come to test. But it's something that will be seen.

I want to get a little bit out of here and into the indoles. I may not get very far into them. The basic neurotransmitter . . . I'll draw the neurotransmitter, the phenethylamine—this is called a CNS transmitter primarily. The companion neurotransmitter in the brain is a material called "serotonin."

Remember when I talked about the phenethylamines, how you have an aromatic system and then about two carbons away from it, a certain distance away, you have a strong base? So, you have a weak base, a sort of a pi nebulous-type of base, the aromatic system, then so many angstrom units, a strong base. That is the vade mecum of the CNS-active compound. In the case of dopamine or mescaline or all of the materials we are talking about, the weak base is this aromatic system, the 2-carbon removal to a strong base that is an amine, strong amine, and its electron pair.

In serotonin, same thing. You have an aromatic system. In this case the aromatic system is called "indole." The name of the ring, this ring, in chemistry, it is known as "benzene." And in the system, you have the separation by that magic; so many angstroms to a strong base, the amine. Go through the alkaloid world. Go through all the drugs in the *PDR*. Work through the Merck Index. Look for things that are centrally active. You will generally find an aromatic system or a big diffuse basic system and a separation to a strong base that is built in there.

When we get into the next lecture, into LSD, it is a great big monstrous molecule that takes up half the blackboard, and you find in the middle of it an aromatic system, there's your nitrogen. It's in there. It's a nice little tool. It's a little signature that says, "I'm going to be a CNS active, maybe a stimulant, maybe hallucinogenic, somewhere in there." If you find the separation is only one carbon, you'll find a lot of these systems. If you find a great big aromatic system that's got a pi-basicity and a one-carbon separation from the strong base, you're usually in the things that are delusional, anesthetic, disruptive. Not constructive, but disruptive. I've mentioned PCP, I've mentioned ketamine. These materials are of this ilk. They have this general separation. So the nitrogen is close. The two bases are fairly close together, and you get into a disruptive anesthetic, delusional, confusional, cognitively disturbing, an amnesia-generating world. If you get a little further separation, you usually get into the stimulant, into the reinforcing world. If you get further separation yet, if you have any activity at all, you generally get into an anticholinergic activity, and it generally goes after the muscles but does not have a central activity. Very few compounds with a 3-carbon separation are centrally active. Two-carbon generally stimulant; one-carbon

LECTURE 19 ~ *Psychedelics II*

generally disruptive.

So you have this material, serotonin. Okay, or 5-HT, a very nice common term that is used as a nomenclature for the 5-hydroxy, in the five position, tryptamine.

Yes?

STUDENT: What is the indole?

SASHA: Indole is this part of the molecule. It's that aromatic system, as benzene is this part of the molecule.

ANN: Can you give an example of what kind of things are indoles?

SASHA: Serotonin, tryptamines, all of the hallucinogenics I intend to talk about for the next 14 minutes.

I wrote this on the board because there's no good way of organizing this portion. [Referring to blackboard.] Amazon, Orinoco. The main problem in this whole area of all the hallucinogenics—and here you have things that are rapé, paricá snuffs, cohoba, ayahuasca—you have things that have plant names. I have three typical plant names. Countries in which they occur, rivers they are associated with, tribes that use them: it is a terrible hodgepodge. You'd like to say, "Here is the organization that will allow you to get it and take it apart and be able to know it perfectly forever." But you'll find that this tribe has taken that plant and that plant and mixed them, this person over here has taken the residues of a snuff package and found the chemicals that are in it, this person over here has gone to the plant and found the chemicals that are in the plant, not realizing that two plants have gone into one snuff. This person has grown up in this area where they use several plants, but they have learned from another tribe over there that has learned from something different.

There is no pattern of consistency. And where are the chemicals in this mess? It would be nice to say this chemical is found in this tribe in that thing from that plant. It doesn't work. So the best you can do is get the picture of all the tribes, of all the areas of where something is, or what's used, how they're used, and the chemicals that are in there, and not try to say, "This snuff has this compound used by this tribe from this plant," because it doesn't quite go that way.

THE NATURE OF DRUGS

All of this world is in the area where the materials are natively found. Let's go back to South America again. If you remember, I made this monster. It should be a more pointed than that. Here's South America, and the general pattern I've given—I started my three little babies up there and kind of came down without dividing it—remember this old routine? It's a nice one to know. Here's Ecuador, because the equator is there. And so you divide that up, you divide these into fourths. These are, in essence, the west coast parts of South America. Your little Guyanas, and of course you have Venezuela, you have Colombia, you have Ecuador, you have Peru, and you have Chile.

The areas that we're talking about, where these are used, are up in this northern and largely western part of South America. The Andes comes right upside—this gets kind of fuzzy—and cuts through in its own way there. This is the umbilical cord that goes to Central America.

The area that is down in here is largely the Amazon drainage area, the upper Amazon, the whip of the upper Amazon. You're up in this area. The Amazon dumps into the Atlantic over here, comes through Brazil, and basically feeds out of the highlands that are on the eastern edge of the Andes, but are in Peru and in Ecuador and in Colombia. This is the Amazon area, and this is the area where a bunch of this drug issue is of interest. The second major river in South America is the Orinoco, which drains out into Venezuela, and comes from these extremely high highlands, the eastern slope of the Andes. This is the treasure house of the jungle lore of active drugs that include a lot of the ones I want to talk about.

Where the Orinoco dumps, you have Trinidad as sort of a benchmark of the West Indies, and this entire area of the West Indies into Central America is the area that is very heavily invested with the snuffs. So you have Colombia as a country, Brazil as a country, the Amazon as a major river, Orinoco as a river further north and draining from Colombia through Venezuela into the West Indies, into the waters inside of the Antilles. And the names of the snuffs vary there. Every one has a local name. The major compound in many of these materials is like a modified serotonin. This is a compound that is called dimethyltryptamine. This is DMT. It is probably the prototype of all of the drugs. It is found in many plants throughout the world. It is a very broadly based alkaloid. It is responsible for many of the neurological

LECTURE 19 ~ *Psychedelics II*

problems that are found amongst animals that graze on various grasses and various plants. Some of the sheep staggers are ascribed to DMT. It is not an orally active compound in most animals, and it's not orally active in humans.

DMT, I brought an example if you want to see an innocuous white crystalline solid. DMT, as with many of the indole hallucinogenics, with the exception of the psilocybin and psilocin group of the mushrooms, has to

be taken parenterally. It has to go up the nose, up the behind, into the blood, into the muscle, somehow, because once it gets into the tummy, it is destroyed. It is not active orally. The active dose of DMT is probably in the order of 50 to 100 milligrams. It is smoked in our culture; it is blown in as a snuff in the South American cultures, characteristically, through little hollow bird bones. You have a bird bone with a hollow center, you put the snuff in the center, someone puts it into his nostril, and someone else blows, and it goes up the nostril and gets launched.

Yes?

ANN: In the movie *The Emerald Forest*, I think, they showed the natives blowing into a long bird bone.

SASHA: Epená, yagé, cohoba snuffs were the primary snuffs that are involved. I have not seen the movie, but I've heard reports of it. Quite authentic in its descriptions.

STUDENT: They described it as a rock or something.

SASHA: Shouldn't be.

STUDENT: Do you remember that? Green crystals, in the movie?

ANN: No, I don't remember that part.

THE NATURE OF DRUGS

LECTURE 19 ~ *Psychedelics II*

SASHA: *Virola* comes from the bark of a tree, the ayahuasca—I'm not going to talk about ayahuasca until later—it comes from bark. *Psychotria* is the leaf of a tree. Only in *Anadenanthera* do you get into seeds in snuff in South America.

ANN: Do you remember it as being blown in the nostril?

STUDENT: Right.

SASHA: DMT, and I'm going to put one more basic compound in this area, and that is the compound 5-methoxy-DMT. I don't know how to abbreviate them, but 5-methoxy-DMT is quite common. These two are found almost always together. But you can't say that they're together in the plant because the snuffs that have been used have often been a combination of different plants. In the snuffs you find them together. Often in plants you'll find one or the other, but in many snuffs they are together. DMT is 50 to 100 milligrams, 5-methoxy-DMT is 5 to 10 milligrams. This also must be parentally administered. It must go up the nose or into the muscle or be smoked. Orally it is not active.

An interesting complication gets in here. There is a third plant that I was not going to talk about until the end of the hour. *Banisteriopsis*, it gives rise to a material commonly known by the name of yagé or ayahuasca, which is a material from a plant that used to be known as *Banisteria*. Its genus has always changed name thanks to the botanists wanting to take credit, and now it's called *Banisteriopsis*, or the alkaloid has been called telepathine. It is a carboline. Its common chemical name is "harmine." It is a material that has a reputation of being psychoactive, but its main activity in the body is to promote the activity of other things that would normally be destroyed. It is a monoamine oxidase inhibitor in its own right.

In the native use, you have really native psychopharmacology. You take this plant that contains materials that are only active as snuffs and combine it with this other plant that doesn't have much activity, except that it makes materials that are not orally active, orally active. You may take the snuffs and eat them if you combine them in this mixture. So in essence, the combination of a drug that is destroyed in the body by an enzyme and a plant that has the inhibitor of that enzyme has made a combination that is orally

THE NATURE OF DRUGS

LECTURE 19 ~ Psychedelics II

active. And that combination is broadly used, largely in the Orinoco area, this area of South America.

Okay, I want to get into *Banisteriopsis* alkaloids a little bit later. But I want to spend a little bit more time on the tryptamine areas. 5-methoxy-dimethyltryptamine. I mentioned dimethyltryptamine. Very fast onset, very short lived. 5-methoxy-dimethyltryptamine, extraordinarily fast onset. If we're dealing with a minute or so with DMT, we're dealing with a fraction of a minute with 5-methoxy-DMT. You smoke it, inhale it, and you may be able to get a second toke and you may not. It is that kind of a very, very rapid onset. You are fully into a different type of altered place for a matter of a few minutes. Then you realize that it's on the wane. And often with 5-methoxy-DMT, within the hour, you are completely where you started. And with a very vivid memory, but with no residues.

A compound that lies in this general classification—I'm going to continue with the eraser job, I'm almost out of time today—is a material with an oxygen in the four position instead of the five. This basic orientation, 4-hydroxy-dimethyltryptamine, this specific compound is called psilocin. The phosphate ester of it is called psilocybin. And these were first uncovered in a mushroom genus that is called *Psilocybe*.

I write these out because they are often messed up in terms. Psilocin, psilocybin, *Psilocybe* is the usual pronunciation. The genus gave rise to the names of these two materials. These materials—the phosphate ester being psilocybin, the hydroxy being psilocin—are orally active. For what reason they are orally active, it's been speculated that maybe there's an internal salt

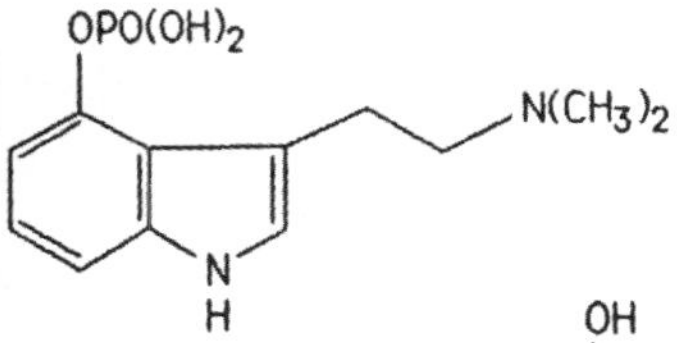

PSILOCYBIN

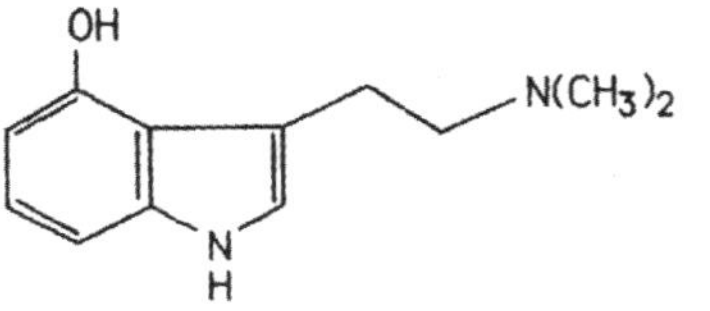

PSILOCIN

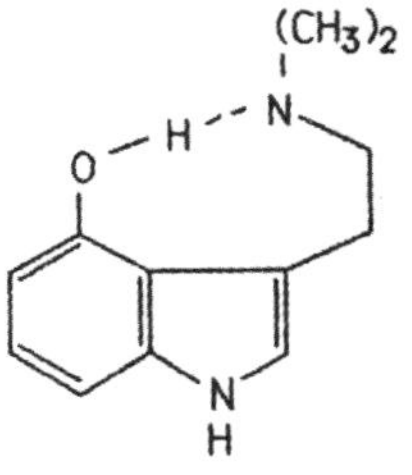

PSILOCIN
(internally bonded)

THE NATURE OF DRUGS

LECTURE 19 ~ Psychedelics II

arrangement that allow some sort of a loss of the extreme hydrophilicity that bars entry into the CNS. The truth is, it is not known why they are orally active, but they are.

I brought in the show-and-tell. We're almost out of the hour. I will bring them next time. See them if you want, they're fragile, be careful—here are some of the *Panaeolus* mushrooms and some of the *Psilocybe* mushrooms that have been gathered in the field, to show you what they look like. This one with the long sticks is an example of a cultured *Psilocybe cubensis*. And I've also brought in a sample of dimethyltryptamine and 5-methoxy-dimethyltryptamine.

The *Psilocybe* mushrooms are in the North American portion. Their origins are in Oaxaca, in Mexico, and was thought originally to be only in this little area of Mexico. They are now found to be throughout the world. All these various species and their alkaloids had not been used throughout the world, now they are. But interestingly, it's only in two or three little spots that the native cultures had discovered its activity, and the primary spot that has been the source of it coming to Western medicine and Western anthropology is in the upper northwestern portion of the state of Oaxaca in southern Mexico.

I'm out of time. We'll get into something else next hour. Be careful in handling; there are pictures in the in the mushroom book of how they look fresh, and these are the dried samples. They are fragile. Handle them with care, please. These were gathered, by the way, in Oregon, in a little town of Siletz, where cows graze freely and mushrooms grow easily. [Laughter.]

LECTURE 20
April 9, 1987

Psychedelics III

SASHA: Just waiting for a few more people to wander in.

ANN: How many people come from the East Bay?

SASHA: How many people are here from the East Bay this morning? Did you come by way of the Richmond Bridge? I did. How long did it take to get across the big one?

STUDENT: I came really early.

SASHA: Okay, you got ahead of it. Two Muni buses going in the same direction exercised the skills they learned in driving school and managed to collide.

STUDENT: On the bridge?

STUDENT: Going in the same direction?

SASHA: On the bridge. In the same direction, yes. Thirty people hurt, traffic, unbelievable. Still being jammed up. I came over the Richmond Bridge and across the Golden Gate Bridge. I think I probably got here earlier than I would have—so if people are not here, they will wander in and hear sad tales about traffic.

I want to spend a little more time on indoles, the second of the two categories. And then I want to get into an area that is only kind of related to drugs specifically, and more to the concept of drug usage. By the way, I should have started the show-and-tell earlier. Someone asked about where these mushrooms were found, and these were the ones I collected in Siletz. I mentioned the name. This is of another group of *Psilocybe semilanceata*. Then I brought in two Sandoz vials that contain LSD and psilocybin in the form that they made them available, when they were still making them available in this country for experimental use.

Also, what I did start doing is putting some of the words I use up on the board. People ask, "What was the word you used?" I wrote out a few that I'll be getting to in today's lecture. I always have trouble with *i*'s and *u*'s. It's *i* before *u* except before *q*, but in this case it's an exception to the rule, so Ololiuqui is spelled this way, I think. I'll get into these terms as I go through the hour. Is the minute hand about right, by the way?

Student: Yes.

Sasha: Okay, I'll observe that. I mentioned in last hour the indoles. We've talked about the phenethylamine, where the indole world is the companion. Phenethylamine is in dopamine, norepinephrine/epinephrine, adrenaline/noradrenaline, chemically oriented. Not pharmacologically oriented, but chemically oriented, those neurotransmitters. The neurotransmitter for the indoles is the indole neurotransmitter serotonin, or 5-HT as it is often abbreviated: 5-hydroxy-tryptamine.

I had mentioned dimethyltryptamine, DMT, and 5-methoxy-DMT. There is no trivial name for these materials. I mentioned psilocybin and psilocin. I'll talk a little bit more about them in the cultural sense this hour. Another compound that is often bounced around in the same area is a compound called "bufotenine" or "bufotenin."

It is literally 5-hydroxy-serotonin. It is a very close analog. I mean, it's 5-hydroxydimethyltryptamine. It is N,N-dimethyl-serotonin. It's an amalgamation, you might say chemically, of this compound and this compound. The two are being put together. It is a compound that does occur in the body, but in extremely small amounts. It is a compound that is found in many, many plants. It is one of the few alkaloids that bridges both the plant and the animal kingdom. For those who bounce around in the Latin world, "bufo" stands for toad. And it is a component that's found in the excrement of toads. So it's one of the strange compounds that is an alkaloid—it contains a very basic nitrogen, and it's found in both worlds. It's found in many of the plant sources of the snuffs—cohoba, yopo, epená—the various snuffs I had been mentioning in the last hour.

STUDENT: It is correct to call it an alkaloid then?

SASHA: It's correct to call it an alkaloid, because it's found in the plant world and contains the basic nitrogen. Although the definition of alkaloid sometimes can be very rigid and sometimes very loose, I would comfortably call it an alkaloid, even though it's found in animals. It's found in humans, and there are other alkaloids that are found in humans.

It was well bounced around. The idea was called "toad venom," and as a matter of fact, there is a toad—a genus *Bufo* toad—that is found in Arizona and New Mexico, the venom of which contains this material and it is actually collected and dried and smoked. So, if anyone says, "I understand you were spending last Saturday smoking a toad," don't be too startled if you find he is from the Southwest. However, bufotenine itself is a strange material. It is classified legally as a hallucinogenic compound. It is probably inactive. If you give large amounts, it is called hallucinogenic, because of work that was done in the 1950s in which two physicians injected two prisoners who were in remission . . . that's not the way you call it. What is a term with an *r* that means you committed a crime and you got out and you're not going to do it again? Or you did do it again?

ANN: Recidivism.

SASHA: Rehabilitation and recidivism. The prisoners were trying to undo this kind of thing by getting in the good graces of the people in the narcotics hospital in Lexington, Kentucky. They were injected intravenously with the bufotenine in 1, 2, 4, 8, and 16 milligram amounts. They just left the needle in and dribbled in more and more and more.

At the highest levels, one said, "Oh, this is causing an interesting thing. The field of vision is becoming yellowish. I'm seeing things as if through a yellow filter." The other said, "I think I am going to be ill."

"And the yellow?"

"No."

"Are you feeling ill?"

"No."

At which point the one who thought he was going to be ill was not ill. The one who saw things the yellow color felt a slight pounding in her head, felt a slight incipient headache. They stopped the injection at 16 milligrams, shut down the entire experiment, and reported in *Science* that the material was a hallucinogenic because it caused color visions.

Well, if you look at the report rather closely, it did not cause color visions. It caused what for all the world might have been a crashing intraocular pressure. The eye was being distorted by some sort of a pressure—there was a headache, there was a blood pressure rise, and there was a strange discoloration, which is not uncommon for a color formation from intraocular pressure to be seen in early stages of glaucoma.

I mean, how many people have looked at a moon on a clear night and seen rainbows around it when there is no reason? Aha! Be careful: you shouldn't. Because there is no reason to see rainbows around the moon unless there is moisture up there in the air. If it's a clear night, there shouldn't be. And if you see it all the time, or you look at a bright streetlight and you see rainbows, be attentive. This could very well be a structure within the eye that is misbehaving and you're getting pressure and you're getting refraction in the eye.

You've had it? How long has it been since you've had intraocular pressure measured?

STUDENT: How long have I been seeing this?

SASHA: No, how long has it been since you've had the intraocular pressure measured inside of your eye? Ever?

STUDENT: No.

SASHA: How many people have never had their intraocular pressure measured? I won't raise my hand. Whoa, never! That's half the class. Of those who raised their hands, how many people have had eye exams? No one. Okay. Next time you get an eye exam, have your intraocular pressure measured.

STUDENT: Do they do this with all eye exams?

SASHA: A good competent person will automatically do it, like a good physician on a checkup will measure your blood pressure, it's automatic. They have a neat little thing. They put a little whatsit against the white of the eye, and they say, "Now, this is not going to hurt," and then suddenly there's this little *bing*, and it's like a little hammer that comes down and hits the white of the eye. It's an extremely small thing and you're not aware of it, your reflex happens, too late, it's all done.

STUDENT: Kind of like that air thing?

SASHA: It can be done with air too. What they do, they deform the white of the eye, just for a moment. But they measure how long it takes to start to deform, and how long it takes to completely deform, and how far it deforms. By some sort of a magic little mathematical game, they can tell what the pressure is inside the eye. No entry, no pressure, no pain. It's almost humorous. But have it done because the syndrome of glaucoma comes from a sustained intraocular pressure. It can be relieved when it is seen developing by merely letting more drainage occur from the eye, letting the pressure be reduced. But once glaucoma sets in, you are headed toward blindness and there is really no easy way out. It is one of the few really preventable mutilations and damages that can be detected by a very simple test.

Anyway, if you see those rainbows up there, and you always see them, and bright lights always have them and you get these color refractions, be attentive. You may well have intraocular pressure, and it can be relieved at an early stage and keep you from having permanent eye damage.

So, is this material a hallucinogenic? No, I don't think so. Two people in an environment such as that can hardly make what we call a statistically, soundly available sample. Nonetheless, this report came out in great time. They were framing the drug bill just before 1970, and it went on the list as a hallucinogenic because their report was included thusly. So, you'll see it as another indolic hallucinogenic, and that was dutifully taken up in every textbook that is now used in medical school. You have a generation of physicians who believe that because of the textbooks they studied, bufotenine is hallucinogenic. I don't think it is. I don't think there's evidence to support it. But it's one of these interesting little things that once it gets said and gets reinforced, it becomes gospel on the basis of frequent repetition.

What other compounds are in there? There are many variants of the tryptamines. I'll just give a hand-waving picture. There is a dimethyl at the end of the chain. The diethyl is active as are diethyl psilocybin and diethyl psilocin. They have been used in clinical studies, primarily in the work of Hanscarl Leuner, who was at Göttingen in Germany, and had done a great deal of the early seminal work in the using—and the publishing of the use of—drugs in psychotherapy. The longer chain materials and all of the psilocin analogs are orally active. All the short chain materials and DMT and 5-methoxy-DMT and whatever activity bufotenine has are not orally active.

In fact, I had a personal friend, a psychiatrist of strange taste, who wanted to test this bufotenine thing in his own way, and I had a little vial with a certain amount of the bufotenine oxalate in my collection. He said, "I'm really anxious to." I said, "Okay, try it." He snorted it. The medical term is "insufflated it"—he took it up the nose, which is called "snorting" in most circles. And he went through my entire vial of about a gram of bufotenine oxalate and got neither oxalate poisoning nor anything from the drug. So, I think there is one living example of the material not being active, at least by that route in that amount.

Okay, I wanted to get a bit into another subject today. I think the term is commonly used. It was part of the glib use of the '60s, but it's not really understood by many people. A lot of drug effects depend upon set and setting. The term "set and setting." "Did you have a good set and setting?"

"Nah."

"Did you have a good set and setting?"

"I had a pretty good set and setting."

What does it mean? There are terms that are the jargon of psychology and, in some ways, of this form of psychopharmacology. They're not easily understood. I want to take the terms apart.

The term "set" is very explicit. It is what you have in mind that a drug experiment will bring. It is pure preparation. It is your viewing of what's

LECTURE 20 ~ *Psychedelics III*

going to occur. You have an idea that you're going to have a good experience—that is your set. You have an idea that it's going to be a bummer, and you kind of wish you didn't do it, except your current girlfriend says, "Hey, you know, this is just the thing to do, and I'm going to do it anyway. Aren't you going to join me?"

"Yeah."

So you'll do it. You'll have a negative set going in. A lot of people around you have said this has been very whatever-it-is, and you have that as your image of how it's going to be. That is your set in that instance.

The setting is quite literally the environment in which you do it. And I'm going to talk about both of these, to some extent to try to take another view toward the area of psychedelic drug use. Both will profoundly influence the type of experience that is had. It can profoundly influence your physical response to the drug. Any drug going into you, or into anyone, does not produce the same thing each time, because you are not the same person, or that person is not the same person, and the environment and the set—what's expected— and the setting—where the event will occur—is not the same every time.

So, to say something is the property of this drug, it's going to have this type of response in humans? Be careful. "They say this is what you can expect. This is the dosage." Well, dosage is a little closer to it, because you'll find you can get dosage by taking 100 people and getting 100 responsive answers and finding what their average dose was and say that's the average dose. But is it going to be a good or a bad thing? Is it going to produce this or that effect? That's not built into the drug exclusively. It's built in the drug and its relationship to you.

Think back a couple or three lectures ago when I was talking about the use of peyote in America and in northern Mexico by the Huichol and the Tarahumara culture. I talked then about the plant, how the plant was gathered, how the plant was used, how it was taken. I talked a little bit about the other components that are in the cactus and the general orientation of that entire thing from the drug point of view. Let me have you review it from the cultural point of view. You're dealing here with a group of 5, 10, or 30 people, and they are bonded together. I mentioned that the use of tobacco is very, very great in the peyote culture. Tobacco is one of the shared drugs, and tobacco goes around the entire group. The peyote is one of the shared drugs, and it goes around the entire group. You are in a brethren-type bonding of people, and as a consequence, the drug is almost incidental to the set and the setting in this case, the setting in which it is used.

They're going into it not with the knowledge that they're going to have a drug experience. They're going in with the knowledge that they are going to be able to address questions to themselves; they are going to interact with people they know well; they are going to reinforce that bonding between them and others in a cultural unit. If you did the whole thing without the drug, you would probably have the same consequence. You have this in many social situations where there is not necessarily a drug to be used. So, that is the set and setting. So you say, "What is the action of peyote?" What is a scientific, a clinical evaluation of the action of peyote? You will not duplicate that demonstration of its action.

Let's go back to the Mazatec. This is the Indian tribe and the language that is used in the area of Oaxaca. Oaxaca is a major state just south of Mexico City, in the southern part of Mexico, just where it begins to bend out to become the Yucatan. And this is the area where the Mazatec have used the Teonanácatl, which I did spell out for you. Teonanácatl—the flesh of the gods. Here again you have a religious-cultural use of a drug. I'll describe it. This is from Maria Sabina, who is now dead, but was the shaman who introduced Gordon Wasson to mushrooms. He is also now dead and was the first person to have deduced that this mushroom culture must have existed, and if it had existed, it still exists. This is back in the late '40s, early '50s. He and Hofmann from Sandoz and Heim from the Sorbonne, the University of Paris, went into this area of Mexico and found and met Maria Sabina, who showed them the entire cultural experience and its use, in both religious and social things.

Her comment is, "There is a world beyond ours, a world that is far away, nearby and invisible. And there is where God lives and where the dead live, and the spirits and the saints; the world where everything has already happened and everything is known. That world talks. It has a language of its own. I report what it says. The sacred mushroom takes me by the hand and brings me to the world where everything is known. It is they—the sacred mushrooms, that speak in a way that I can understand. I ask them and they

LECTURE 20 ~ *Psychedelics III*

answer me. When I return from the trip that I have taken with them, I tell you what they have told me and have shown me."

In this particular culture, it is a talking from this world. I've mentioned other cultures where different things occur. But in this case, it is a going into this world. Here is a shaman who will go and find answers to questions. You say, "Well, you can do a lot better with a T3, T4, and a liver panel." Yes, you can talk about health, but these people do not have the systems for running liver panels, nor do they have the cultural background that invests them in the certitude of that way to determine a person's health and future. They have other ways, and in their culture, these other ways have proven to be as valid. There are instances, of course, where they would not be useful. We will say in our culture with great positiveness, there are cases where there are unhealthy situations that cannot be resolved by this type of inquiry.

They, on the other hand, will say, "Yes, but there are instances in which we have resolved questions that you have not been able to resolve," and they are right also. So, you must be open to all of this cultural aspect. I'll show you the anti-cultural issue. This I gleaned out of a Spanish missionary report in 1656, about the very same culture.

"They," talking of the Mazatec in this case, "possess another method of intoxication which sharpened their cruelty, for if they use certain small toadstools, they would see a thousand visions, and especially snakes. They call these mushrooms in their language 'teonanácatl,' which means God's flesh, or of the devil whom they worshipped. And in this wise, was that bitter victual by their cruel God, they were houseled.[21]"

Namely, it was a complete viewing of the same thing from the point of view of a culture that had another religion and another approach to medical problems. I might say, the Spanish in the 17th century had a practice of medicine which I would almost want to be put totally in the hands of the Mazatec as opposed to the Spanish at that time, if I had a serious physical problem.

Another interesting thing I had found. Those who have dabbled with the New Testament—I imagine many have, but I don't want to take a show of hands—the Gospel According to St. John starts with this phrase: "In

---

[21] "Houseled" means "administered the Eucharist to."

the beginning was the Word and the Word was with God and God was the Word." Interestingly, the entire essay of the chant that goes in the Mazatec, in conjunction with the use of the mushroom in ceremony, the beginning of the testimony is "In the beginning was the Word and the Word became Flesh." Directly out of the Mazatec. A very interesting parallel. It did not, as far as I know, come from the Christian ethic, because at that time they were still using a language that had not been fully known until about 100 years ago, in being understood or being translated.

The ceremony always occurs at night. It is believed that if you use the God's flesh, the teonanácatl mushroom, during the daytime, you'll go mad. It is always at night, and it's in the dark. The mushrooms are gathered from the day before to two nights before, at which point a pilgrimage is made out in the area with the morning, first signs of morning and the first appropriate breeze. I had mentioned the idea of a certain physical purification in the peyote rite; there's a certain natural environment that's very necessary in the teonanácatl gathering, and that is of a certain air, of a certain temperature at certain conditions, but it also is a very social and cultural thing.

The mushrooms are gathered, they're brought in, they're allowed to dry a slight amount the evening that they're used. They're always put over a copal fire. You'll find a great many references in the anthropological texts to copal, which is the burning of a somewhat resinified, almost petrified, tree sap that burns with a very characteristic sweetish smell. As with the peyote going over the cedar fire in the peyote religious and social ceremonies, the mushrooms are always put over the smoke of the fire in the night experience of the Mazatec. And they're always eaten by twos. They are eaten by a family. They can be eaten in groups. They can be eaten by the shaman him- or herself. It is very much of a together experience. Again, the point is to address questions, to find out answers that come, they say, from the mouth of the mushroom or from the mushroom, but actually come from within yourself. The answers are all in there. It's an access to that form of the unconscious.

Another culture. Again, this is all in the set and setting approach, which drug is almost incidental, it's the seed of the culture. The culture of ayahuasca. The drug I was talking about just at the end of the last hour. The material in it is a substituted harmine; you'll find beta-carbolines. Ayahuasca

LECTURE 20 ~ *Psychedelics III*

is found in the western highlands of the Orinoco. Even more so over the ridges into the western headwaters in Peru and northern Peru. Largely in northern Peru on the eastern slope of the Andes and in the Amazon grange, each into western Brazil.

Do you know the concept of *watershed*? This is something that is kind of a neat little thing. They say that the Rocky Mountains are the watershed of the United States. The term "watershed," are you familiar with it? It's a high point that if you were to drain a cup of water on one side, it will go into one ocean, and drain a cup of water on the other side, it will go into the other ocean. A very interesting thing. I never quite came home to the Andes as the watershed of South America, it is almost on the western coast. There is very little drainage as you have in the western United States. The Sierras are not the watershed in the United States. The Rockies are the watershed. The Sierras draining east still have the water going into the Pacific.

But in Canada, you have a third ocean called the Arctic Ocean. Ever realize that you have, therefore, a whole new pair of watersheds? That which separates the Arctic from the Pacific and that which separates the Arctic from the Atlantic. Hence you have three watersheds in North America, because you have three oceans. And if you have three watersheds, you have a tripoint—the point in which you can stand and spit in each of three directions and have what little bacteria to get rid of go to each of three oceans. Any idea where it is? It's in central Canada. It's the headwaters of the Athabaska, which is found in Jasper, a big national park, north of Montana. But I discovered that I never realized that there has to be a third ocean. There has to be a whole new drainage in the Athabaska, and it is one of the major drainages within the Yukon. Of course, it's the one that drains north too.

Anyway, enough of watersheds and the Andes. This is the northern part of Peru. There is a group of Indians, the Jivaro, who are users of ayahuasca. Here's another cultural group that has an entirely different type of attitude toward the religion-social-pharmacological view of drug usage—set and setting again. To them, the natural world is totally unreal.

Plug yourself into this. They see the natural world around them as being unreal. All of the interactions they have with this world, and their origins, their fates, their health, and their sickness, is all due to the supernatural. It's all about the supernatural. It has nothing to do with the world out there.

THE NATURE OF DRUGS

And the only way you address the supernatural is with ayahuasca. So they are stoned all the time. They are. They use it Monday, Tuesday, Wednesday, Thursday, Friday. One in four of the male members of the Jivaro are shamans, and probably more are who won't admit it. Not so many women, but when a woman becomes a shaman, she's an especially powerful shaman.

So there is a great investment of this in their whole life. From birth to death, it is an investment in interacting with, understanding, and coming to peace with the supernatural that controls their entire life. What we see as the real nature—trees, birds, flowers, food, the works—is all incidental to the idea of conducting your life and surviving in this very complex world. Their complex world. Our world is complex in a different way. But they've got their complex world. So here you have a drug that is invested in an entire living process. They will occasionally take a pharmacological break from this rather harsh material, and in essence, wallow in an unreal world, which we would call the straight world.

I'll tell you, anthropologists who come into this area to study them come out changed anthropologists. To really get to know a group of people and be accepted by them is to become, in a sense, them. There is an absolute need of identification at the personal level, at the speaking level, at the emotional level, with a group to understand them, because they do this to understand you.

I often offend people by saying, "I really believe that to understand a murderer, you must murder." People say, "I don't have to murder. I know what goes through that guy's head." You don't know what goes through that guy's head. You have to be in a position where you are provoked to the point that you would kill, that will actually bring out that part of you. I don't know how many of you have killed. Some possibly have. But you've got to be able to realize that each and every one of you has that capacity in you to kill, and you think, "Not me. I'm into pacifism of the Quaker religion," the whole thing. No! You have got that built into you, but you've got it under good control. Maybe. You have incidents like, "Oh, that son-of-a-bitch cut in front of me, I'll get him!"

Occasionally it can come out and be expressed in bits of anger. We have little straw men we put up to defuse this anger, so we go and punch it out in the punching bag or shout at the umpire in baseball or whatever it is, ways of diffusing that anger. Beware of the person who doesn't have a target to defuse the anger. The anger is part of you. You've got it. I mean, there are things that really will provoke it. Take a mother with a newborn and find that newborn caught underneath a car that may run over it, you'll find that mother will go and actually kill, I mean, actually will if necessary, to keep that child from being hurt. There is that kind of instinct. It is there.

Well, these are people who get rid of this by means of ayahuasca, which is a very intense, strong body experience. I mentioned nausea. I mentioned the diarrhea. I don't think it's just the bacteria in that area that leads to the bowel problems that the anthropologists there come back with. What little they have left harbor the biggest collection of worms and parasites known. It's a really rich area to spend a year in.

I know two people who have spent a year with the Jivaro. One of them I knew a bit before, the other I didn't. One is a man, a shaman in New York who calls himself Michael Harner. The other is a woman anthropologist from the southern part of the state whose name is [Marlene] Dobkin de Rios. Neither of them came out exactly the same person that went in. They interacted and they learned and they saw that aspect of it and to report on it they became, in part, that. You do that every time you deal with a new culture. You deal with a new experience, a new environment. You go in there, and the only way you learn it is to become part of it, and to become part of it, you also change. There is no neutral, objective observer in any of this area. To observe, you participate. A little editorial; we're gonna go aside.

I want to get into another cultural area that had the use of a drug in a different way, the Greek Eleusis. There are the so-called Eleusinian mysteries that have always been bounced around. Eleusis is a little town about 15 miles west of Athens. Kykeon is a drink that was used in the Eleusinian mysteries. It was a mystery school. The concept of a mystery school was not a school as we know it, or like the "Mystery House" down in San Jose or wherever it is, where what it is is not known. A "mystery school" comes from the Greek word which means to shut the mouth and shut the eyes. Not seen, not spoken about. Private. Secret. Covert.

ANN: Occult.

SASHA: Occult in the literal sense. *Occult* means it is not viewable at first glance. Maybe you take it apart and look and you can find what is in there, but it is occult. This persisted for hundreds of years, and no one knew what was going on inside of this large chapel thing in Eleusis because no one could talk about it. To talk about it would result in whatever you do to people who talk about things that are occult and you're not supposed to talk about them. And yet people came from the entire eastern end of the Mediterranean for whatever went on in there. There are songs. There was the thought that it occurred at night. There was a drink involved. The entire thing culminated in an orgy of dancing and singing and whooping and hollering the following day in conjunction with the sun coming out.

Someone said the sunbeam struck the so-and-so, so you know there were a lot of visual things involved. The mystery has been kind of unraveled into some sensible form by the work, again, of Gordon Wasson[22], who is the person that brought the Mazatec teonanácatl into Western culture. He unraveled some of the few records that had been found about this drink that was used, and one of the key components of it very likely was ergot—a material I want to talk a little bit about.

Ergot, the black horn. I forget its name in German. Often it's called a spur. It grows classically on rye, as a fungal thing on the rye; the whoozziwhats with seeds and the stalk sitting up there. But some rye has a little black spur system that's on it. This is known as "ergot," called "black rye." It was thought—up to about 150 years ago—that the black spur was part of the rye. They looked at it; it tasted like rye; it could cook with rye. Some had it; some didn't.

But there was a series of plagues. It's hard to tell exactly how many, because a lot of the different types of plagues were only recorded by certain observations. Of course, no one knew how to give a differential diagnosis of a plague. But the plague, "St. Anthony's Fire," the holy fire, swept through the western Asian area, Eastern European, Central European, Western European, periodically. A kind of connection was made with the rye. Yet this black growth, the spur on the rye, was not connected with this specific malaise. The ergot grows on a number of things other than rye. It grows on almost

---

[22] R. Gordon Wasson popularized this idea, but credit for the proposal that the *kykeon* may have been based on ergot belongs to Carl A. P. Ruck and Danny Staples who brought this to Wasson's attention.

LECTURE 20 ~ *Psychedelics III*

any grain. It grows on some grasses; the fescue can handle it. And it has been implicated in certain miscarriages and bad damage in grazing animals. Yet this is a fungus, as indeed the teonanácatl of Mexico is a fungus, a mushroom in that case. In the case of rye, it's a more base fungus, but still fungal growth. The fungus is known as *Claviceps purpurea*. Its common name is "ergot." Yes?

STUDENT: What was St. Anthony's Fire?

SASHA: St. Anthony's Fire is named for St. Anthony, who was made a saint shortly after he died back in the 1000s. There was a chapel, I believe in eastern France. St. Anthony's chapel was named for him, and he was sainted because he was associated with the chapel that handled this religious fire-damning. People would come to it with this burning. There's a common name given to the ergot poisoning; it's called "holy fire." It was expressed by putrefaction; it caused gangrene. It would close down circulation to the hands and the legs, the hands more often. You would have cramping of the fingers and cramping of the feet, and circulation was cut down, and the stench was unbelievable. The tissue would rot, it would fall out—dead tissue without yet causing death. Death would follow. So you had infections that would come in from the gangrenous rot, and there was extreme burning.

There was a term, "formication," which is a neat word if you like strange words. From "formica," "form-," which means "ants," and "formication" is a medical term for the feeling that ants are crawling underneath your skin. Ants or animals underneath the skin. You can't see it, but actually the visual input from the ergot can cause a fair amount of visual distortion. You can't actually see that the skin is wiggling, but you have the awareness that there's something underneath your skin. This is in the early stages of it. Burning, intense burning, very painful, and, in essence, a rotting of the tissue from the gangrene.

This is largely the western expression of the ergotism; the eastern was convulsive or what was commonly called—instead of necrotic—called "convulsive ergotism." In this you would actually have epileptic-form convulsions. You would have muscular spasm. In some areas, you'd have both the tissue rot and the convulsion. It would often pass. It would recur a number of days later. Not always fatal. But very often there was fatality, especially from the necrotic effects.

Necrosis: necrotic. Nice word in here. There are two terms. I don't think we're going to get into either form of spider, but things that cause damage: "lytic" for lysis. "Histolytic" is a nice term because "histology"—the study of tissue. "Histo"—dealing with tissue. "Lytic" is to lyse, to take apart, to dismember. You are lysing tissue when you are destroying tissue. The bite of the *Loxosceles reclusa*, the brown fiddle spider.

STUDENT: The violin spider.

SASHA: The violin spider, the fiddleback spider. *Loxosceles*. It is a spider in the western states. It's one of our two poisonous spiders in California. We have the black widow in this area and the *Loxosceles* in Southern California. Brown spider, small body, legs unduly long for a small spider, and if you look very carefully—I would not suggest it—you'll find on the light brown part a black or very dark brown figure on the thorax that has been said to look like a violin, a sort of ill-shaped hourglass type of figure. The black spider with red or orange underneath is known as black widow. This is the other comment in this area. It is characterized by extremely strong webs; webs that are strong and resilient and will spring back far beyond the strength you'd expect a web to have.

ANN: Which is it?

SASHA: The black widow, that's what is in our area, the black widow. I've never seen a *Loxosceles* here.

ANN: I don't believe they can live this far north.

SASHA: They said the same thing about *Anopheles* mosquitoes. I suspect they're in the area or coming into the area. They've been reported north of Santa Barbara. Just a matter of time. By the time they get here, you may very well have the same thing. By the time the killer bees get here, they are so interbred with so many other types that they have lost the signature that makes their approach a panicky thing.

But the fiddle spider is a histolytic poison. You get a bite, and it actually erodes the tissue; you go right down to the bone in a little volcano cone of eroded tissue. It's a hard thing to control because the only way you keep it

LECTURE 20 ~ *Psychedelics III*

from continuing the damage is to remove the tissue, at which point you've damaged and mutilated the person.

The black widow spider is a neurolytic poison. It goes after the nerves. It causes nerve destruction—I've been stung once that I know of, and probably twice. You have this very sore, very difficult handling of the muscles, and you have a slight swelling but a very intense neuropathy, nervous irritation and pain. Neither is considered particularly lethal. Both are very disruptive.

STUDENT: In small kids they are.

SASHA: Small kids, same thing with many of these types of things. You give an amount of venom into a small organism, the concentration is too high.

STUDENT: Also, there is this one that just in from the coast of Chile. It's in Los Angeles County. *Loxosceles laeta*. It has a lot more venom.

SASHA: I don't know the spider.

STUDENT: It's similar, it's just a different species.

SASHA: It's not indigenous to here; it was brought in?

STUDENT: The only place you'll find it is in California, in Los Angeles County.

SASHA: Not aware of that one.

STUDENT: I heard ergot being connected with the witches of Salem?

SASHA: Not valid. Better to look into the henbane and the datura group—what grew on the east coast—the jimsonweed. That's the usual thing that's associated with that, where you get the illusion, the confusion of the datura. The ergot probably, and the argument has been made that it participated through some fescue use in the brewing of this broth that was used in the Eleusis. It's a speculation. You're not going to find a record of it that can be pinned down. But it was known, has been known for a long time, that these types of compounds in modest amounts do give rise to a lot of mental changes, a lot of visual changes, often lights and sparkling, sensory changes and interpretive changes. Those that survived the St. Anthony's Fire survived it with mixed memories—the pain and agony were bad, but the mental effects are recalled with fond memory.

The ergot alkaloids—ergonovine, ergometrine, ergotamine, about 10 or 12 major alkaloids—are present in ergot. Each of them, by the way, matched with a very close ally. Almost all of them have an iso counterpart. They are in pairs. Some of the isolates are mixtures of alkaloids and their pairs—different ergots, different concentrations. And some of them are extraordinarily effective as vasoconstrictors, and these are the ones that lead to the coolness, the itching, the discomfort. I think one way or the other, when your circulation closes down, each person has his own way of knowing it. Coldness of the extremities is probably one of the first marks of this. Some of the very earliest LSD work—all this is working up to the discussion of LSD—is an outgrowth of the ergot studies. Some of the crude LSD preparations that were made available on the illicit market early in the time of the Haight-Ashbury were largely ergot with only a little bit of LSD, and the chilliness, the feeling of coldness . . . [The tape dropped the rest of the comment.]

Ergot has been used as a postpartum snapper-together of the uterus, or used in predelivery for uterine muscle control to cause the uterus to begin closing down and begin to apply pressure of smooth muscle to expel the fetus. Or in the case of a clear birth, with the afterbirth having been cleared, to close up the uterus so that you do not have it open to infection, and, in essence, to terminate that aspect of the internals of the childbirth. So, if you have a pregnant person or animal that is not at the labor stage and at the partum stage, then contraction of the uterus will destroy the developing embryo. And so yes, miscarriages and uterine damage, offspring damage, are common.

Yes?

STUDENT: Is that given sometimes during pregnancy or during childbirth to the women?

SASHA: Yes, very much so. It's one of the—

STUDENT: Are they actually getting high?

SASHA: Yes, there is a certain amount of a psychological thing that

LECTURE 20 ~ *Psychedelics III*

comes along with it, often very much confused with the psychological turmoil that is going on at the time of childbirth anyway. And so, it's often overlooked, but it's a very real contribution. That is one of the minor arguments that can be made against the use of LSD in pregnancy when there is an uncertain purity, because you are mucking around with a material that comes from a source that can carry quite a bit of smooth muscle contractive capability and hence, uterine contraction. Other things are used in the area, I mentioned other drugs.

There is a drug known as Sansert. Its chemical name is "methysergide." It is used as a prophylactic for migraine. I know one person, a woman of about 30, 32, who had migraines and had tried a number of prophylactics and was not handling them too well. It was suggested that she use Sansert. From the drug point of view, she was a totally naive person. Had never used drugs. As far as I know, did not even smoke or use coffee at any point. At about the third or fourth day, she said, "You know, this is making me feel funny." She was quite open about the fact that she had migraine seizures and she was taking a prophylactic against migraine. "Feeling quite funny."

"How are you feeling funny?"

"As if I didn't know if I was walking straight or not." I love probing, because she clearly was getting an LSD effect from this medication, and Sansert will produce an LSD effect. But she had no idea what LSD was, so it was a rare opportunity to get a naive voicing of what was being interpreted by a person who had no background to do that interpreting. And these were her phrases: difficulty in walking straight, thought process perfectly fine but almost lightheaded and not liking it. Thoroughly not liking it. Very often—in fact, more often than not—when a person is administered a psychedelic drug—a drug in this area—and is not familiar with that type of thing and is not told that the drug is there, you get a person who is poisoned. These materials are poisonous, and I consider it to be totally irresponsible to give a drug to a person without that person being aware of set and setting, being aware of what's going to occur, what the drug is, and what might be expected to come out of it. It is irresponsible in my eyes for anyone to give a drug to another person without their knowledge.

This gets into a very interesting outgrowth of the set and setting argument. I'm going to get back to LSD for the moment. Let me continue that

THE NATURE OF DRUGS

argument, because I've heard people say, "Well, we cannot ever really condone peyote, ayahuasca, ololiuqui, or *Psilocybe* use until we have studied it scientifically, until we know what it really does to the human person." So you say, "Well, let's not look at the Mazatecs in Oaxaca. Let's not look at the Jivaro in northern Peru. Let's not delve around in the Native American Church and snoop around teepees or hogans in the central, southern, western part of the United States. Let's go to the laboratory and do it scientifically and look at exactly how it's done in the laboratory."

Part of the early work on LSD was done by a person named Dr. Sidney Cohen, at the VA Hospital in Los Angeles. Some early work on mescaline was done by a psychopharmacologist named Ron Siegel, also associated with the University of California, Los Angeles. Quite a bit of work was done on the East Coast by people at Princeton, people at Yale, people at Harvard. It was done in Illinois; Hartford; Carbondale, Illinois—a lot of the early research work. It was done clinically. The work done in Los Angeles was done in a clean room—in a sensorily isolated room. It was done with a physician with a stethoscope and with a white jacket and a blood pressure device at hand. According to Siegel, "I would never conduct an experiment with mescaline unless I had at hand—not in sight because it may not be conducive to a good experiment—both straitjackets and IV drips in case we have a real medical problem."

So a lot of this early study work that was done in the medical field was done in this type of environment, which is hardly the group-togetherness, around the fire in the middle of the night, all knowing that they're going to go into the same area and have it as a bonding experience. It was done by one person, the volunteer, and the one physician, the experimenter, with him or her watching him or her (whatever the sexes might be in any order). What comes out of it is a report such as, "There was an element of panic in the early portion of it. It developed into a blend of paranoia, loss of self-identification, and their affect was largely negative. There was a rise of X millimeters of blood pressure. There's a certain amount of urinary this, a certain amount of eye that." You know, anything that can be measured, probed, opened up, looked into, or developed was done and reported.

In the studies that were done by Cohen, in a report to Macy Foundation about 15 years ago, he was amazed that he had almost a 90 percent record

LECTURE 20 ~ *Psychedelics III*

of people having completed the experiment who chose not to repeat the experiment with LSD. A 90 percent denial of a repetition.

A group was studied using LSD or Sansert, and psilocybin, psilocin, and three or four of the amides of LSD that were analogous to LSD. By the way, Sansert—as opposed to LSD—takes 10 to 20 milligrams, and so it is several prophylactic dosages. It nonetheless is a prescription drug and has been used by some psychiatrists in place of LSD. All these experiments were done in his living room with volunteers he knew personally, with music going and in a social environment, and his batting average was about 85 percent of the subjects chose to repeat the experiment. So, I think there is a contribution to the experiment, again, by both set and setting.

I mentioned briefly the Lexington hospital. This was an outgrowth of something I had mentioned in the development of the law that addiction was a crime. In the passage of the Harrison Narcotics Act in 1914, there would be no interference with the practice of medicine. But addiction—which comes from, in legal eyes, the use of heroin and cocaine and the other things that were handled under that law—became a legal problem, not a medical problem. Hence the treatment of an addict is a criminal event and not a medical event. Hence a physician who treats an addict is committing a crime. This is the way they, in essence, turned the entire law to make the medical community separated from narcotics addiction. Well, it produced a situation in which people who were criminals but needed medical help couldn't get it, because they couldn't find a physician who would face the harassment that came with treating narcotic addicts.

So, they developed what are called "farms," narcotic farms. There was one in the Midwest near Fort Worth, Texas, and the major one that still exists is in Lexington, Kentucky, where you put people who are criminals, but their crime is the use of drugs and being addicted to a drug. So, you have a captive population of prisoners, and they have been the raw material for a great deal of the human work that's been done on hallucinogenics and on narcotics.

There was a wide-open hearing. I think it should be widely published. It was published by the federal government and scattered to whoever is an archivist that collects these sort of things. Hearings were conducted into the way in which these studies were done in the prison there, and it was a rather interesting reward system. Stop and think what kind of a volunteer you have who is (*a*) drug-oriented; (*b*) a prisoner in a situation where they want to get out—most prisoners choose to get out. So they can buy goodwill by volunteering for an experiment in which a drug is used. They're being called upon because of drug experience. They will buy goodwill, which will eventually lead to an early release, and rewards, by the way, for volunteering for drug experiments. The experiments were often designed in such a way that the person who is being worked on, the subject or volunteer—I use the term loosely—knew what drug was being used and what was expected.

There are patients in VA hospitals who are there quite consistently and often have no choice. They do not choose to go out in the real world, because they have a family, an environment, in the VA hospital that supports them and feeds them, and they are very willing to cooperate with physicians.

"Just tell me what you want, doc. You want to experiment on me. Go right ahead. Here's my arm. You tell me what you want to find. I'll find it."

It's a very cooperative population. It is in this hospital too.

"We wonder which of these narcotics leads to a higher euphoria. Let's try one today and one tomorrow."

Yet they're both pretty good, and you get this kind of a reinforcement of preconceived notions. It is the opposite of a double-blind experiment. Plus, the rather, to me, outrageous thing, and this was brought out in the hearings, arguments were made as to whether this was ethical. The person said, "There's no law against it, and it seems to be a proper way of running it." The rewards for successful participation and volunteering were drugs. They were given chits that could be turned in at the pharmacy for drugs of choice. If it was a long experiment, they got four chits, and it was worth two shots of heroin; a small experiment, two chits worth one shot of heroin and one Demerol or whatever they want. They can take the chits and turn them in for drugs. So, in essence, the concept of these being prisoners on drug crimes, they are in an environment in which drugs were available, where they could be earned, and where they were being experimented with.

This is the Lexington Hospital in Lexington, Kentucky. It still exists today. This is not done today, I presume. I've not been there. But the testimony was, "Yes, we did this." Harry Isbell is probably one of the major names associated as being an MD who was in charge of research there. "We

LECTURE 20 ~ *Psychedelics III*

did it. It is neither unethical nor is it illegal." Yes?

STUDENT: What was the name?

SASHA: Harris Isbell. Now retired. If you are a newspaper man, get an interview with him, it would be a fascinating story to get.

So what are you going to learn about what the effect is of peyote, of mescaline, on humans? Of psilocybin on humans? Of the teonanácatl? Of the ayahuasca? Of these various drugs on humans? You can go into the culture, you can go into the area where they're used—and they're getting harder to find because of our acculturalization of these—and find out how they are used and what they're like. This is the world of the anthropologist. The physician doesn't go out there in the culture and expose himself to various parasites and stinging things and biting things and into an area where he uses the drugs too. No.

You will not find that area of research from the medical point of view. The medical point of view is the white-walled clinic with the IV drip and the sensory isolation. In this country, or in the medical environment, the so-called scientific method has nothing to do with the drug as it has culturally been used, but they will measure the blood pressure changes and the dilation of what all.

Where do you get this information? "Well, we'll synthesize the two together." But they are two different experiences. They are two different worlds, and they are not to be synthesized. At least no one has succeeded in bringing them together. You have the entire morning *Chronicle*—how many people read the morning *Chronicle*?—four pages on the 20-year review of the Haight-Ashbury. Read that from the point of view of the anthropologist and the sociologist—not from the point of view of the physician where every drug under the sun was being used, and everyone was stoned and using too much meth and LSD and scatter, scatter, scatter. Read from the point of view of the generation of love, the love-in, the social aspect and the cultural aspect and the reinforcing-of-one-another aspect. It is an interesting survey of a very interesting time that played a great deal of influence on our cultural development.

So anyway, the ergot is this collection of alkaloids. I'm not going to draw the structure. It's a great big four-ring monster with bumps going out right and left and double bonds in and out and aromatic rings. But ergot is a family of materials that have in common a ring system that is known as the erg—not ergine, argh! I've forgotten the name of the parent ring system. But the parent compound underneath it is called "lysergic acid." Lysergic acid, a white, easily discolored, light-sensitive material, again, in this pair—lysergic, isolysergic. Lysergic acid and isolysergic acid are called epimers of one another chemically. They're due to the inversion of one little hydrogen at one position. And so, they're very closely allied and they're often found in mixture. You synthesize, you usually get them both in some appropriate ratio.

Lysergic acid is not found in nature in its free form, but it is found as many amides, and these amides are called the "ergot alkaloids." It can be just plain ammonia, ergine; it can be a hydroxyethyl up there; but most of them are amino acids, usually di- and tripeptide, mostly tripeptide amino acids. A very complex amino acid glump that is connected with an amide to this big lysergic acid glump. The combination together are called "ergot alkaloids." Here are the materials that cause the vasoconstriction. Here are the materials that cause the smooth muscle contraction. The materials that cause the capillary constriction that makes them effective in certain headache situations—migraine or prophylaxis migraine.

Now, if you discard the amino acid collection and just take the lysergic acid moiety itself, then there are amides that have been made into lysergic acid, some of them by biological systems. There is a *paspali*, an organism, a culture of fungus that will put on a hydroxyethylamide in a broth, and that will produce over a milligram of ergot alkaloid—of this particular alkaloid—per milliliter. This is a gram per liter. And in a 22-liter broth this produces 22 or more grams per cycle, and you cycle in a matter of minutes. So the ergot alkaloids and the ergine does not depend upon going in the fields of eastern Bavaria and picking little black things off of rye. It goes in big vats, and Italy is one of the prime producers of these materials, where there are drug houses in Italy that produce them by the ton, and it does not take many tons to produce many dosages of LSD, as it is active in a matter of micrograms. So, in essence, you're dealing with an immense amount of material that can make an extremely potent substance.

Of these amides that have hooked on, the most famous is the diethyl,

LECTURE 20 ~ *Psychedelics III*

and hence the D of LSD stands for diethylamide. L-S-D: lysergic acid diethylamide. Acid in German is *säure*. Our English word is based on that—sour. The sourness of acid, *säure* in German. LSD is *lysergicsäurediethylamid*.

It is amongst the most potent of those that are known. It has its companion, isoLSD, which is not active. The pure material is free of iso. The crude material when made illicitly has both materials present. It is active in man at probably as little as 10 or 20 micrograms. And the typical dose range is between maybe 50 to 150 micrograms. How much is a microgram? The dust that is on my hand from having done that to the chalk is probably 100 micrograms. I had one instance of a good friend of mine who I worked with industrially who had weighed out a sample of LSD for me but had not realized it was a sample of LSD. He became quite curious because he had always heard that LSD was odorless, colorless, and tasteless. He came in one time after having weighed out 3.278 or whatever milligrams, he said, "It may be odorless, and it may be colorless, but it's not tasteless. It's slightly bitter."

"How do you know slightly bitter?"

"Well, there was a little dust left on the spatula." He thought he was weighing psilocybin, because I had two vials. You could see the vials are identical in appearance, but he had not bothered reading the label because he was quite sure what it was, and he tasted the spatula. Well, how much solid do you get from the dust that's on the spatula? I have no idea.

He said, "Well, it's slightly bitter." I can tell you now, quite honestly, that LSD is slightly bitter in one person's experience. So we tried to deduce how much he had taken and we guessed probably a few tens or a few hundreds of micrograms. Well, a few tens would be an interesting but very benign experience probably, correctly handled. A few hundred can be a very busy experience. But I'm familiar with the territory. He trusted me totally, and I said, well, we're going to have a busy day.

We were in a very big industrial laboratory near Antioch, and we spent a very busy day. That's exactly how you handle an experience like that, where a person is totally naive—in an area where they do not know where they are going. They know that you know where you are, they know where you are, and they trust you. And you go into that territory, and that's how you handle overdosages in this case. If you have the time and the patience in an

THE NATURE OF DRUGS

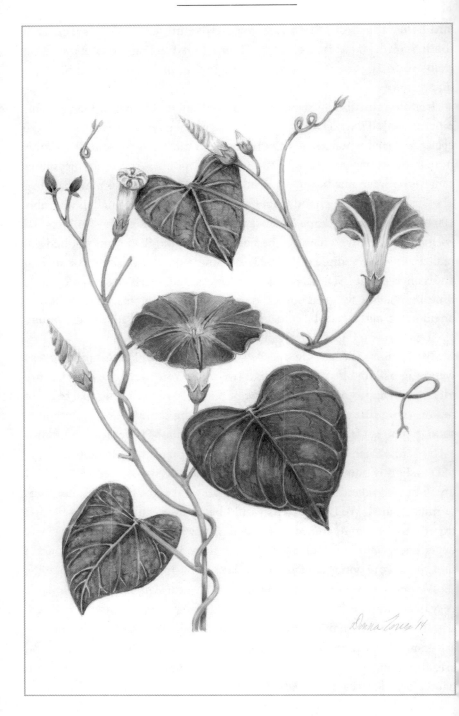

LECTURE 20 ~ *Psychedelics III*

emergency room, a person comes in stoned to the eyeballs, they are crocked to the gills: "Where are you?"

"God, I don't know."

Next thing you see, he's gazing out there, following things falling on the ceiling. Join them! Don't tell them lies; they'll see through it. The cognitive structure is totally intact. They just don't know where they're going psychologically, nor do you. Don't say, "You will be all right." How do you know if they will be all right? You don't know. So don't say it will be all right. Just say, "I've been there. And let me tell you, it's got some rough spots and got some interesting spots." Join them, walk. Be with the person. Get into that territory. Let them begin relating back to you and walk with it. Then tide them over until the body takes care of it, metabolically the liver chews this up and pees it out. The level of the drug drops and the level of anxiety and paranoia often can go very quickly with this kind of very close association and togetherness.

That's how you handle this experience. If you are in a busy hospital, you shove them full of Haldol. Okay. They're not bothering you. They're sitting over there with glazy eyes kind of looking at nothing, maybe somewhat comatose. Yet out of your way, and they'll get over it. The body will still grind up whatever it is and pee out the metabolites with Haldol present or not. But the person is in a really interesting, fascinating, frightening—it can be very anxiety producing—paranoid possibly, psychological place. Don't lock him up. Don't stick them in a straitjacket. Don't stick them on an IV drip. What are you going to put in the IV drip? You don't know. Haldol? It's going to help you. It's not going to help the patient.

Talk to that person. Be with them. Spend the time with them. And that, I think, is an outgrowth of the paramedical experience that came out of those days in the Haight-Ashbury of 20 years ago, in which a lot of people got into bad places, and other people who had been in bad places knew it, and they took their hand and they walked and they talked. That is the quintessence of how you treat a bad experience in this general area.

So, LSD is a very complex thing. It is a white, colorless, odorless, not quite tasteless, crystalline solid. Very, very fragile. You expose it to light, it will decompose. Photosensitive. You expose it to moisture and light, it will decompose because moisture adds to the double bond. There is

THE NATURE OF DRUGS

LECTURE 20 ~ Psychedelics III

a material called *lumi*, L-U-M-I, Lumi-LSD, that is a process of adding water or alcohol to the double bond that is this little chemical spot in the molecule that destroys its activity. If it isomerizes, it destroys. If you put it out in the light, it will very rapidly oxidize from the air, photochemical oxidation from the light and the air. Any moisture and it will go in three entirely different directions. I would say, a little bit on a moist day in the open light, white crystalline LSD put out into the open sunlight, you would probably not have any LSD left in probably a matter of minutes. That kind of sensitivity. If it's in solution, it may well be seconds when exposed to UV. Put it in the UV light in a spectrophotometer as a water solution of LSD, and you can watch the spectrum change as you're running the spectrum. Very, very sensitive.

The products of it are inactive.

ANN: How do you take it if you can't mix it with water?

SASHA: If you mix it with water, you don't stand out in the sunlight.

ANN: Water is okay?

SASHA: No, ordinary water is not all right. LSD is only dealing with 20, 30, 40, 50 micrograms of material, and water contains chlorine. Chlorine is a superb reactant with the double bond and with the 2-position of the indole. There is a trivial amount of chlorine in water, but there's a trivial amount of LSD in the water too, and the chlorine will inactivate the LSD. You cannot make up a water solution with tap water. You will destroy some, certainly, of your LSD just from the chemical exposure.

I mentioned ololiuqui. I couldn't find my collection of seeds at home. It is the native name of seeds of the *Ipomoea* genus and of the *Rivea* genus, and now probably extends to the *Argyreia*, morning glories. There was quite a flap about 25 years ago. "MGS," morning glory seeds. It became just the thing. *Ipomoea*. Everyone was turning out Pearly Gates for some reason. The botanical name of one of the principal *Ipomoea tricolor* variants was Pearly Gates, which was a natural for the Haight-Ashbury days of the late 1960s. *Ipomoea tricolor, Ipomoea violacea*. These are the red-on-white morning glory or the white-with-purple-around-them morning glories. Go up on the beach and look back behind you in the hills. You will see morning glory

THE NATURE OF DRUGS

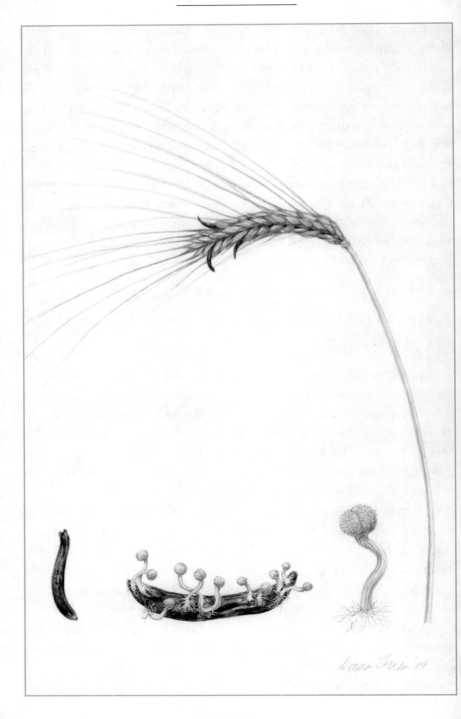

LECTURE 20 ~ *Psychedelics III*

plants—little bell-shaped blossoms, five petals in a bell. Very small.

Who has not seen a morning glory? No hands or maybe one or two. Okay, morning glory, little bell-shaped blossom. Five pointed blossoms, a little seed with three divisions in the seed. The pod contains three seeds. They bloom one day. Morning glories you see blooming today were buds yesterday and they will be seed pods tomorrow. Every day a new batch of blossoms. The seed pods contain these three seeds, which are called in the native usage, again in Oaxaca, ololiuqui, *Rivea corymbosa*. There are a number of Spanish names that I do not know, basically "black seed." They are small, grain-of-rice-like size. Fifty, 100, 200 must be chewed up, eaten, and swallowed. Very difficult on the tummy. You get upset. You can feel that you are carrying a load. You are carrying a load. You're carrying a couple of hundred ground-up, chewed-up morning glory seeds.

The components are ergots. It is another source of ergot, but for the first time in a higher plant. Not a fungal source, but actually from the *Convolvulaceae*, a higher plant source. This gave a kind of a closing-of-the-circle for Hofmann, who had first made LSD in 1938 and discovered its activity in 1943 and was involved in finding the active component of ololiuqui and discovered—having gone through psilocybin and the Mazatec Indian experiments and all of the *Psilocybe* mushroom and that fungal area—had completed the circle back to the morning glory that contains the ergot compounds, very similar to those that are found in the fungal *Claviceps*.

It's the hour. I don't know where I intended to go; that's where I got to. I may go on a little bit more with LSD if there is any more interest in that as to its variants and the type of action. And we'll take one week vacation. Have a wild week, and Tuesday, a week-and-a-half hence, we'll continue with whatever is the title of the lecture for next week, which I don't know. Good.

LECTURE 21

April 21, 1987

Prescriptionals

SASHA: Okay, 2, 4, 6, 8, 10, 12, 14, 16, 17—and they rally.

We have two more lectures in the general area of drugs—this lecture and the next lecture—and then we get into the serious stuff. We'll get into the more general, broad things. We're seeing the end of the course. It's getting brighter at the end of the tunnel. I think you're aware of that as much as I.

This is lecture 20, 21, something like that. I have lecture number six written up and ready to go. Number seven is half done. If I can get my stuff together, I will have it done on Thursday. I will never catch up. But that brings up a point that I mentioned last hour, about the final.

Let me go back. You already had your midterm. Luckily you all did very well on the midterm. I was very happy to see that. [Laughter.] About the final—the final will be an essay. This will be a written essay. It will be a couple or three pages; it doesn't have to be long. You can get it done ahead of time. In fact, do it ahead of time so that you can hand it in on the last lecture of the course.

The last lecture will be on Tuesday the 19th of May. We're now in April—30 days, so we have about four weeks to go. On that day, hand in your final. Do it ahead of time. You can do it now if you want, but that's absurd. Do it a few days ahead of time and write it out. It will be on a subject that the course has dealt with. As I mentioned before, and I'll rephrase it now, try to find something with which you disagree with me.

I'm sure that if I have not offended each of you at one level or another, at one time or another, I've not done a good job. That's part of what I'm trying to do. Find something that—when I said it—you cringed and said, "Mama wouldn't like that" or, "I don't go along with that" or, "That's contrary to my philosophy" or, "I learned something different elsewhere" or, "I read in a book that so-and-so was saying something opposite to that." Present something that is contrary to what I've been saying, and defend another side. Defend your side. When you say, "I don't think that marigolds are gold because I read in an encyclopedia that they are blue, because they found a blue dye in them and they say they're blue." Okay, give a citation of the encyclopedia, if that is the source you want to use. Do something to try to get me to change my lecture—which I'll do if I buy that. "By golly, I was wrong. Marigolds *are* blue. I was misinformed. I've learned quite a bit about it." In my next lecture or when I write out the notes, marigolds are going to be blue, at which point you get an A. If it turns out that marigolds are gold, and you misread the word because it was a translation from Sanskrit, etc., I'll say, "No! I'm going to hold that marigolds are really gold and not blue." I'll give you a B. You tried, but you didn't convince me to change my lecture. [Laughter.]

Don't give me the false compliment that everything I said is so sterling correct that you can't find one hole in it somewhere. I'll give you some holes.

ANN: You said before that if we didn't counter your argument on something, that we could add something that you don't know much about?

SASHA: Add something that I don't know. I want to learn.

ANN: So, it doesn't have to be disagreement.

SASHA: It doesn't have to be something I said wrong. It could be something I should have put in and didn't. Give me something I can learn from. If I learn from it, A. If I don't learn from it and it's already in there and you didn't read the notes, fine, B. And if you don't hand in an essay, that's an F. That's fair enough.

So a C will be the average curve. A bunch of As and Bs and maybe two Fs. It's an unusual way of grading, possibly. But it's very—

STUDENT: The essay doesn't have to follow a format?

SASHA: Oh, good heavens, no! I would like to have it typed unless you have very good handwriting, because I'm not very good at reading script. And English is the preferred language. That's it. Actually, no format. If you can do a nice typed little page, that's marvelous. Don't make it 10 pages long. There are 30 of you and one of me.

LECTURE 21 ~ *Presciptionals*

ANN: What's the latest deadline?

SASHA: There is no latest deadline. There is a time that it's due, which is the day of the last lecture. Okay, any questions? I think it's going to be one of the few courses with all As. [Laughter.]

STUDENT: You can make a claim about being a great teacher.

ANN: Yeah!

SASHA: Okay, we have this lecture and Thursday's lecture, in which I think Thursday was assigned to marijuana. I spell it with an *h* and a *j*. It's an interesting story in its own right. I was supposed to get into prescription drugs in this lecture. It occurred to me when I was coming back from Lone Pine, where I spent the last three or four days, that I had never even discussed endorphins and enkephalins, brain neuropeptides that are like morphine but are generated in the body. They are a big chapter. Hadn't even thought about it when I was setting up the course. So I want it in here. If I'm going to have neuropeptides in here, that means I've got to have polypeptides, that means I've got to have amino acids, which means I've got to have DNA, which means I'm going to get into one of my threesies discussions, which is going to be far away from either LSD or from prescription drugs or marijuana. I have a feeling that for the next two lectures I have a bunch of notes and no way of connecting them together. So I'm going to start going, and I'm going to bounce around as the mood and the questions change my course, and then we start next week with something outside of drugs—I can't remember what it was.

Okay, one thing on the show-and-tell, prescription drugs—oh, no, let's go back even further. LSD, we were just finishing up with LSD in the last hour. That was a week and a half ago, which I can barely remember either. Someone said, "Are you going to talk more about LSD?" And I said, "Yes, because I haven't said much about it." And I can't really remember much of what I had said, but let me give some background of the material, a little bit about LSD.

LSD, a good way of remembering—the German word for acid is *säure*, s-ä-u-r-e, from which our English word "sour" comes from, the taste of lemons due to citric acid and other acids. Lemons are sour. The German word for acid is *säure*. Lysergic acid diethylamide is LSD. It would have been LAD if it had been in this country.

It is not a natural compound. This is one of those little myths that has come in periodically: "This is natural LSD as opposed to synthetic LSD." It is not a natural compound. It does not occur in nature. It has never been observed in nature. I did talk a little about the ergot and ergotism that has given rise to that entire area of pharmacology, but ergot itself is the major source of the starting material for LSD. All LSD is synthetic. All LSD is made from lysergic acid. It is converted from an acid to an amide. All lysergic acid comes from natural sources. The principal natural source is the ergot alkaloid, which I had mentioned as being the spur on the various grasses, fescues, and grains that are found largely in the southeastern European and central European area.

Lysergic acid is tied in with ergot alkaloids, and you usually isolate alkaloids from these fungus sources—the source is a fungus. The alkaloids ergonovine, ergometrine, ergotamine—there are a variety of different names, and all have a medical use, but they can be hydrolyzed to lysergic acid, which in turn can be chemically converted into LSD. LSD was first synthesized in 1938 by a person by the name of Hofmann in Switzerland at a company. Someone said, "I have never heard of that company." It's a well-known one. I'll write it up here—Sandoz. It's a major pharmaceutical house in Switzerland. I once had an opportunity to talk to the research director of Sandoz, and I asked him a question I'm sure he had been asked by others.

He said, "LSD has been quite a notorious development from Sandoz."

"If you had to do it all over again, if you saw something like LSD and you knew it was going to become notorious the way it has been, and it had that type of action, that type of metamessage, would you do it all over again? If you had to do it all over again, would you?" And he thought for a moment and smiled. I'm sure he had the answer all prepared.

He said, "Yes. We would. Because although LSD has brought a lot of strange reputation to Sandoz, and has brought some administrative grief, before LSD no one had ever heard of Sandoz. Now they have." And Sandoz has become one of the major—probably one of the half-dozen major pharmaceutical houses in the country—in the world. It's Swiss by country, but

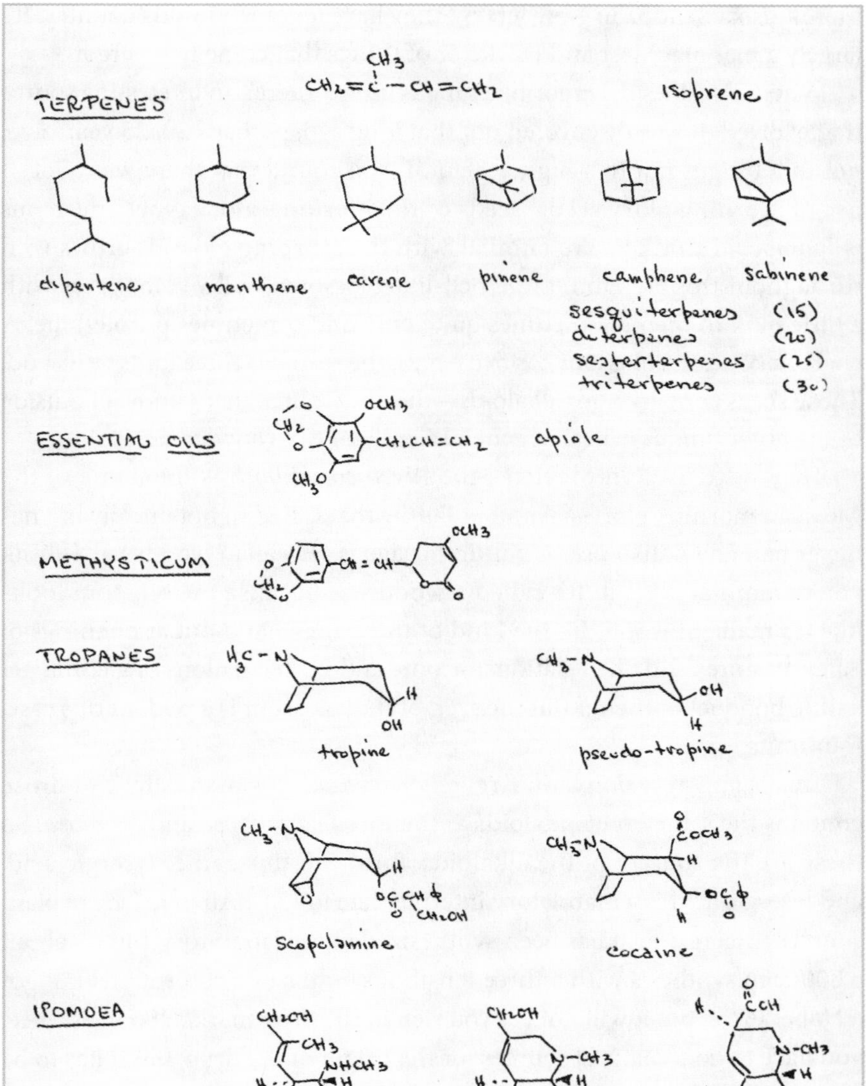

LECTURE 21 ~ *Presciptionals*

TERPENES

$$CH_2=\overset{CH_3}{\underset{|}{C}}-CH=CH_2 \qquad \text{isoprene}$$

dipentene   menthene   carene   pinene   camphene   sabinene

sesquiterpenes (15)
diterpenes (20)
sesterterpenes (25)
triterpenes (30)

ESSENTIAL OILS   apiole

METHYSTICUM

TROPANES   tropine   pseudo-tropine

scopolamine   cocaine

IPOMOEA   chano-   elymo-   lyserg-

A. muscaria   ibotenic acid   muscimol

there's also a Sandoz in New Jersey. They have a branch in this country. It's largely a manufacturer and marketer of things that come from ergot.

Where else does the ergot-like fungus grow? There's an interesting source that evolved. It was discovered not that long ago, perhaps 20, 25 years ago, not in a fungus but in a higher plant. It was found that there was a lot of use of morning glory seeds, seeds of the morning-glory type. The genus is *Ipomoea*. I think you're familiar with the morning glory. It grows wild throughout this area too. Little bell-like blossoms, spiky sometimes with a little pink in them. Sometimes quite pink and sometimes purple-blue. A characteristic: the blossom has five petals, the seed has three units to its pod. Those seeds contain ergot alkaloids—the only source that I know of outside of the ergot fungus is in that general family of *Convolvulaceae*.

They were used in Mexico—the *Rivea corymbosa* is the name of the Mexican morning glories. Another family that is tied in botanically but has never had any native use or cultural usage is Hawaiian woodrose. I think you're familiar with it. It's called a woodrose because the blossom looks like it's made of wood. It's the kind of thing they hand out at openings of Safeway stores and Shell stations. "Come and get ten gallons of gas and get a little bouquet of them." But the seed of that blossom is a pod, in this case, four units.

This is an *Argyreia* plant: *Argyreia nervosa*. Hawaiian baby woodrose contains the same ergot alkaloids as the *Rivea corymbosa* and *Ipomoea*. So these are the sources of the alkaloids. But from this comes lysergic acid, and lysergic acid is a mandatory intermediate to LSD. All of these are plant sources. Lysergic acid has been synthesized in the laboratory, but it's about a 500-step synthesis with a three ten-thousandths of 1 percent yield. It got a Nobel Prize, but it will not get you rich in the drug market. So in essence, you have to go to natural sources for the origin of the drug, but it has to be made in the laboratory.

LSD is an extremely potent compound. It is often said to be the most potent psychedelic. It is not. There are materials that are more potent. It is active in humans in a few tens of micrograms. When trying to estimate the weight of a microgram, it's a hard thing to estimate, or to look at. A little bit of dust of chalk will be a few micrograms. Typical dosage would be somewhere between 50 and perhaps 200 micrograms. You're dealing with something on

LECTURE 21 ~ *Presciptionals*

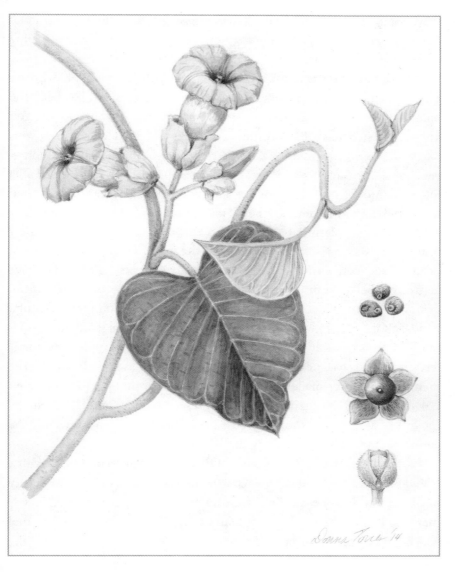

the order of a tenth of a milligram, which is a ten-thousandth of a gram. There are 28 grams to an ounce. How many people have a gestalt of what a gram is? Most. How many people have no idea how much a gram is?

Okay, a couple, three. Go put a teaspoon into a saltshaker and bring out the tip of the teaspoon full of salt, just the amount you get on the tip, the amount you could balance on your fingernail in a little pyramid of salt. It's maybe a gram. Of that, if you had ten thousand grains of salt in that gram of salt, each grain would be 100 micrograms. That would be the kind of feeling of the LSD amount that would be active.

LSD has an extremely negative reputation with many people who are not that familiar with its personal use in the drug scene. It has been the focus—it has been the projection—of everything that's been negative. You know, "Psychedelics are interesting, but the trouble is once you start with psychedelics, you get into LSD." Yahh! And you get this negative response, yet people who have used LSD realize it is another drug in this general area for study and for investigation. It is a Schedule I drug. It was broadly used in the 1960s and was put into the 1970 drug law, placed into Schedule I.

It is not particularly lethal, not particularly dangerous physically. I know of no death that has come directly from the use of LSD. There have been deaths associated with LSD usage, but usually they are traumas that have come from behavior or from actions that were not under conscious control or under cognitive decision processes, such as if a person jumps out of a window. One of the most famous ones was Art Linkletter's daughter—that put him on the crusade against drugs in general, the death of his daughter—in which she came out of a window, presumably as a result of LSD. The mention was never made that there was a six-month separation between LSD usage and the suicide. But he was so hurt by that, and rightly so, by the loss of a daughter who was in the drug-usage scene, that he made the connection that LSD is the thing that started her on her downfall.

As a matter of fact, an outgrowth of that was an interaction I had with a friend who was the head of the Department of Psychiatry at the University of Chicago at the time, a fellow by the name of Freedman. Daniel X. Freedman. He is now the head of psychiatry at UCLA. But he was on a panel which Art Linkletter was on, and it was one of these roving panels telling the facts about drugs, this and that, trying to dissuade people from drug usage. And Linkletter was a very—he is a stage person, he is a public person. He's a good speaker, he tells a good story. And he was telling about the trauma, the negatives of drug use.

At this particular panel, Danny Freedman—who is kind of a wiry psychiatrist, a curious ally in a strange, interesting, and a very maverick way—was listening, sitting right alongside Art Linkletter. They were getting into a

LECTURE 21 ~ *Presciptionals*

discussion, and another person was climbing on what Linkletter had said, and was giving the message more and more clearly that the use of drugs was destructive, this, that and the other, and Linkletter saw his entire pattern being warped and turned around and being presented to the audience with an extreme stance that was not his original intent at all. And the next thing, someone in the political area was climbing up on this particular presentation, and the whole thing moved over into politics. And Linkletter sort of had this little strange light go on over his head, so to speak.

He turned to Danny Freedman and said, "Danny, do you think there's some element of politics or some form of administrative policy that has woven its way into this drug education program?" Danny, bless his heart, turned to Art and said, "It never occurred to me. Tell me, tell me why you think so?" [Laughter.] That was the last panel that Art Linkletter was ever on. He has dropped out of that scene entirely now.

But there is a blending of politics and drugs, which I think we've touched on before, and we will again. I'm going to go back and forth, especially as we get toward the end of the entire lecture series. It is about things more political, more social, more environmental, all of them in their own ways destined to somehow influence your behavior. And that, of course, is the heart of what drugs are. They are there to influence behavior, but so is radioactivity and so is—it's astonishing—fundamentalism. One way or the other, it's all there to influence behavior, and that's what I want to get into.

So, I want to wind up this whole area, LSD and prescription drugs. Prescription drugs, we're going to get toward that. I came across a thing I should have mentioned if I didn't in the very first part of the hour—I have about seven things to touch. I spent about 10 minutes trying to get an arrow going from this to this to this as a logical way of getting through the seven things. Every time I do it, I've left one thing out. When I go to the one thing I've left out, I've messed up the pattern, so I have no way of getting through these next two lectures in a rational way. So I'm going to bounce.

Okay, prescription drugs. I found a thing in the newspaper just three days ago, on the dollars that were spent in 1987 on prescription drugs. On the top five drugs, and I think this says a great deal about us, drugs, our country, our culture, and where we are going. The first of the five drugs—this is wholesale shipments, by the way. I'm sure that not many pharmacists

THE NATURE OF DRUGS

sell these things for wholesale. Almost $400 million value—Tagamet, the big in thing for ulcers right now. Three hundred seventy million dollars—Inderal, for hypertension, one of the biggest beta blockers for the treatment of high blood pressure. Number three, Zantac—$290 million wholesale, ulcers. Number four, Dyazide—$280 million, a diuretic, something that makes urine flow more, gets rid of fluids by urine flow. And number five, Tenormin for hypertension—$250 million wholesale. The five leading drugs, every single one associated with stress—two for ulcers, two for hypertension, and one is a diuretic, which deals with hypertension. We're dealing with blood pressure. It's kind of an interesting thing. There is a wholesale of over a billion dollars of drugs, all dealing with blood pressure.

STUDENT: You said this was in 1987?

SASHA: 1987.

STUDENT: So far?

SASHA: How in the world can they do that? How can they know it is top selling? We read copyright 1987, Universal Press Syndicate. Source: Pharmaceutical Data Service. I don't know. I can't unravel it. I wrote "the dollars in '87," but obviously as this is '87, it can't be. Let's make it '86. I'm probably a liar, but the music is about there. Maybe it is so far in '87. I don't know the answer.

Yes?

STUDENT: Does it go by the cost of what it is to sell?

SASHA: Wholesale cost.

STUDENT: But that's not saying how many prescriptions are actually given. Because I know that Tagamet costs a lot more than Zantac. So if you're just going by cost—

SASHA: I'm sure the costs are probably five times this. But that's the kind of thing—

STUDENT: But what I'm saying is that you're not really saying how many prescriptions were given out.

LECTURE 21 ~ *Presciptionals*

SASHA: How many pills? No. How many prescriptions? No. How much is used? No. This is the money in pocket. I don't know the validity. I just found the paper and I thought it a good thing to file under prescriptions. The prescription role of drug usage came in legally through an amendment to the Food and Drug Act. It is, in essence, to protect people from misuse of drugs and invest the knowledge of how drugs are to be used in the hands of a physician. Triple prescriptions are clearly for keeping from misuse of narcotics and things that are "high abuse potential" and only relevant to their medical utility. But the number of things sold on prescription, the abuse of prescriptions both from the user's point of view and from the physician's point of view, is monstrous.

I came across this, just by chance a few days ago—talking about drugs—let's talk for a moment about Ritalin. Ritalin is a stimulant. Ritalin is one of the major stimulants of use for what's called "minimal brain syndrome," MBS. What is it called now? There is a new term now for misbehaving kids.

STUDENT: Hyperactive.

SASHA: Hyperactive kids, but it's ADD, something or other—

STUDENT: Attention Deficit Disorder.

SASHA: Attention Deficit Disorder. A kid misbehaves, in a sense he gets too active, and he is unruly. "How do you keep Junior quiet?" The school wants to send him home. How do you keep him quiet? Well, they have created a syndrome. Syndromes don't come on trees. They are created. "You have a disease that is a bothersome thing." We look upon it as a disturbing thing. Let's say a small child is too happy, he does not pay attention. He is unruly. He is a nuisance. It has been found that stimulants in small children have a paradoxical response. They tend to quiet the child, and Ritalin is one of the in things in this whole area right now.

I found a report on Ritalin, that about three times the amount that's sold in New York is sold in Alabama, and 50 percent more of all that is sold in Alabama is sold in Atlanta, in the suburbs. They made a little inquiry in Atlanta, and they found that the parents have been sold a bill of goods to pressure the teachers into getting happy-and-willing physicians to prescribe

THE NATURE OF DRUGS

Ritalin to students. The amount that is used in the student population in Atlanta is unbelievable.

STUDENT: I have a friend who was hyperactive as a child, and he swears that the reason he is a drug addict now is that he was constantly being given drugs from the time that he was a little boy. And he has a problem with it now. He's been through detox several times, and he swears that it started because of his parents.

SASHA: I totally buy it. I totally buy that. Something will modify your behavior. You're going to do better. Shut up and take the drug. You get into the habit of feeding stuff into your mouth as a drug to produce an effect. You're building on early into this process of associating drug use with an issue, and drug use is a pattern that is deeply invested in.

Ritalin—I came across a reference, I didn't get the name. One MD, I'd love to know his name and where it is. I'm sure he's no longer in this role. I believe the date was the early 1980s. One MD handled 2.5 percent of the national output of Ritalin. One MD.

I found some interesting things on methaqualone, which is another thing in talking about stress. This developed a whole process in New York, which has now been stopped. But boy, for a while, there was no way of stopping it. They were called "stress clinics," in which you go to a stress clinic, you get 30 to 60 methaqualones, 300 milligram tablets. Methaqualone, as I mentioned, are Quaaludes, "ludes," the 714, sopors. They're still very much in demand. Ten minutes in a stress clinic and you get a "script," which is a prescription that goes to a pharmacy, and you get it filled for your methaqualone or Quaaludes. It costs approximately $100, a typical price for your 10 minutes for the script. Ninety percent of all the methaqualone that was sold in New York came through one clinic.

Interestingly, this was a little bit more difficult to show, but of all the prescriptions that were made by that one clinic for methaqualone, 50 percent of them were filled at one pharmacy. It turned out there was $1.50 per hoozie coming back to the pharmacist; there was $100 for 10 minutes coming to the physicians; they were putting ads in the newspapers for physicians who had licenses to practice to be consultants at their stress clinics. You say that drug abuse is something outside of the medical community? That is about

LECTURE 21 ~ *Presciptionals*

as flagrant an abuse of the prescribing license as you can have. Now they re-relocated methaqualone from Schedule V, minimum control, what have you, to Schedule II. That was pretty much stopped by the need for triple prescriptions. But methaqualone was a broadly abused drug.

I was on a trial case in San Francisco in which a methaqualone factory had been under surveillance. I learned a great deal about the skill, by the way, of both the narcotics officers of the government and of the state in staking out and watching an operation.

This was down a little alley in San Francisco. There's one that goes along beside the *Chronicle*; Minna, I think it is. There's one the next street over, I don't know its name. There was a warehouse, and apparently they had this thing under surveillance for 72 hours. A bunch of really wily people were in that warehouse, and they were making methaqualone. They had a pill press. They were there making buckets of the stuff and putting it in the pill press; they were running for 48 of the 72 hours that they were under surveillance. They never tumbled to the fact they were under surveillance.

And I was quite impressed with the skill of what was going on. Someone would leave to be picked up—they had it tied in with a house out in the ocean area—someone else would leave to go down the freeway. They had a policeman pull them over because of a faulty taillight, check the license plate, and say, "Get your taillight fixed." They had it all set up for the surveillance, so they knew where everyone was going, when they went, who they contacted. They actually observed a robbery in process, because outside of this warehouse one fellow parked his car and some guy came with a giant bat, at two o'clock in the morning, broke into the car, took out the stereo, portable thing with big speakers from the back of the car, and went off with it. All this being dutifully recorded. [Laughter.] Seventy-two hours of intensive high surveillance of a group who are clearly up to their ears in illegal action; they were making methaqualone, and eventually they were arrested and were convicted. And they were discussing in the session that was recorded, whether they should report to the police the theft from the car. You know, they decided that it was not cool!

Methaqualone right in downtown San Francisco. How many labs like this are there, where illegal things are done? I'd never really thought much about illegal labs. There are no such things as illegal labs. This is one of the interesting things you'll find the papers: "an illegal lab was busted on the corner of such and such in suburbia." There are no illegal labs. There are laboratories everywhere, but no legality. Labs are needed; labs do not have to be licensed. Labs do not need a permit. Labs do not have to have the authority to be. You may have a laboratory; there is no legal or illegal lab. There are private laboratories. There are public laboratories. There are university laboratories. There are clandestine laboratories. There are illicit laboratories, laboratories that probably wouldn't fall into the approval of the society if they knew of it. But there are no illegal laboratories. There are laboratories in which illegal things are done, and where people may be breaking the law or doing things that are not in accord with law; these are called illicit laboratories. A clandestine laboratory is a laboratory that you don't want anyone to know about. A private laboratory is a laboratory where you don't want anyone coming in. These are all different and valid adjectives, but illegal does not apply to laboratories. Okay, that was another aside of something you will find commonly misused in the newspaper.

Prescription drugs: I brought in a collection of pharmaceuticals. I was really inspired quite a while ago. I received a letter from a person in Sweden ... let's go back even a little further. Sweden has nationalized the pharmacy industry. There are no private pharmacies in Sweden. They're all taken over into this benign socialism, under government control, owned by the government who will sell from the pharmacy and control the whole thing. Well, there was an embarrassment, because suddenly, the government found a source of income that had been going into the pharmacists' pockets. Mountains of money. And they are not supposed to make money in this way and are only supposed to break even. So they had this big increasing amount of money and had to spend it on areas related to drugs and drug use, what have you, so they began putting on conferences.

So I got called. They said, "Will you come over to a conference in Sweden and present a paper at a meeting on marijuana?" That ties in with prescriptions, marijuana. "Come over and give a talk, give a paper on your research on marijuana." I said, "Sure, I'd love to! But I can't afford to get over to Sweden. I don't have that kind of money." It was a critical mistake. Never say, "Yes, but..." The answer you should say is no. I didn't. I said, "Yes, but..." So, the next thing, in return mail they said, "Here is a round-trip,

LECTURE 21 ~ *Presciptionals*

first class ticket on SAS Airlines, and we're putting you up at such and such place where the conference is being held for five days. We're expecting you and your paper in Sweden in about six weeks" (or whatever the time was).

I really hadn't done any work on marijuana research at all. [Laughter.] So I spent five weeks in the lab. What can I do to make a reasonable presentation? I would love to go to Sweden, it's a nice thing to do. So what I did, I took the whole area I had just been working on in amphetamine chemistry and I modified the chain over here [gesturing]. If you take the molecule, turn it upside down in this way, it looks like the side chain on marijuana. So I'm going to make a blend of marijuana and amphetamine and see what comes out. So I thought I'd mix it up in the laboratory and try it out. Nothing was there!

It was an interesting compound. I worked out the chemistry, but there was no action. Very often in pharmacology and medicinal chemistry, you find that you can take this very neat little active type of molecule that does whatever this is, and take another neat little type of molecule that does that, and you say, "I'm going to combine those two neat little kinds of molecules, and it's going to do both this and that." And what it usually does is neither this nor that. The track record of combining active things, different types of actions to produce something rather sparkling, has almost always produced a dud. It has not worked well. In this case, it didn't work at all well. But this will all tie in.

So I got to Sweden, I gave my paper, and I met a very interesting person named Andreas Maahly, m-a-a-h-l-y.

ANN: M-a-e-h-l-y.

SASHA: M-a-e? Really? Thank you. "My name," he said, "is Andreas Maehly"—he didn't spell it right. [Laughter.]

Have I told this story before? No? Okay, no. "My name is Andreas Maehly. I've always admired anyone who has his own laboratory." He is a 55-, 60-year-old person, gray suit, vest, well-polished shoes, clean shaven, you know, established really. He said, "I have my own laboratory. I'd like to have you see it. I'm very proud of it." I'm not about to wander into some—not illegal of course, but clandestine, private or what have you, laboratory in Stockholm. But I don't want to get myself into a strange spot.

"Well, I don't know."

"But I would be very honored if you visited. We could drop by the Karolinska on the way and meet some mutual friends."

He must be okay if he has mutual friends at the Karolinska Institute; we did stop by there, and then we went down to his laboratory. To my amazement here was this entire block in downtown Stockholm with a two-story building that filled the entire block. Two armed guards at the front door allowed us into this darkened laboratory. He was the Hoover of the FBI of Sweden!

He was the head of the narcotics laboratory of Sweden. Good god, I'd never seen this type of equipment, golly! It was dark, and every now and then you'd run into a guard who then disappeared. We wandered through there. Many things impressed me, but one of the things that impressed me: We went into a room, he pulled out this drawer, and the drawer was filled with a little set of about 10 by 10, 100 tablets in a little grid, and another 100 and another 100 and another 100. He pulled out another drawer, there's another 500 tablets and the room had drawers all the way down this wall, that wall, and all the way around the room. He said, "In this room, arranged by color, shape, and—" one more parameter, I forget which. "There are 77,000 drugs that are legally available or have been legally available in Sweden. We have one of each."

I love it. That's my kind of thing. I love it. There is a white round one, a white square one, white hexagonal ones, pink ones, and yellow ones. And so, if they have a seizure, they can go through this little key. You can go right down the line, and by golly, there it is. If they want to get a reference sample of a drug, they go to the thing, pull it out, break off a piece of the pill, make an extract for GC and a peak. It's a beautiful reference collection. I said, "I want that."

So, I came back, full of fire. I began clearing out detail residues in doctors' offices. Physicians have a strange difficulty, very often. I don't know if you know the term "detail man." I saw some nodding. These are the people who come through Sandoz, Parke-Davis, you name it. They say, "You've got to try this. It's the newest thing we have for hypertension." They dump this mountain of drugs on the physician. You go to a private physician's office and open that closet drawer, you see this cascade of things that have been left by detail

LECTURE 21 ~ *Presciptionals*

men. They hand them out: "Try them out and send a report." Sometimes they are known pharmaceuticals, sometimes they're still at phase II or phase III level of exploratory use. Sometimes the physician sells them happily or they give them to the pharmacy they co-own, but that's another story.

So I get these different things and started the collection. This is typical of a box. I have about seven of these at home—that's 700—and I said, "If Sweden has 77,000, how many do we have in this country?" I sort of gave up and I stopped at that point. I've kept records on all of them. For a while I did narrow it down to only drugs that are illegal, that have been sold in the illegal market for what is called in the trade "ballistics." It's a nice term. It's a term used in forensics. It's when you get a tablet and the tablet is whomped on a machine that can make a lot of loose stuff into a hard tablet. The whomper leaves little marks, like this whomper has a little burr on that side and a smaller burr on this side. So you look at this thing under a microscope, you see a little scratch and a small scratch and you say, "Oh yes, from the ballistics, this was made on machine 43R. The last time we knew of 43R, it was in Newark, New Jersey. We seized it in an illicit MDA lab. We've lost track of it. Now we found a tablet of illegal methaqualone. This comes from the same machine."

But a lot of physicians have a lot of goop that they don't really know what it is. They don't want to throw it in the garbage because a lot of people look in physicians' garbage for these very things. And so, I served a role: "If you'll give it to me, I assure you it will not get into naughty hands." "Take it!"

It's ballistics. You know, with a rifle, you shoot a gun, the bullet goes whirling down the thing. You take a look at the whirling, you shoot a trial bullet and put it in a comparative microscope. You get the same little scars, you say, "Aha! That bullet came out of this gun." You begin tying bullets to guns by ballistics, that's the original use. But you can tie tablets to whomping machines, whatever they're called, by the same tool, by the same sort of concept, and in crime it is called "ballistics." And you will find there are literally hundreds of whomping machines that are known by code and by name and where they were, and where they were last used. You get strange tablets from a strange place, and you'll identify which whomping machine it came out of. And it's used as a way of tying illegal drug preparation and sales and so forth together.

THE NATURE OF DRUGS

So I began collecting. I have a couple of flats of illegal tablets of LSD from here and there, and tablets of this, tablets of that, so that I could eventually begin tying this together. It's not worth it. I abandoned the entire process. But I thought I'd bring one of them as a show-and-tell of prescription drugs. This is from the prescriptional chapter of things.

Also, in the prescription areas: *PDR*. How many people have never heard a *PDR*? Oh good, more than half, you have now. *Physicians' Desk Reference*, *PDR*. If you want to know, it's one of the nicest and one of the often misused sources of information. It comes out every year. You'll find it at our local library. It has lots of colorful pictures of what pills and tablets look like. You'll find it in used bookstores, often with some area ripped out. If a drug is prescriptively available, it is in there. It has information about the size of the tablets, the color of the tablets, the manufacturer's description, how it is to be used, how it is to be prescribed. Beware, the information in the *PDR* is written by the company that makes the drug, so you are not only going to get rather colorful things about how good it is, you will get what is known to be bad about it.

They take every single thing that has ever appeared in the literature that has been negative and they jam it all in there too, so that they can never be accused of not having warned the physician about the negative aspects of this drug. So legally they cannot be attacked quite as easily. But of course, the poor patient who has had a runny nose and has been prescribed something or other, and her cousin, niece, or someone that has a *PDR* goes and finds all these horrible things that can happen and suddenly she comes down with all the horrible things. That's why physicians never put *Your Health Today* in their waiting rooms. They will put *People* but not *Your Health Today*, because sure enough, people will browse into this medical book, read about the latest disease that's hit the trade, the latest thing that is associated with some malady, they'll read into their own self something that is going on, and they'll present with symptoms that they never had when they came into that office. It can really mess up the diagnosis.

There are valid problems with prescription drugs. All drugs have a certain percentage of negative effects. All drugs have a certain percentage of people who respond badly to them. These are accepted as being part of the trade in prescription drugs. Some are notorious for being individually

LECTURE 21 ~ *Presciptionals*

handled or not handled well: antihistamines, certain antibiotics, certain treatments. In fact, we have two treatments for ulcer. It may be competitive, and it may be that one person is sensitive to one drug and not sensitive to another. So there often is a great deal of redundancy in the treatment, in the prescriptional area. A lot of these side effects are disturbing, but they are tolerated. Sometimes antibiotics have rather a high percentage of side effects. Chloramphenicol has some nasty side effects. It's a dangerous drug to use. But there are certain bacterial and fungal problems which it is very specifically effective on, very difficult diseases that are a problem. And so, you use the very dangerous ones with very severe problems. You use the very benign ones with trivial problems, but you use any drug knowing that the drug has a certain probability of negative effects, and you titrate the risk-benefit ratio and choose the drug accordingly.

This is one of the reasons that there is no allowance for any negative effects in the illicit psychedelic, or nonmedical, or should I say, the paramedical, usage of drugs. If there is no virtue in the eye of the person who's making the risk-benefit ratio, if there's no benefit, no risk can be tolerated. It's a very logical conclusion of a ratio in which the denominator is zero. If you had some virtue, you could tolerate some risk. If you had a lot of virtue, then tolerate a lot of risk. No virtue, no risk. Therefore, if there is any negative associated with the drug—as I said, all drugs have negatives associated with them—if there is any negative, then the drug obviously has no justification for being used under any circumstances. That is the monolithic philosophy that has brought the drug laws into their present sense. And what led to prescriptions.

Is the minute hand about right?

ANN: Yeah.

SASHA: Prescription drugs. Oh boy! How many of you have taken biochemistry? One, two, three—how many people have not? Let's get a tally. I guess that's about everybody. Those who have taken biochemistry—amino acids. Familiar? How many people are not familiar with amino acids? Ah, good. More than half.

I want to start by using a concept that I love, because it's an eccentric word—I call it "threesies." This, by the way, is the nicest poster I have in the

THE NATURE OF DRUGS

whole discussion of creationism versus evolution, whereby basically either you've got a lot to explain about evolution or a little bit for the creation side. The fact that we have DNA—the concept of DNA, I think anyone who reads a newspaper is aware of DNA. It's something funny that's way down in there, in a very basic cell that carries the information of what's going to happen. It is a source of genetic information. Humans have a DNA that's, you know, billions of somethings long and that carries all the information of whether it's going to be five fingers or four fingers, or whether it's going to be smart or dull, or black or white, or tall or short, or male or female. The whole cheese is built into that DNA. Everyone has their unique DNA. There are no two the same. Identical twins would make an issue. But generally, there are no two that are the same.

But this DNA is a large sequence of things that are called "nucleotides"—a big word—that are derived from nucleosides, that are derived from nucleic acids like pyrimidine. You can forget all that, I'm going to get a handout. I will have it if you want it. I did give one handout that has a little bit of it, one of the handouts had nothing to do with the lecture that's for today.

All of these things come in threes. That's why I call them the "threesies." You have a nucleic acid, DNA: desoxyribose nucleic acid. Ribonucleic acid: RNA is another companion, which is a little bit further away from the gene information, but it's more practical to be used. They come in threes, and what the threes do, they tell you what amino acid is going to be used in what peptide and what protein.

About 5 percent, maybe it's 1 or 10, a very small percentage of the DNA is actually dictated for protein, and the rest of it has functions that are really not well known. They're not known at all. They call it nonsense, but I don't see our packing around a lot of DNA as nonsense; it's for reasons we don't know. But for the little bit that is used for proteins, you have three of these in a row. The so-called triplet code that dictates this DNA says, "We're going to take this amino acid, that amino acid, then that amino acid, and somehow we're encoding a protein." A bunch of amino acids hooked together to form a protein based on the knowledge of the gene.

Proteins are the building structures of the body. They are all the enzymes. Things that do things to the body are proteins. All the little messengers that convert hormones, release hormones, and most of the hormones are

LECTURE 21 ~ *Presciptionals*

proteins. They're all built from this basic structure of amino acids. There are many amino acids, and all the enzymes and proteins and hormones and messengers and transfer-thises-and-thats, are, in essence, expressed as sequences of amino acids.

The heart of the amino acid is the ability to build a chain of molecules—amino acids hooking onto amino acids—that carry a structure that is not based on the molecule itself. As we've talked about with drugs . . . here's lysergic acid, LSD. It's got this bond and a ring and a thing sticking out—it has a physical shape. It fits in the receptor because the receptor is the opposite of its physical shape. It fits in and does the job. Amino acids are loose wobbly things. When you get a chain-linked thing, you can pull it straight, you can crunch it together. What is the shape of a piece of chain that is 100 links long? Whatever shape you put it in. It doesn't have an intrinsic shape. A molecule has an intrinsic shape. The atoms are attached, they do their vibrating thing, but they are more or less shaped in the form of the molecule. A chain doesn't have a shape.

But the amino acids, besides forming the links of the chain, have arms that go out. There are many different amino acids. Some of them have negative charges; some have positive charges. Some are very water loving, some are very water hating. If you took these amines, these chains, and attach two each of the links of the chain—a little arm that had a plus or a minus property or had a water-liking or water-disliking property—then the chain could condense upon itself and take a different shape. So that which loved water was near something that was like water. And that which was something over here that was positive was near something that was negative. There's a molecule that will take a physical shape due to these very subtle interactions. By changing one link in the chain from one thing to something very near to it, you're going to change the shape just a little bit.

So this is the shape that these polypeptides and proteins have. If you pull them out straight and let them relax in water, they'll reform this shape. That is the shape of the receptor where they are active. We have many of these peptides in the body. I'm going to get back to DNA. I may never get back to threesies. I'd like to, but I'm off into peptides.

We have many of these peptides in the body. One of the major prescription drugs I wanted to talk about was a drug that is going to be a major thing to watch. How many people have heard of the morning-after pill? Okay, fine, diethylstilbestrol (DES). It has now become really a nasty thing, because it has been shown pretty convincingly that if you do not abort, if you do not succeed with the DES, and you have a child and the child is female, and the female child goes up to maturity, the chances of cervical or vaginal cancer are much greater in children of mothers who have used diethylstilbestrol as the morning-after pill.

So it is in the area of female hormones. The whole approach to contraception deals with mucking around with female hormones, and believe me, there is no simple description of what goes on in this whole process. Would you give me a one-sentence description of how a contraceptive pill works? No, there's no way you're going to get into that simply. You have two large areas of hormones that are in the female process. You have the estrogenic hormones; you have the ones that are correlated, progesterone itself. Let me approach this from the physical point of view.

Point to your forehead and point just in front of the ear, into your head, follow where your fingers are going. You're going to go and hit something called the pituitary gland. It is a major unit in the head. How many people have heard of the pituitary? Okay, that's where it is, in there. I had mentioned when I was talking about the portal system, getting from the gut to the liver, the hepatic portal. I said it was one of the portal systems of the body. Those who were attentive would of course ask, "Where is there another?" There is another in the pituitary. It's a very neat little system.

How many people have taken embryology? Only one or two. Okay. Embryology. When you build out the body, you first lay out from a cell, two cells, four cells, lots of cells, big piles of cells, you are in a process that's called "building the embryo." The embryo is a microscopic development of the individual that will then grow to become a fetus, then the newborn child. We have embryonic development where you're laying cells in different ways to form the shape of what you will eventually find. Then you have the "fetal development," which is the development of that shape and growth of that shape. It is during the embryonic development that you get into such things as teratogenicity. I think I mentioned this in the very first lecture, the laying down of something wrong in the cell division. This is one of the fundamentally difficult problems. Another drug I should mention in

LECTURE 21 ~ *Presciptionals*

the prescription area—I think I did—thalidomide. That was the one that precipitated one of the tragedies. It was a very safe sedative hypnotic. It was safe in all people except pregnant women.

If it was taken at the point of teratogenic sensitivity in the pregnant woman, it would cause the loss of the development of the arms, a process known as phocomelia, or the flipper-arm construction. This teratogenicity is a fault in the first exposure, perhaps in the fourth, fifth, or sixth week of pregnancy, when you have not got a fetus yet. You're still developing in the embryo stage. The cells are developing into that form. The cells, the developing embryo, can be very sensitive to certain drugs. The difficulty is you don't know which. You often cannot tell by animal experiments which drugs will be dangerous in humans. Thalidomide's danger cannot be shown in any animal other than a rabbit. Even in that, it's very uncertain. That may not be exactly specific, but there are certain animals in which it will not show that potential, and it was tragic that it had to be shown in humans.

This is why I'm a very strong advocate—if you're pregnant or think you're pregnant and you are in that first trimester of pregnancy, don't use any drugs. Stay away from alcohol. Stay away from caffeine. Stay away from anything but very ordinary food because you are very sensitive to this sort of distortion of that development.

What can happen in development? One thing that happens in development, and this is tied back to the pituitary—take your fist and say this is the gut that is developing, this embryo. Take your fist and say this is the mouth area of that gut, and let your thumb be what's called a "diverticulum," a little thing that's coming up on the mouth area. Everyone take your right hand and make a fist with the fingernails down. And take your little third finger down. This is nothing symbolic, just happens to be a good measure. [Laughter.] This is the brain that's developing, and this is a little stalk that's coming down on the brain. Again, a diverticulum out of the third ventricle, the middle ventricle in the brain.

The brain is developing, and the gut is developing, and then of course they are developing so the roof of the mouth is close to the bottom of the brain. They develop in the same place, and these two touch like these two nails. The diverticulum from the brain actually is grey material. It's called neural material. The gut material is tissue, is gland material, it's called adenoid tissue, and the two touch. A very tight transfer of the fingernail from the gut to the fingernail of the brain. That area is known as the hypophysis or the pituitary.

It goes down out of the hypothalamus, which is really the brain's brain, the thing that runs the brain as well as running the body. And that little organ—about the size of a pea, maybe weighs a gram—is located there. It's held by a stalk that has two portions. There is the front portion called the anterior pituitary, which is the one that came from the gut, and it is the source of all these marvelous little polypeptide hormones: growth hormone, lactating hormone, follicle-stimulating hormone, ACTH that goes and moves stuff in the adrenals. All these things come from that front end, the anterior end, the tissue end. "Adrenal hypophysis" is the medical term. Hypophysis is a fancy name for pituitary. It is fed by a little portal system from the hypothalamus, which is the thing above it. You may have six hormones coming out of the pituitary, but you've got six of these other hormones that are releasing hormones of those six hormones, all of them polypeptides.

In the back of the pituitary you have the posterior, the rear. The posterior of the pituitary is called the neurohypophysis because it's attached to brain. There you have oxytocin, something that affects the tension of the uterus; you have vasopressin, which is a pressure agent and causes you to build up blood pressure and retain urine. These are polypeptides. That little organ that does everything is right in front of the pineal, which has its own world too. That little organ is really the controlling organ of almost all the functional things in the body other than insulin itself, in terms of the pancreas, which is also a polypeptide. All of these agents come out of that little thumbnail-sized organ up there, the pituitary. It's a hard thing to get to. If you have problems with pituitary cancer, you've got a difficult problem because you can't just go opening holes in there. It's at the very center of things.

How do you treat a pituitary cancer? Well, we're going to get into this later on and talk about radiation, which I hold to be a drug. In essence, you can put a person in a field where the focus of the radiation effectively is on the pituitary and destroys the part of the pituitary that may be cancerous. And you know something is going wrong because you suddenly have a flood of these peptide hormones that are completely out of balance.

LECTURE 21 ~ *Presciptionals*

These are the hormones that do indeed control the entire menstrual cycle. This is where you are mucking around with contraceptives. What you do, you have the estrogenic hormones in the uterus, you have a lining—endometrial lining—and that is prepared by the estrus hormones.

In the ovary, you have a development of what's called the Graafian follicle, out of which the egg will come. That developing follicle is encouraged by the estrogenic hormones. It, in turn, will become what is called a corpus luteum once the egg has been thrown out and goes wandering down the fallopians. It may be fertilized, maybe not. Once it's gone, what's left in the ovary is called the corpus luteum. The "yellow body," literally. A yellow body, and that's a source of progesterone. The progesterone then tells the uterus, "Hey, you've got a nice vascularization going on there. You've got nice new spongy tissue. You're going to be getting an egg pretty soon. It may or may not be fertile, but we'll see."

And if the corpus luteum drops away, falls away, you get a little bit of follicle-stimulating hormones. A new follicle coming up over here—the other ovary or possibly the same ovary. That yellow body, the corpus luteum, disappears. Progesterone sort of drops off. The uterus sort of loses endometrial lining. You go into the menstrual period. That is the dumping of all that business that wasn't used.

If you get implantation and you begin getting the development of a demand from the uterus itself, you can maintain progesterone from the uterus, which suppresses more ovulation and maintains the level of the progesterone. Your pregnancy tests become positive, all that sort of thing goes on. What this drug does—RU-486 is a progesterone antagonist. It boosts up progesterone. And so it is as if you were not pregnant, and if there is an egg about ready to implant, the endometrial lining falls, the period starts and the whole washing out system: out goes what may have been a fertile egg. So it is not a day-after contraceptive. It is a month-after contraceptive.

Now the question comes up: in the balance of pharmacology and politics and ethics and society and everything else, is this drug going to be marketed in this country? It will be marketed as of July in France. It does not have a trivial name yet. It does have one name, it's a French name, which I'm not going to give you. I can give it if you want it. Why should I not give it to you? The French name for RU-486 is Mifepristone. M-i-f-e-p-r-i-s-t-o-n-e.

Look at where this is going to go. You have on this side the right to live; you have on this side choice. This side is calling it an abortion pill. You are destroying, you're killing—a new form of killing. On this other side, you don't even know if there's a fertile egg or not. You are merely instituting a normal period. You're going to have a conflict of philosophy, of belief. I'm sure if you were to take a private tally of this class you would find a complete range of opinions, and all valid. Anyone who says the whole area is free of greys, that's mistaken. The whole area is grey; there is no simple answer to it.

But look at it from the point of view of the pharmaceutical house. How much does it cost to put a drug on the market? It used to be $25 million. It's now, if you were following the research that developed the drug, probably closer to $200 million. There are no such things, nor will there ever be any such things again, as small pharmaceutical houses. The giants that we have are the ones we will have for a long time. Unless they're sued out of existence. [Laughter.]

Look at the track record of the Dalkon Shield. They're still settling lawsuits over the Dalkon Shield. Is a company going to leap into a situation with a hormone that's going to dump the whole uterus endometrial lining just like that, and may be used every month as an automatic thing to start a period? It could be almost a routine thing. That's a pattern that's actually built-in. Use it as a post-facto contraceptive. Is it effective? Yes, it's effective.

The latest study that I saw was published in France—published in this country from France. They took cases of 100 women who were chosen as being possibly pregnant and not wanting to be pregnant. Not a terribly difficult inventory of population to find. They took the 100 people and divided them roughly in thirds. One third they gave 200 milligrams a day for four days. The second group, 400 milligrams a day for four days, and the third group 800 milligrams spaced out over two days. So roughly a gram per person. Of the hundred, every one of them started menstruating within four days of starting the dose.

Of the hundred that started menstruating, 85 were by means of straightforward assays determined not to be pregnant. The 15 that may or may not have been went through a cervical examination. That was part of the study. So there was no pregnancy as a result. Eighty-five out of a hundred is pretty good odds. There were no difficulties with lightheadedness, dizziness, high

LECTURE 21 ~ *Presciptionals*

blood pressure, the usual sort things you look for. So it looks like it's going to be an effective drug. It is going to be a very fascinating thing to watch. Its fate, its development, what company will choose to pick it up and take the virtue. Selling a pill a month is not like selling a pill a day. They don't get the return on it. So there's not the monetary return. All companies thrive on what's called "chronic medication." You find something and, if you're lucky, they'll take it for the rest of their life. Meanwhile you cash in on what comes.

Five minutes ago someone had a question. I saw a hand almost come up in this area. Yes?

STUDENT: What would the rate of pregnancy have been if the pills hadn't been applied at all? Was there a projection on that?

SASHA: Not in the article. One missed period. I don't know what the structure was. I frankly don't know. The spontaneous miscarriage and loss of pregnancy probably runs 50 percent.

STUDENT: I mean, so you're saying they weren't even sure whether or not they were pregnant to start with.

SASHA: They don't know. They were primarily looking for the effectiveness in triggering the period. Each was done within 10 days of a missed period, and in every case, a period started within four days. So that was 100 out of 100, which was most unusual. There's another hand bouncing over here.

STUDENT: This is about teratogenicity.

SASHA: Yes.

STUDENT: I know a young person who was conceived while his parents were tripping on Ecstasy. They also did the same the following weekend. Is that too early to have any effect?

SASHA: For conception it's too early. What is the actual time of sensitivity? Usually, in most experimental animals you can, with a fair amount of abandon, go in and observe the changes. It's the time of the folding of the neural crest. You have the whole cells of the body. At one point the developing embryo tends to fold upon itself. That folding upon itself brings down what will eventually become the spinal column and the brain. That is development of the neurons, of the neural crest is the term. It is during this short period, usually about 36 to 48 hours in the development of the embryo, at the time of the closing of the brain stem and the brain. That is the time that the development of the organism is sensitive to drugs. I would say that in humans there is probably a 48-hour period of sensitivity.

When it occurs is not known accurately. It's probably a few weeks after conception. So it may conceivably be a time when you don't even know that you're pregnant. Because you may have a period that is not a true period, but a little bit of a partial endometrial slough. Slough—who can spell *slough*? It's a neat word. S-l-o-u-g-h. I found to my surprise, there is a legitimate word "s-l-u-f-f." There is a word *slough* [pronounced "slew"], which is where you go up to the delta country and get your rowboat stuck. There is a "slough," which is the endometrial things coming out during the period. There is also "s-l-u-f-f." The only place I know it used is the game of bridge. You throw a card in. You sluff a card. That is spelled s-l-u-f-f.

Questions? We're almost there. Yes?

STUDENT: What about the American Indians that I think use pennyroyal and tansy as a tool to induce menses and abortion?

SASHA: In a case of amenorrhea, to induce the menstrual cycle. Yes, pennyroyal has been used and contains terpenes that are very toxic. I would make a blanket statement. I think there is no plant probably on the face of the Earth that has not been used by some culture.

STUDENT: Have they done any studies on its effects on fetuses?

SASHA: Yes, it induces abortion. Okay, we'll go on next time.

LECTURE 22
April 23, 1987

Cannabis

SASHA: Okay! A number of things. One: someone in the class gave me a handout from *Playboy*. There are a few cartoons in it also but it had an article on drugs and the espousement, or the nonespousement, by the editorial staff of *Playboy* 15 years later, after they published this big thing. I tried to make copies and it didn't come out very nicely. I finally made the right size copy in the middle of the triptych—I still don't care for that word—and it still didn't come out well. Someone handed me this new one, and I tried copying it and it came out worse. So I don't know what to do about that. But I read the article and it had a nice point in it that I like, and I'm going to kind of stick it into my thought process—a distinction between addiction and physical dependency.

Remember way back we talked about dependency? Physical—the body gets depending upon it, the metabolism changes in the liver, things handle drugs differently, something goes wrong if you don't get the drug. That's the concept of physical dependency. As opposed to psychological dependency, where you want to do it, you choose to. It feels good. You tend to reinforce those things that give you a reward. I argue that the term "addiction" really has gone out of medical usage because it carries the feeling of something illegal, something criminal, an onus, a mark of Cain, something wrong about the thing being addictive. Whereas physical dependency and psychological dependency are nicer words.

This *Playboy* article did bring up a nice use of the term "addiction" with our life. It separates physical dependency—which is more or less the euphemism for what used to be called "addiction"—from something that's a property of a drug. The drug leads to a certain change in the body or a change in the psyche that leads to a dependency upon its usage. And says addiction is a property of the person, and it suddenly makes a kind of sense.

THE NATURE OF DRUGS

There is a term in psychology known as the "addictive personality." God knows we all have a little bit of it sneaking around inside of us somewhere. If you find that your relationship with a drug is such that you cannot break the stride of using it, it could very well be that it doesn't matter which drug it is. That's been found quite often by people who are dependent on—addicted to, whatever term you want to use—Drug A. You get into an abstinence, you've got to learn to say no, you do all kinds of things—lick the ashtray, whatever one does to try to break that awful dependency on whatever the drug is. If you can get the person away from that dependency, they go onto Drug B. They don't care about cigarettes. They finally gave up cigarettes, turning up this way for coffee. You get them off coffee, next thing they're into cocaine. Get them off cocaine, they choose smack. I don't know why. It's almost as if the personality has that compulsion. And I like the idea of getting the addiction on to the person. "I am an addict. I don't care what I'm addicted to. I'm an addict." As opposed to the drug carrying the responsibility of changing the body, blaming the physical dependency on the drug. It's a nice balance. I'm going to play with it. I'm not sure I totally buy it. But it's a nice way of looking at it that allows addictions, which had been sort of discarded, wasted, to be pulled out as a word and see if it might have some value.

Okay, today I was going to get into my threesies argument. I'd love to get into a threesies argument of the genetic code. I love the concept of "add a time in Guano City." We're talking about mnemonics—"add a time in Guano City" is my outgrowth of Peru, which if you're in Peru, the Guano Islands are off Peru. If you "add a time," you have adenine, thymine in Guano City, which is guanidine, cytidine. It's a cute way of getting the various purines and pyrimidines of genetic code into a pattern that you can remember.

But I do want to go back to one thing. I was called on by a student about three weeks ago when I gave my north-to-south things being alphabetical. I said I discovered that the passes in California are in alphabetical order from north to south. One student came up with a little smile and said, "You forgot about Carson Pass," which, of course, lies south of Donner, and *c* is before *d*. And of course, north is Tioga. You can't change where the pass is, but the *c* is out of order. So this is a total aside from drugs, but I want to redeem myself. I had a marvelous opportunity of going down to Lone Pine during the week's vacation, and of all things, I went over Carson Pass. I was really

LECTURE 22 ~ Cannabis

delighted when I found, at the top of the pass, it said not "Carson Pass"; it said "Kit Carson Pass." [Laughter.] So this has nothing to do with drugs. You have Donner that goes across that way, way up there. If you go just south of Donner, you get Echo.

How many people have gone over the Sierra by car? Everyone has. Okay, you know your passes. So you have Donner, you have Echo. Then you get into what I call the KLM Complex. I think there's an airline called KLM, isn't there? Consider that KLM complex. You have Kit Carson, you have Luther, and you have Monitor, and they're all mulched together. They don't all go across the Sierra, they kind of go across one and then across the other [gesturing]. You will cross this when you go across that when you're into there. You have to cross two of them to get across. So they're not really passes in the simple sense. They are part of a triumvirate. Then the rest, of course, is like falling off a log.

Everyone knows about Sonora, and of course the one at the bottom, the first one south of Whitney is Walker Pass. So I'm redeemed with north to south in alphabetical order.

STUDENT: Ha!

SASHA: What do you mean, "Ha?" [Laughter.]

Back into serious things like drugs. I wanted to do a little bit of catch up. Today I'm going to talk about marijuana. I also want to use today as a follow up on the other drugs we've been talking about that there may be questions on. This is going to end this little cycle, and I want to get into the cycle of other things that influence behavior, which I will call drugs for convenience, like radioactivity and pollutants, and environmental things and pesticides, and nerve gases and biological warfare and anthrax. All these nice little things that are really drugs—urine testing and what all—things that influence behavior.

I saw a thing in the newspaper just this morning. Apparently, drug use is down from X percent to Y percent in the military now. Having gotten down largely because of the urine testing program and the surveillance program that has been enforced upon people in the military, which I think is a noble thing. It makes the military surely run more smoothly.

THE NATURE OF DRUGS

STUDENT: Alcoholism is up?

SASHA: Alcoholism is up. Addictive behavior, exactly that. You have a person not wanting to get into drugs, they can get off this drug to go to that one. I think you've seen that. The major drug I think in junior high schools now has gone over to alcohol because of efforts to restrict other drugs, and alcohol is still there. I'm sure you're going to find the smoking situation is quite the same philosophy.

So that's the last third. I'm going to wind up this third in these various areas. I've brought in some marijuana plants. To most of you, I'm sure it is a familiar geometry, to some of you. How many people have smoked marijuana? Okay, I'm not going to ask the opposite. [Laughter.] I've brought in male and female plants. This is an experiment I did at home as an introduction to marijuana: three males, three females. It was an experiment that I made nominally to see if seeds from different places did indeed carry within them the genotypic knowledge of where they came from. "This is excellent stuff"—stuff from the northern plains of Peru or wherever it is. God knows where all.

So I got seeds from different sources. But the experiment was not really to see if seeds from different sources produce different kinds of plants. I grew them all in the same environment, kind of a not-too-fertile place with a certain amount of eastern exposure, and got seeds from Mississippi, which is the place that the United States has its marijuana farm. I don't know if you're familiar with the fact that there is an entire marijuana enclave in Mississippi, run by a group called NIDA, the National Institute on Drug Abuse. They have, in connection with the University of Mississippi, an immense marijuana farm. That is where the government raises all marijuana that is going to be used in research, that is going to be used for forensic purposes, that's going to be used for genetic analysis. They have fences around it that you wouldn't believe. There are cyclone fences with big—what do you call the stuff that has razor blades on it and wire—

STUDENT: Barbed wire?

SASHA: Not barbed. Concertina wire. Right, concertina wire across the top. They say it's to keep the deer out. [Laughter.] Anyway, my experiment

LECTURE 22 ~ *Cannabis*

was not to raise marijuana to see if indeed the seeds from Mississippi, from Afghanistan, and from the DEA—which they said came from Asia—produced plants of a different nature. My experiment was to see how easy or how difficult it would be to get through the administrative headaches of raising marijuana. That was real research. It's like a person doing a certain amount of research so they can write an interesting article, how they can get around and through the editorial process to write the article. The research is almost incidental; it's taken care of. I can get the seeds with a certain amount of difficulty. I wanted to learn the operation of the administrative and judicial system in how you do this thing. The whole process. First of all, to raise marijuana you've got to get permission. It's nice to say, "I didn't ask for permission, I was never refused permission." But when you get into things that are class A1 felonies, you got to be a little cautious about being innocent with the law. So I got permission. I got a research license from the DEA.

STUDENT: To get a research license with DEA, don't you have to have a particular qualification? Don't you need to be a doctor or—

SASHA: Nope.

STUDENT: No?

SASHA: The answer to that question is no. You do not have to be a doctor.

STUDENT: What kind of qualification do you need to obtain a permit with the DEA, though?

SASHA: No qualifications were required. I think an individual can go in, knock on the door, say, "Here's five dollars. Please give me a license for Schedule I research." They're going to say, "We'll give a license to your institution." I'm not part of the institution. That's exactly what I ran into. Did I tell the story? I think I brought my license in one time.

STUDENT: Yeah.

SASHA: I have the license. But it was a matter of diligence. They said, "We are not allowed to," and I said, "On the contrary, you are required to." But I said it nicely. Soon I had a license that said, "Schedule I, II, IIIA, IIIB,

THE NATURE OF DRUGS

IV, and V." I still don't know about IIIA and IIIB, why they're different, but there are six entries on there, for analysis. I also got a little green slip of paper that says, "Schedule I, II, IIIA, IIIB, IV, and V, research" and put it on the wall. So that process was taken care of.

How do you get marijuana seeds from Mississippi? Well, that was going through NIDA. NIDA says, "We are not allowed to distribute them because they are under the control of the IRS." Well, that was a new one for me. The IRS, you know, taxes, W2s, all that sort of thing. But marijuana seeds? Well, it turns out, this is back at the time—I mentioned this was probably '71? I have it on my code here. But this would be prior to 1970 that I went through this, because I mentioned that up until the establishment of the—no! The DEA was 1972, so it was 1971.

Before the establishment of the Drug Enforcement Administration that was empowered by the Controlled Substances Act, which was the public law or whatever back in the 1970s—prior to that, all the narcotics control and the operation of what was then the Bureau of Narcotics and Dangerous Drugs was under the Department of the Treasury. I think I mentioned this very early. The earliest laws on drugs were fiscal laws. "You may do as you wish, but to deal with the drugs, you've got to have a license." To get a license you've got to pay $50 or $2 or whatever it is. Once you have the license, you can do what you want. If what you want is illegal, and you don't apply for the license, they'll nail you for not having a license. When it's illegal, if you apply for license, they won't give you a license since it's illegal, and there's a lot of this going on. But all that permission was licensing and paying a fee and having something on the wall. That is all administered not by the Department of Justice, but by the Department of the Treasury, and the Department of the Treasury is the mother of the Internal Revenue Service. So all of this was controlled by the IRS.

So I wrote to the IRS. "Dear Mr. S., I would like a pound [laughter] of marijuana seeds from Mississippi."

"A pound of marijuana? What are you doing?"

"I've got permission from the authorities to have a Schedule I drug. I need then, with that permission, to go through you to get it."

They said, "You've got to have revenue stamps. You've got to have an application form, and you've got to have revenue stamps."

LECTURE 22 ~ Cannabis

TREASURY DEPARTMENT — DUPLICATE Value Two (2) Cents
UNITED STATES INTERNAL REVENUE

ORDER FORM FOR MARIHUANA, OR COMPOUNDS,
MANUFACTURES, SALTS, DERIVATIVES, MIXTURES,
OR PREPARATIONS UNDER THE MARIHUANA TAX ACT OF 1937.

| MARIHUANA ORDER FORM NUMBER | A4097 | DATE ISSUED BY COLLECTOR: | January 19, 1971 |

TO: National Institute of Mental Health
5454 Wisconsin Avenue
Chevy Chase, Maryland 20203

Sir:

Application having been presented and transfer tax in the amount of $36.00 having been paid, as evidenced by transfer tax stamps affixed to the original hereof in accordance with the provisions of the Marihuana Tax Act of 1937 and regulations issued thereunder, you are authorized, in so far as the provisions of that Act and the regulations issued thereunder are concerned, to transfer to _Alexander T. Shulgin_ to be delivered to him in person or consigned to him at _1483 Shulgin Road, Lafayette, California_ a quantity of marihuana not to exceed _36_ ounces in the form of the following products:

ITEM	NAME OF PRODUCT OR PREPARATION	QUANTITY
1	Crude Marijuana Plant (1 Kg)	35.27
2		
3		
4		

Signed _Bennie Carter, Chief Narcotics Stamps_, Collector,
By _Western Service Center, Ogden Utah_

NOTE: Not valid to authorize a transfer of marihuana unless signed by the collector and the full amount of transfer stamps indicated above are affixed to the original copy.

TO BE RETAINED BY THE TRANSFEREE FOR A PERIOD OF TWO YEARS.

I said, "So sell me some revenue stamps."
"They cost a dollar apiece."
"Send me 25."

And by golly, they did it. I have—where is it? In my files at home I have, I think, one of the few examples in hand of IRS revenue stamps, which are US revenue stamps, overprinted, that say, "Marijuana Act of 1937." Anyone

THE NATURE OF DRUGS

TREASURY DEPARTMENT — DUPLICATE (Value Two (2) Cents)
UNITED STATES INTERNAL REVENUE

ORDER FORM FOR MARIHUANA, OR COMPOUNDS,
MANUFACTURES, SALTS, DERIVATIVES, MIXTURES,
OR PREPARATIONS UNDER THE MARIHUANA TAX ACT OF 1937.

MARIHUANA ORDER FORM NUMBER: **A4099**
DATE ISSUED BY COLLECTOR: *April 27* 19 **71**

TO: *National Institute of Mental Health, 5454 Wisconsin Avenue, Chevy Chase Maryland, 20203*

Sir:

Application having been presented and transfer tax in the amount of $ *4.00* having been paid, as evidenced by transfer tax stamps affixed to the original hereof in accordance with the provisions of the Marihuana Tax Act of 1937 and regulations issued thereunder, you are authorized, in so far as the provisions of that Act and the regulations issued thereunder are concerned, to transfer to *Alexander T. Shulgin* to be delivered to him in person or consigned to him at *1483 Shulgin Road, Lafayette, California, 94549*, a quantity of marihuana not to exceed _____ ounces in the form of the following products:

ITEM	NAME OF PRODUCT OR PREPARATION	QUANTITY
1	*Marijuana Extracts and seeds*	*4*
2		
3		
4		

Signed: *Benwin Carta, Chief Narcotic Unit*, Collector,
By *Western Service Center Ogden, Utah*.

NOTE: Not valid to authorize a transfer of marihuana unless signed by the collector and the full amount of transfer stamps indicated above are affixed to the original copy.

TO BE RETAINED BY THE TRANSFEREE FOR A PERIOD OF TWO YEARS.

here collect stamps? A philatelist? That's one of the reasons I did it. To find these things that are not in the stamp catalog. I believe I'm one of three individuals who actually applied for and got marijuana stamps, and I got this beautiful thing with all these stamps on it. I had to turn some in to get the marijuana, because it was an order form that had to have revenue stamps on it. But I saved a few of them just as a collector's item. I turned

LECTURE 22 ~ *Cannabis*

TREASURY DEPARTMENT
ORIGINAL — Value Two (2) Cents
UNITED STATES INTERNAL REVENUE

ORDER FORM FOR MARIHUANA, OR COMPOUNDS, MANUFACTURES, SALTS, DERIVATIVES, MIXTURES, OR PREPARATIONS UNDER THE MARIHUANA TAX ACT OF 1937.

MARIHUANA ORDER FORM NUMBER: A4099
DATE ISSUED BY COLLECTOR: April 27, 1971

TO: *National Institute of Mental Health, 5454 Wisconsin Avenue, Chevy Chase, Maryland, 20203*

Sir:
Application having been presented and transfer tax in the amount of $4.00 having been paid, as evidenced by transfer tax stamps affixed to the original hereof in accordance with the provisions of the Marihuana Tax Act of 1937 and regulations issued thereunder, you are authorized, in so far as the provisions of that Act and the regulations issued thereunder are concerned, to transfer to *Alexander T. Shulgin* to be delivered to him in person or consigned to him at *1483 Shulgin Road, Lafayette, California, 94549*, a quantity of marihuana not to exceed _____ ounces in the form of the following products:

ITEM	NAME OF PRODUCT OR PREPARATION	QUANTITY
1	Marijuana Extracts and seeds	4
2		
3		
4		

Signed: *Bennie Carter, Chief Narcotic Unit*, Collector,
By *Western Service Center Ogden, Utah*.

NOTE: Not valid to authorize a transfer of marihuana unless signed by the collector and the full amount of transfer stamps indicated above are affixed to the original copy.

TO BE RETAINED BY THE TRANSFEROR FOR A PERIOD OF TWO YEARS.

them in, and sure enough, by return mail, UPS I think, came packages that contained a pound of seeds and a pound of marijuana. So that was the first hurdle of how to get the seeds. Then I had to get the plant material. Of course, you're dealing with research in the area of scheduled drugs, so all the research must be approved. You've got to apply—what's called a protocol, and get it approved by the Feds and by the State. The state laws

THE NATURE OF DRUGS

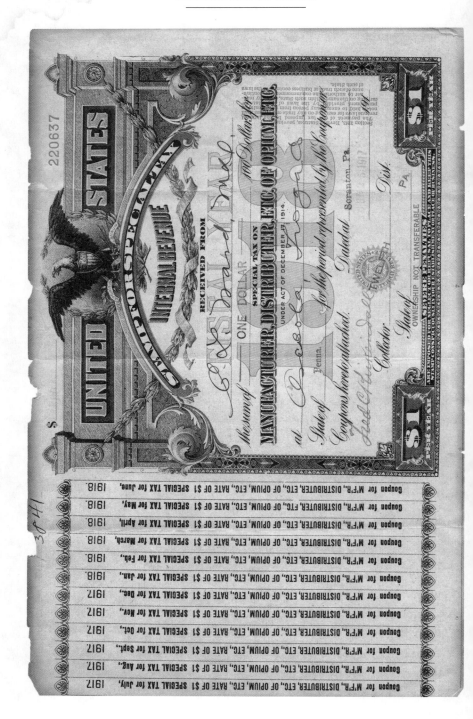

LECTURE 22 ~ *Cannabis*

are very often written in very close parallel to the federal laws, often with the same wording.

The state has, I think I mentioned before, the Research Advisory Panel. I think I did, yes? No? In any case, I will do it again. There is a group in California known as the Research Advisory Panel, which is a group of 12 appointed people who are in the position of saying, "Yep, that sounds like a good research project. We approve." Or, "It sounds like a no-good research project. We don't approve." Which is okay because you're dealing with things that are social crimes, things that have evil properties and all that. But to me at some basic level, as a person who has mucked around in the scientific arena for about 30 or 40 years, I don't like people saying, "We approve of your research; we disapprove of your research." I kind of like to look at a stone and say, "I wonder what kind of bugs crawl out from under stones?" And I lift the stone. If I get stung by a bug, okay!

I like to do it myself in my own way. I think that's really a basic, intrinsic, seminal portion of research. But in this area, this is not a freedom that was quite left intact. So there is a Research Advisory Panel. In California it's known as the California Research Advisory Panel, which was amusing in its own way. [Laughter.] I'm not making this up! It's right out of state law. They had this whole thing of letterheads and all that. Some rag—I know who, but I'm not going to say who—spotted this thing and started calling it the you-know-what panel (CRAP). And they immediately said, "No, this is the Research Advisory Panel of the state of California," which is fine.

They meet every three months, and they approve research or they disapprove research. This technically covers all research dealing in any way with scheduled drugs. It got so awkwardly complex that they have backed away, and it is now research dealing with scheduled drugs in a medical context, because people on the panel are medical people. You'll have someone over here who wants to do something with cocaine—there are a lot of people doing a lot of research on cocaine. Pick up *Index Medicus* or pick up the Current Contents and look under cocaine. With that many papers produced per week, I would say that probably the production of papers which involve cocaine are largely animal studies, or biochemical studies, or in vitro studies, or evaluation of certain things, largely in biochemical systems. I would comfortably say there are 5,000 papers a year published involving cocaine.

THE NATURE OF DRUGS

You are not going to have the Research Advisory Panel or the state of California have time to get to all of them. Let's say you have maybe 10 percent of those applications going in saying, "We wish to see if cocaine binds to protein in the blood. We intend to take a little bit of blood, we intend to add a little cocaine, we intend to shake them up in a little warm spot for a while, separate them, analyze that which is soluble, that which is not." A trivial question asking a trivial point about a trivial thing to get a trivial answer to put one number in the table somewhere. I can't see a board of 12 professionals sitting down for three months discussing whether this experiment really should be approved or not. So that just kind of slips through by being ignored. Dealing with human experimentation or experimentation with a medical flavor, the board still meets.

So, I applied to this board. I said, "I'd like to plant marijuana seeds from three different places in the world, side by side under ordinary conditions, and see if there's an appreciable difference in the THC content." THC. I got ahead of you this time and drew it on the board. THC, tetrahydrocannabinol. Tetra, four; hydro for water; cannabinol, which is the single word for marijuana. The genus of marijuana is *Cannabis*, c-a-n-n-a-b-i-s, and many of the plant compounds in marijuana, active or not, are called cannaba-something-or-other. It gets to be quite a wild thing of words, so they abbreviate with the letters, and THC is one of the most common combinations of letters of the cannabis plant. Tetrahydrocannabinol. I asked them if there was a different amount of THC that was present as a function

of the origin of the seed. Mainly, does the seed carry with it the knowledge of what has been produced independent of the environment in which it is raised? I didn't know the answer; I wasn't particularly interested. But I was very curious about how one could go about going through all this paperwork, and I found out. I got permission to do this experiment. They didn't

LECTURE 22 ~ *Cannabis*

quite know how not to give me permission to do the experiment. I found out later how they could achieve not giving me permission, but at that time it was still the Age of Innocence back in the early 1970s.

I got permission to do it, raised six lines of marijuana. These are samples of three males and females from each of the three different seed sources. I forget what the THC content was, but it didn't vary by more than 20 percent over each of the plants. If you go toward the drier, stonier, hotter area of the planting with marijuana, the material contained an increased amount of THC. The plants were scrawnier, much more tortuous in their appearance, and contained increased amounts of THC.

THC is in the resin of the female plant. It is not just in the resin of the female plant, it is in the leaf, the stem, not in the root to my knowledge. I did not look at the roots. I did look at the stem and scrapings of seeds. It's on seed only by mechanical transfer. If you wash the seed, it's not inside the seed. It's on the leaf above and below, and inside it's largely in the resin that is encasing the female seed. The harsher the conditions of the marijuana plant, the more resin is produced to protect the seed from the harsh conditions: ultraviolet light, heat, air, temperature, all these things are intrinsically destructive to the life process, and the plant will put out whatever is necessary to protect the seed. Of course, what it puts out is the resin that contains THC. So in essence, you do get a stronger, more rich THC variant as you get more strange conditions in which it is being raised.

STUDENT: So you grew these seeds in two different settings?

SASHA: No, in one setting, but the setting was a continuum from rich, fairly good soil to a fairly dry, rocky area. I drew parallel lines in three rows, each row having different seeds. So it was kind of a two-dimensional plot. I did find out a marvelous thing in the process. I remember reading—I read everything I could find. At that time, the '70s, there was a lot of stuff being published in paperback on how to do this. The term "sinsemilla" did not exist at that time. Everything was still raised as the entire plant.

There is an argument that plants must always be hung upside down to be cured. I think people may have read that somewhere. I thought, "Why should plants have to be hung upside down?" You know, the idea of draining the resin through the tip or something. I tried to get some rationalization.

I found out it's very simple. That's the beautiful thing about experimenting. You can't hang plants any other way. I tied a rope around the roots, and there they are. They're upside down. You know that idea that you have to? You have no choice. Try hanging a plant right side up sometime! There's just no way of doing it. So a lot of these little things fell into a logical place.

Okay, so that was the process. I learned that. I closed up the whole shop and didn't reapply for another thing until the following year. But they were onto me and they didn't give me permission, so I had to give up my license.

Yes?

STUDENT: Do you consider it more difficult now to obtain permission to experiment on a Schedule I drug?

SASHA: Yes.

STUDENT: And would it be much more difficult to obtain permission from the board or whoever you had to apply through if you're not a doctor or not a professional scientist and don't have a lot of experience in the field?

SASHA: I believe so. I haven't actually tried firsthand to do it. I could probe to find out, but I feel that would be what I would find.

Yes?

STUDENT: What did they say when you went the next year?

SASHA: Oh, I applied for some other inquiry. These were actually trivial questions, and I applied for another trivial question. They said, "We don't think it has social or scientific merit." I responded perhaps a bit brusquely, saying, "I think your role is not to determine social or scientific merit, but to determine safety to the public" or something like that. They came back very snippy on that and said, "We are evaluating all aspects of the research; tell us more about how you intend to run it before you leave." They wanted the position of the bathroom in my laboratory, the square footage in front of the fireplace, where I stored my drugs. Marvelous things. They are interesting points, but they have no bearing on my request.

Finally, the whole thing that was the end of my research license with the federal government—they gave it only on the basis that I had state approval—the board said, "Because we have not approved this research

LECTURE 22 ~ *Cannabis*

project, we recommend that the federal government withdraw your research license. You have no legal basis for your research." Then they sent a carbon copy of that to the federal government with the state's recommendation that the license be withdrawn. Fortunately, I have friends in interesting places, and I got notified that this little request was going in. I just beat them to it and wrote to the federal government and asked them to withdraw my license because I don't have any research to do right now, and I didn't. They let me withdraw my license, and it never was reported.

STUDENT: That's just for the marijuana license?

SASHA: No! That's a research license in scheduled drugs. The Federal license for research is a DEA license for that type of department, a Bureau of Narcotics and Dangerous Drugs license. The IRS had a separate licensure, but that was for buying, creating, and selling marijuana. That was what marijuana required, but that's through the IRS.

Yes?

STUDENT: So you don't have a research license?

SASHA: I do not have a research license. But at the time I applied for the research license, I also applied for an analytical license, which is a separate type of license. The state never knew I had that, so they never asked for it to be withdrawn. I still have that. That's why I can bring in stuff like pot and not have to worry about it.

Yes?

STUDENT: Well, you'd already gotten these seeds and this pot from Mississippi.

SASHA: From the federal government.

STUDENT: You didn't have to give them back?

SASHA: No. Because that it was presumably consumed in the course of the experiment. [Laughter.] The word "consumption" is not used here as a verb. [Laughter.] This is an interesting idea, an interesting aside.

You bring up a point that's worth comment. In licensure, there are roughly five types of federal licenses for scheduled drugs. There is a research license, there's an analytical license, there's an importation license, there is a manufacturing license, and I forgot what the fifth one is. All of them require exact inventories except analytical licenses. If you are in a drugstore and you are carrying methadone and Demerol and god knows what all down the shelf, you have to make an inventory every six months. The inventory book is that thick. You have to go over there and check off 42 bottles of how many pills. Check! The whole thing has to be done periodically. I think it's every six months. It has to be available for inspection every six months or a year. You have to send in the old copy to get the new copy that allows you to go on for another six months. It is a monstrous amount of paperwork.

I have a friend over the VA hospital here in San Francisco, the one over near the golf course, near the Golden Gate Bridge. He heard someone say to someone that their methadone in the methadone clinic apparently had gotten misplaced. Somehow this word got to the DEA. They had a surprise compliance visit, and they popped in: "Show us your inventory. Count the pills and see where your methadone is and make an inventory right now." And by golly, they were scurrying around, and it turned out that they were at the bottom of the safe or something.

Inventory on pharmacies, on physicians who have inventory of scheduled drugs, and manufacturing, by golly—you better believe if you're going to be a manufacturer of methadone, you're going to keep a complete inventory of what you made, where it is, who it went to, paperwork, what have you. But with an analytical license, there is no inventory. Their argument is when you run something through the GC, you're never going to get it back. If you can do analysis over here, you're never going to get it back. There's no way of keeping inventory because the material is consumed—I'm using the word now in the analytical sense—it is consumed in the course of doing the very thing you have the license for, hence there's no requirement whatsoever of an inventory. It's an interesting little turn.

Another interesting aspect is that in the area of analytical licensure, the DEA is very much in the position of making advice, giving its opinion and its judgment as to how things should be stored, where it should be stored, the security of this, that, and the other. But the law says they are in the position of giving advice. They are not in the position of requiring.

I had a compliance visit about three or four years ago from the DEA.

LECTURE 22 ~ *Cannabis*

They have every right to compliance, as I've got a license. They come out and make sure everything is where it should be. They said, "We would strongly advise that you have a four-hour safe." They name safes by the hour, not by the size or by the strength. I think it's for fire purposes. When you have a four-hour safe, that means it's going to be in a blazing inferno for four hours before it burns through; a two hour-safe, a one-hour safe. They recommended that I have a four-hour safe and that it be bolted to a concrete system of at least such and such a square tonnage or footage or something or other and that it be located in a way that was out of sight or something. They had a whole bunch of things for which I thanked them profusely, because I had no knowledge of how to mount a safe on concrete for four hours or anything like that. When they were gone, after I thanked them for their marvelous advice, I ignored it. Because there's no requirement that you need to have a safe and that it needs to be four hours and needs to be on concrete. They, in their judgement, say what is good advice, and I respect that.

STUDENT: The continuance of your license or having the renewal revoked had nothing to do with the points of that advice? It's just advice?

SASHA: No, I have an analytical license. A little aside and then I will answer your question. You must remember that if you are the authority establishment—this applies on the big scale for congressmen and administrators, the IRS and the WPAs and whoever is in their place, big operations to small people in big positions, from academic administrators to people who direct traffic—wherever you find the feeling of authority, you'll find that authority is what it is because the authority has been empowered. If a comment is made by a person in authority, that comment will be correct. I don't care if it turns out the moon is made of green cheese and is up there backwards. That comment is correct, because if someone else came up and said, "I can prove you're wrong," that person isn't an authority anymore. The person who can prove he is wrong is an authority so to speak. That's the way you get to the top of the Brazilian army—I use Brazil, but believe me, we have some aspect of that in this country.

When you are an authority, you are in a position of being right. If you are shown to be wrong, well, what are you going to do? You have been put in a face-losing situation. You have been shown up. You made the wrong decision. You can either back away gracefully and say, "Okay, it's time to retire. Maybe I'll go into a consulting business." Or you can fight back and say, "You are wrong and I will show it." They'll make up facts to be correct. This is the way things are in spite of the fact that in the journals and newspapers and such, on facts about drugs—I'll talk about LSD. I'll talk about blindness.

How many people have heard that LSD will make you go blind? Okay, that's another generation. How many people have heard that LSD will break chromosomes? That's still maintained. And on and on. We'll talk about all these things. They're not factual, but they were stated by people in a position of authority. "LSD makes you go blind." I've lost your question. I'll get back to it somehow.

"LSD makes you go blind" was reported by the head of the Department of Public Health in the state of Pennsylvania in the early 1970s. He said that he had an actual report of five people who—under LSD—sat down and, without protecting their eyes from the sun, looked at the sun with sufficient duration and openness, and they were blinded. Three of them permanently and two of them temporarily. It was in the newspapers throughout the country: LSD makes you go blind.

They didn't say that you had to sit and keep your eyes open while looking at the sun for a period of time. That's incidental. You wouldn't have been sitting there looking at the sun for a period of time if you weren't stoned on LSD. It goes with the territory. [Laughter.]

So, if you go back to three days later, on page 23 of the *New York Times* you'll find a reprimand being given to the head of the Department of Health of Pennsylvania for having put out a false story. There were no such five people. There was no one sitting under a tree on LSD looking at the sun. But it made a doggone good anti-LSD story, and it was perpetrated as "LSD will cause you to go blind." But it was absolutely false. That person got caught in the fib, and instead of losing a power as a structure, he probably lost his power. I would imagine he was not much longer the head of Department of Health for the state of Pennsylvania. But this is the kind of position you're in. If you make a statement, it is a correct statement. If someone wants to challenge it, they have to challenge a dogmatic statement of fact. There's a lot of this in politics.

LECTURE 22 ~ *Cannabis*

Back to your question. Yes. I was visited by compliance of the DEA and they found that everything was in order. So what went on the books? "Everything was in order." You think they are going to be wrong? No one's going make any challenge along the idea that things aren't in order, because they visited and made a dogmatic statement from a position of power. "That laboratory is in order." And it's fine. Don't fight the system—use the system. If you can find a way in which they will give you a clean bill of health or say everything's okay, fine! Congratulate yourself and make sure that's the way it goes, because if they say it the other way, you're in trouble forever. I will have another compliance visit, probably every couple or three years someone will drop by, and they'll drop by to make sure that nothing was wrong with the first report. And by golly, they're not going to find anything wrong with the first report, because if they do and it's really, really wrong, then the first report was in error, and someone in power has made a mistake and doesn't have their strength anymore.

So all you have to do is marginally go on. Don't go through red lights shouting at the top of your voice at two o'clock in the morning at 80 miles an hour. Don't bring attention to yourself. Don't, on the other hand, give up all your freedom and become a slave to the system. Just be sensible in a very reasonable way. Don't offend a person who is in power. You know, people say, "How is it that you don't go up there and write a real scathing article in *Harper's* about the president's new drug amendment that just passed? It's absolutely disruptive and really stops all research in the area." Sure it does. But I'm going to quietly do my research. I'm going to continue. I love looking under stones for bugs and I'm going to continue looking under stones for bugs. If I write an article in *Harper's*, I have just rubbed the whole nervous system of the other people who passed this law, and I don't want to do that. I just want to do my thing. So I'm going to stay in a very good, obedient place, out of sight.

A friend of mine once gave me a very good bit of advice. He said that if you are a threat to the system, but you're not known to be a threat, you'll be left alone. If you're not a threat to the system but you're thought to be a threat to the system, they'll get you. It's kind of a weird little philosophy, but it has its merit. It's something to consider when you're considering driving 80 miles an hour down through a red light at two o'clock, stoned out of your mind. Be careful! Slow down and stop at the stop light, and you'll be able to continue.

Okay, back to where we were. More questions on that?

STUDENT: Marijuana.

SASHA: Marijuana, thank you. Spelled with a *j* or an *h*. I spell it with a *j*. I will tell you, a little bit of bureaucracy and what it does is marvelous. For a long while, marijuana was spelled with an *h*, as the older generation—especially if they never used it and never condoned its use and can't quite understand why this younger generation wants to go out and blow pot all the time—managed to pronounce it with five syllables. Do you get it? "Mar-ee-hu-a-na." [Laughter.] I don't know how they get that all in there. It's something like the pronunciation of *aluminum*. In English, in America, the metal is pronounced with four syllables, *al-u-min-um*. In England it is pronounced with five syllables. Someone from Britain says it and it comes out *al-u-min-i-um*. You've got five syllables. If you have ever been on the Continent, have someone in a department store in Paris pronounce it in French. It's the only word I know that's pronounced four and a half syllables. It comes out *al-u-mi-ne-uh*. [Laughter.]

Marijuana. It used to be *h* or *j* in the middle of the word. An actual edict came out. It was published and distributed to every subdivision of the law enforcement system. It was sent to me by about five dear friends, five highly placed friends. It was widely distributed, certainly, to all law enforcement groups, all drug enforcement laboratories, all crime laboratories, all police enforcement laboratories, all customs, all FBI, and one more large family; the IRS was involved too, because they were involved in tax stamps. Henceforth, for the sake of consistency and conformity with what is now the accepted practice, "Marijuana shall henceforth be spelled with a *j*." This was a memo that went throughout all the ranks. It shows the importance in its own way of this type of bookkeeping. Marijuana is now legally spelled with a *j*, and all records, of course, have been changed.

The material has been around since the old expression "heavens above." It has been around as probably one of the oldest drugs. Its origin was China, in all likelihood. It was in the earliest of the Chinese pharmacopoeias, comfortably 2,000 years before Christ, and probably much earlier than that. As

LECTURE 22 ~ *Cannabis*

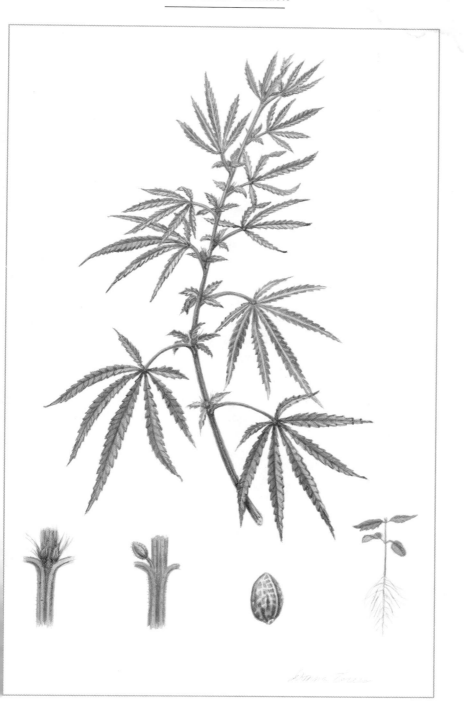

I mentioned in our discussion of tobacco, smoking came into usage from the New World. Cannabis was taken as an ointment or used as a salve, but primarily it was an eaten drug.

Its origins were in China. It moved down from China into India, and from India across the southern part of Asia, Asia Minor, across into Africa. Probably the religious crusades were the primary instrument of bringing it into Europe. It crossed in its flow with opium, which probably has origins in Europe, or one could argue possibly in Switzerland or somewhere in the Arab countries just east of Switzerland, as there are now reports of the areas. Opium moved south into Africa and across Asia Minor, crossing the path of marijuana into India, from India to Southeast Asia, and from there to China. The final part of that move was political.

Yes?

STUDENT: So it didn't grow in the New World in the last few centuries?

SASHA: Not broadly in the New World until probably in the 1600s. There is no record of this being here. God, you would hear people argue, "Marijuana was big in the New World." I find no documentation for it, no believable record that it was not brought here.

Who has taken botany? Someone. I've not, that's why I can't raise my hand. What does the term "dioecious" mean? Botany. "Dioecious." Finally, hands going up.

STUDENT: When the leaves drop—

SASHA: That's "deciduous." "Dioecious?" Can you pull it out?

On the other side of the coin, there is a word for a plant that has both sexes and can switch sex midstream. I kind of feel the word for it is "dioecious," that kind of thing—I know what the plant is. The plant is male or female depending on the conditions and the environment, which is great, so you can have male or female plants, very prone to polyploidy and all kinds of weird, nice chromosomes that give them a lot of weird hybrids, most of which are not fertile.

A great deal of the work was done on marijuana in the 1910s and 1920s by the Department of Agriculture, because they definitely wanted to use it. The alternate name for it is "hemp," and hemp, of course, was the basis of a

LECTURE 22 ~ Cannabis

lot of the trade in the New World, after Europe. The marijuana was raised here in quantity in the early colonies, but primarily for its hemp purposes. I've heard numerous things said about Washington and his marijuana patch, but the truth is, it also happened to be a hemp patch, and I don't know the extent of or even if the plant was used as a drug at all back in the colonial days. I have no way of getting that kind of verification. It was certainly around. Of course, the Indians didn't have their time of cultural assimilation, because it was introduced quite recently from their point of view.

The plant did not really get into serious scientific research until the 1920s or 1930s. There are two names that are really fundamental to the early part of the history. One was Adams, a chemist at Illinois, and one was Todd, who was a chemist in England, I believe at the University of London. They were the primary researchers in chemistry, investigating what it was in marijuana that made it the intoxicant and the poisonous material that it was.

The publications of both spanned the range from about 1935 to about 1940, up to about the time of the beginning of World War II, and it is a fascinating story to follow because the two were at each other's throats. I mean, scientific cooperation, "We'll work together"—oh no! They were at each other's throats, each one claiming this and that, and having the better method, and being the first to discover, and having staked out the territory. Rare at that time, you don't find it at all now as it's carried out at scientific meetings. At that time it was done in the scientific literature. "We, despite the lack of respect to our priorities in this area of inquiry, would like to find a conflict of findings to whoever it was—" Footnote, Todd. Adams writing.

This country largely used the dog ataxia test. If you give extracts of marijuana or marijuana, feed it or give the extract of it to a dog, the dog will lose control of its rear legs. It's a very sensitive response. It was called "ataxia," the lack of motor coordination, and was used as a method of evaluating extracts and evaluating subsequently synthetic variants. In England, Todd used the loss of reflex of the eye of the rabbit. So you have this beautiful battle between dog ataxia, motor ataxia, and rabbit eye areflexia. Neither of them would accept or even tolerate the other's test but would use the test and got surprisingly similar answers, in a very quiet way, indirectly testing for the validity of this type of assay.

A very basic thing when you're working with anything from plant sources is not to say, "I'm going to find out what it is that is active, and I'm going to synthesize it and prove that it's got the structure I think it has, and then I'm going to make something that's going to be more active, or it's going to be active and not have this negative side effect. I'm going to find out what part of the molecule is responsible for the dog's leg doing this or for the eyes getting red or for the intoxication or whatever it is." This whole approach, one looks at it easily to say, "I'm going to go in the laboratory, and I'm going to isolate it and synthesize it and improve upon it," without realizing that when you start with a plant, you don't go to the plant and unzip it and take off the nice white crystalline solid that is the active component. You've got a *gemisch* of unbelievable mixtures of things in that plant.

How are you going to even tell what the active material is? You've got to have an animal test or graduate students, or whatever people use for assays. [Laughter.] You've got to have some way of evaluating where activity is, so you can divide an extract in two portions—a water soluble and water insoluble. Evaluate, evaluate. One of three things will happen: it will be this fraction, it will be that fraction, it will be neither fraction. The third is all of what you actually observe. It is in neither fraction, because it was in the residue that you didn't extract. Or it turns out that they both have to be present to get the action—it is a synergistic or potentiated thing, so they both have to be present. So there's a lot of labor in isolating the active material.

The "Red Oil of Hemp" was what the whole study was called in this country, in Illinois. Hemp was gathered where it grew wild in Kansas and Nebraska, where, by the way, it still grows wild, having been discarded alongside the roadways. It's a bane to farmers because it tends to take over areas they want to cultivate. And of course, the simplest way is to be sure that is known broadly to the authorities as being an illegal drug growing wild in Kansas and Nebraska so that the authorities are responsible for getting it out of there, which saves the cost of the farmer having to clear the waterways and the fields. So there's a lot of politics in the growing of wild hemp in Kansas and Nebraska. The material, by the way, is quite low in THC, for those people who want to do their PhDs in decorated VW buses. [Laughter.]

Anyway, this battle went on. Adams probably did the larger amount of work. There is a term that rings in the annals of marijuana research

LECTURE 22 ~ *Cannabis*

called "Adams 9-Carbon Compound," or sometimes it's called the NCC, the "Nine Carbon Compound." Adams did a monstrous amount of work. In Adams's case he did not actually know the structure, and this is true with Todd, and this is true until the 1960s when an Israeli chemist by the name of Mechoulam unraveled the mystery satisfactorily. They knew there was a double bond in the molecule, but no one knew where, and there was no good way of really pinning it down. So in the chemical part, there was one true double bond somewhere in that ring. Well, it was hard to determine where, but Adams had found you can make a lot of compounds with double bonds easily.

Let me take the structure apart a little bit. How many have heard the word *terpene*? That's not many. One, two. Okay. I'm going to draw a funny little thing over here. It's got kind of nice swing to it. This is a basic structure of a whole family of compounds in botany that are called terpenes. Anytime you cut down a Christmas tree, the thing you smell is terpenes. Turpentine. Extracts of plants and things that, in essence, contain the body of terpenes. Probably things along this line, I would say comfortably, were the first and probably are still a major source of raw chemical materials for doing things, because they come universally in plants in general that are called terpenaceous. They contain terpenes.

There is a nice thing about terpenes. This is a little aside. But if I were to take this portion of the terpene, I would find the following system. This has nothing to do with anything. I have fun with chemistry and I do it just because I enjoy it. You'll find a system of four carbons with one sticking out. Here's one, here's a one, two sticking out, three, four. Here's one, here's two sticking out, three, four. I call it the "one, two bump, three, four" philosophy of biosynthesis.

There are a lot of compounds in the body. You have cholesterol, you have compounds that are fundamental building blocks of most of your steroids and building blocks of many of your complex hormones that have this one, two-bump, three, four origin. This is known as isoprene, and it's a building block of terpenes. Notice this terpene is one, two bump, three, four. One, two bump, three, four. There are six or eight classes of terpenes. There are things that have not two of these, which are terpenes, but three of them, which are called sesquiterpenes.

THE NATURE OF DRUGS

Good word, "sesqui-." The prefix "sesqui-." How many people are familiar with "sesqui-?" Sesqui- is one and a half. It's a nice prefix. One and a half terpenes. If 10 is a terpene, what's 15? Sesquiterpene. How many people collect stamps? I've gone through this before. In 1920-something-or-other, they had a memorial stamp with a great big Liberty Bell on it. Two cents was the postage at the time, and it celebrated the sesquicentennial of the United States of America, the 150th year, because it was one and a half centuries, sesquicentennial. Kind of a neat word. Prefix for one and a half. Prefix for two: di-, bi-, if any of you are Latin or Greek. A prefix for two and a half? Now there is a real treasure. This will get you 50 points at the right time in Scrabble. Prefix for two and a half. Anybody want to volunteer? Sester-. So you have a—in fact, that *was* a question on the final I gave for my criminalistics toxicology course in Berkeley. I said, "In the year 2027, we're going to be celebrating the 250th anniversary of the United States of America. They will probably issue a stamp and we'll have fireworks. What will we call that celebration?" The answer, of course, is sestercentennial. If they didn't know the answer, they weren't following my lecture on terpenes. So that's two and a half times, again. Twenty-five is sesterterpene. Triterpene, 30. Tetraterpene is 40; pentaterpene is 50. There is no prefix for three and a half. So you have sester- for two and a half, sesqui- for one and a half.

Okay. Why am I into this? Yes, I'm into this for this reason. If I were to draw circle around this portion of THC, you see our marvelous little thing again. Here is this entire ring system built into THC. Half of the THC molecule is a terpene.

Of course, in the synthesis of THC, for this terpene, often you go to plant sources for the terpenes which you use. Pulegone from the oil in orange rind—in fact, orange rind is a favorite starting material because it contains citronella, limonene, and a host of nice terpenes, all of which are very reactive. I wrote a short essay once—I never dared to publish it—on how you can make THC synthetically from orange peels and lichen that hangs down from trees; lichens, algae, and fungus all together. The algae are boring. But the fungi are fascinating because they contain beautiful things including this. The lichen includes this portion of the molecule.

ANN: Why didn't you publish it?

LECTURE 22 ~ Cannabis

SASHA: I might someday, but it didn't quite feel right. But you can take this portion of the molecule from a lichen, that portion of a molecule from orange peel, lump them together with $CoCl_2$, which is nice inorganic active catalyst, and out of that you can get THC at a fairly modest but usable amount. In fact, that brings up another whole interesting thing.

I had a situation that dealt with Adams Nine Carbon Compound one time. It's a good time to talk about it. Let me talk about Adams Nine Carbon Compound. This was a double bond here. [Pointing to the drawing on the board.] And it had two more bumps, and it had two methyl groups there. So I have a total of one, two, three, four, five, six, seven, eight, nine carbons with two methyls next to this position. The Adams NCC was the most potent compound that Adams was able to synthesize in imitation of marijuana. It was some 500 times more potent in his dog ataxia study than the synthetic material he made; that was as close to the Red Oil of THC as he knew how to make it. So it's not 500 times more active than marijuana extract. It was 500 times more active than the THC which he was able to synthesize. The most active of them all.

Subsequently, we went into World War II. Of course, part of the World War II urge was to find materials that could be used for chemical warfare purposes. What better than intoxicants, in the way of that—one whole lecture is going to be chemical warfare—those things cause intoxication, stupidness—what do they call it? Stupefaction, I think, is the legally correct term for the actual marijuana kind of thing. A great push was made to make this Nine Carbon Compound that was so highly potent. No one ever tried it in a human, but it certainly did things with the dog. Also, by the way, it did the same with the eye areflexia of the rabbit. And so there was a push to make this material. It never got very far, but it did get quite a bit of interest.

There was a push by the chemical warfare groups to find out how to make it. That was my first introduction to the chemical warfare group. I was working in industry at the time, and this group of people came in for what they call a "brainstorming tour." They wanted to get ideas from people in the chemical industries on how to do things. They said, "Here, we've got an interesting problem." We all went in and sat in this little conference room, and they said, "Let's pretend that you had this." I think they had an ester group down here of some kind. The double bond, I think, was a diene form.

THE NATURE OF DRUGS

Let's say you had something like this. Hanging out of it they had this nine carbon thing, and something sticking out here and something sticking out there. Clearly, if this and that were mixed together in some magic way, you're going to come out with the Adams Nine Carbon Compound. It spoke for itself, as they say in law. There's no way to get out of it. This precursor, that precursor, to this. But this was then a classified compound, because our military wanted to get into it, they wanted to make it, but they didn't want to tell you what they were making, because that would reveal where they were going. So they give the entire design down to the penultimate step, and then left out the final thought. How would you make these two precursors? "For reasons we can't tell you, we're interested in making the precursors."

I said, "Well, I don't know what you're trying to make, of course." [Laughter.] "But let's pretend—" I drew up the structure. "Let's pretend, for the sake of discussion, this structure, and you want, if something like that is what you want to make, you want to make kilo amounts of it, a quantity of it, because you can make small amounts—"

I knew there was none left on Adams's shelf. Adams's whole research facility at the University of Illinois is still there. In fact, now the chemistry building for the University of Illinois is the Adams Building. But apparently none of these things survived the leaving of the graduate students who were working on it. They apparently were the first true human experimenters in the area of the various analogs of synthetic THC, and the vials were empty when they went there to get reference samples. They could not get the original samples. I pointed out the fact. I said, "If these are the kind of things you're after, you've got to realize—" But of course, the real true one they wanted was the material with the double bond. This is where it is in nature, the THC. They wanted this compound. I said, "You've got to realize, a very subtle thing involved in here, you've got a couple of asymmetric centers."

I think the one they had originally had one more asymmetric center out there. "If you want to get what Adams had originally made, this one out there—" Clearly I'm treading right on top of what they wanted—they couldn't acknowledge it. "If you want to get what Adams actually made, you're going to have to make it exactly like Adams made it. Because if you make it any other way, you're going to get a different mixture, and you may get a lot more of what you want, or you may get none of what you want. So

LECTURE 22 ~ *Cannabis*

consider seriously the possibility of not finding a better way of making it, but just making it the way it originally was made—or, entertain the task of making all possibilities of compounds, and try making them separately." And that was kind of the end of the seminar. They went on their way, and they were hired by Arthur D. Little out on the East Coast, but they were from Edgewood Arsenal. No feedback from that whatsoever for about three or four months.

Oh, by the way, one of the other things I had mentioned. Did I talk about the trip in the Atlantic and the Trinidad phone booth? Ha! Okay. I said, "Now what you really should seriously consider is nature's work when nature made marijuana. Nature loves alkaloids." I made this all up, but I said it as fact. I tend to say a lot of things as fact that I make up. [Laughter.] When nature first grew marijuana, it was very bad soil and there wasn't any nitrogen in the soil. Had the whole system been right, nature would obviously put a nitrogen right there, which is where an alkaloid would have it. It's a phenethylamine. That's where nitrogen was wanted. That's the perfect spot for natural nitrogen. But nature not having nitrogen did the best it could with carbon, and it didn't make this compound, but it made the compound with a carbon there instead. Not as active, but active enough. Put nitrogen there, you're going to have a super active compound. Let it go at that.

I got a call about a month later, "Would you please come back to give a seminar in New Jersey?" Sure, why not? I go back to New Jersey, Edgewood Arsenal. That was a wild one—Edgewood Arsenal. I had never been in a military establishment. I was in the navy in the war, but that was not a military establishment. This was the real deal: barbed-wire fences, guards, the whole thing. And here I'm wandering in. I don't have security clearance. How many people have security clearance? My hand is down. One, two, three—okay.

I am a very strong advocate against security clearance, merely for the reason I love shooting my mouth off on whatever I want to say, and you can get a lot of information. You get security clearance here, then you get into confidential, then you get to supersecret, then you wear a green badge, then you wear a red badge, and finally, if you're really good, you wear a purple badge. And whatever it is that you have in the way of clearance, you're getting a lot more information, but you are stopped forever from talking about it, except with other people who wear the same color badge that you have. So you're getting a lot, but you're giving up a lot. People say, "I will give up all of that for getting what you get." I am a firm believer in "I'm not going to give up diddly-squat for anything." Hence, I don't want the clearance.

So I was able to give a seminar. I just went ranging over all kinds of nice structures, and I couldn't watch them closely enough to see where the eyes dilate when I hit something close to the mark, but I wish I could have. I gave a nice seminar. I was never invited back. I talked about a whole bunch of compounds they were working with that were psychoactive, this type of thing, nitrogen in THC, which apparently really rang a bell somewhere. I flew back to the West Coast. That was it.

About three or four months later, my mother died. So to break up the depression of my father, I decided I was going to take him, my now late wife, my son who was at that time was 11 years old, and myself for a year's trip in Europe, which I did. There is a whole story associated with how we got that little trip launched. But the whole heart of this story deals with our being one day out of Trinidad, which is just at the southern end of the Antilles, just off the coast of Venezuela, down the stream from where the Orinoco dumps. You remember our maps. Okay.

I was there. Deck D. I went first class, the cheapest I could possibly go first class. I wanted the good food, but I didn't want to necessarily pay much, so I was on Deck D, but on the first class portion of the ship. I heard this pound, pound, pound on the door, and not remembering I was on the top deck with a two-bunk thing, I stepped off the bunk, hit the bottom, and found my way to the door.

Here was a Mr. Portez, who was a radio man of the Tucson. He said, "Is your name Dr. Shulgin?" I said, "Yeah." He said, "Got a radio message for you." It's dark. He said, "Don't bother turning the light on." What it said was, to give you a whole rundown, namely, they would like to have me phone from Trinidad to Arthur D. Little concerning the synthesis of compounds containing nitrogen that might be of interest to the US military, and so on. I really struck a nerve with this. So I said, "Thank you very much." Ten minutes later, the door again. Portez said, "All right, another message. This one came by RCA." There are two different ways of radio messages getting to ships at sea. Apparently, whoever sent the message used them both. He

LECTURE 22 ~ *Cannabis*

said, "Don't bother reading it. It's word-for-word identical to the last one." I said, "Thank you." By the way, meeting Portez was one of the highlights of that trip. He was a very interesting person. But that's another entire story. [Laughter.]

Anyway, here I was. The message said, "Phone from Trinidad when you get there to Arthur D. Little or Edgewood Arsenal, one or the other, concerning a possible writing of a grant or contract or something." Sure! Well, Trinidad was 95 degrees temperature, 95 percent humidity, and inside of the phone booth in Trinidad was one of the most traumatic experiences of my life, and I was there for about 45 minutes talking long-distance, in there with the call to New Jersey or Boston, I forget which. And the answer was very simple. "We are very interested in having your ideas of how to synthesize a compound that looks like THC but with nitrogen as the synthesizing compound. You'll be arriving in England on the 12th of May. You will be able to send off a special-delivery airmail at that time so that it will arrive to us on the 15th, we have to file our grant application on the 15th on how to make the compound."

Well, here I am on the Tucson, on the high seas of the Atlantic. "Sure, I'll send you how to make it." I go desperately to the research library. There is no research library on a steamship. You know, they have a library. In the library, the only research book I found was a thesaurus. A *Roget's Thesaurus* [laughter] published in 1800-something, which I dearly would love to have stolen. I should have stolen it. It was the only thesaurus I ever saw that had a whole collection, three pages, of these marvelous words like "a pride of lions," "a something of swallows." These little collective words that represent groups of animals. Three pages of these treasures. I've never seen a good text like that since I let it go.

Anyway, they wanted me to tell how to synthesize this compound, and I didn't even have a book that could spell "carbon" correctly for records. So I disappeared in the library. I sort of went back the best I could photographically to everything. I came out with a reasonable flow of how to get there. Mailed it airmail special delivery from England when I arrived there. It went back to the Edgewood Arsenal and to Arthur D. Little both.

Apparently, I don't know how much work they did making the compounds, but the compounds were made. They turned out to be stimulants and also intoxicants with a nitrogen in that position. Somehow the military didn't want them. I actually sold the whole idea to Bristol—not Bristol, some other big pharmaceutical house, I can't remember its name, who realized they had a patent for Arthur D. Little, but my name was on it. They had to buy the patent. So I got a dollar for the patent, and eventually it went into some sort of production thing. They made several drugs associated with it. But it's an interesting circle, because the whole thing went through the military and through Arthur D. Little, which is a big consultant place like Stanford Research is here. Arthur D. Little is on the East Coast. And actually, some drugs went on the market, largely as antinausea drugs or for their stimulant properties. So it's an interesting little turnabout but brought very much to mind the structure of cannabis. And this is a lecture on marijuana. Ha! Okay.

The terpene side, the orange peel side, the lichen side, and the combination of the two produces THC. There's a term "cannabidiol"—"CBD." Cannabidiol is the same molecule with the thing opened up. It has a diol. It is not appreciably active. But, in the course of pyrolysis—pyrolysis is heating something with heat, with fire, until it is lysed, it gets broken apart. You put a flame to sugar, you get a pyrolysis product that is known as caramel. If you put a flame to a plant, you get a pyrolysis product. The burned plant carries the chemicals out as an aerosol in the smoke, and of course, the diol becomes, in part, THC. You have cannabinol, which is a totally aromatic compound. Totally inactive.

They are in the plant. Their term was coined by Ružička, who was a Swiss chemist, and Mechoulam, who was an Israeli chemist who did all the really great work on THC. Cannabinoids. They coined the word "cannabinoid," referring to things in the plant that look like cannabis. There are now about 50 or 60 cannabinoids that are known. The structures have been worked out. The positions of the double bonds have been worked out. Aspects of the configuration have been worked out with great elegance. There has been an immense amount of human work done with THC, with cannabis, under controlled experiments. I am frequently asked the question, "How does it work?" The answers is "I don't know." You ask people who have done this work, "How does it work? Why does it do what it does? Why is it an intoxicant?" They don't know.

LECTURE 22 ~ *Cannabis*

What is going on in the brain that makes THC a CNS-active compound? What causes the peculiar type of intoxication it has? It is not known. The type of intoxication is a very rapid onset, as most of you probably know. A very rapid onset. It causes a form of disruption of the cognitive process that I personally don't like. Some people do, some people don't. What I don't like about it, personally, is what I've heard other people say that they observe—the same thing and they like it, which is fine. To each his own. That is the inability to remember toward the end of a sentence where the sentence started. I don't know how else to express it.

If I'm in one of the experiments in which I'm using this for some reason, and I'm in that intoxicated place, and someone is starting a sentence, I start to think of my response. I've memorized the first four or five words of what I want to say, but then I don't remember. I find that just pisses me off. I don't like it. [Laughter.] It's not my drug. I don't like it. Some people thrive on it. It's definitely a disconnector of cognitive integrity, which is a virtue if you are all wound up cognitively. I happen to like being wound up.

I am going to close up. That clock is right? Okay, that's fine. We'll talk more about it. I'll get back to LSD and whatever the things are for next Tuesday.

LECTURE 23
April 28, 1987

Cannabis and LSD

SASHA: Okay, one, two, four, six, eight—

I want to talk about some radioactive metals, but I'm going to hold off until the next lecture when I get into radioactivity. I mean, what is radioactivity? Do you see it, smell it, taste it? Where does it come from? I want to get into that in a lecture. A lot of it goes into drug problems, just like using chemicals as drugs goes into behavior. Chernobyl broke loose and they evacuated 80,000 people, so there was a behavior aspect to it too, and residual radiation and mutation and all that. I want to get out of the specific drugs and into the more generic classifications. So I'll certainly touch on some of these things. Oh good! After the hour, people wander in.

There was a request after last hour's lecture—which was spent largely on marijuana, and a lot of other things too—to talk a little bit more about marijuana, asking about what hashish is. Also, a question came up about LSD flashbacks. Other points on specific things about drugs before we get into the windup stretch of the course, this is a good time to sort of tighten them up.

Hashish is a common name given to the compacted resin of the marijuana plant. You can see some on the examples I brought in. The leaf of all, by the way, carries this resin. Let's go back over that. The leaf of the marijuana plant is serrated; it is characterized this way, although not uniquely characterized. It carries on it the little beads of resin on the underside, also things called cystoliths.

Cystoliths are the small, apostrophe-shaped containers that carry inside of them a little bit of carbonate. It's an inorganic container that is a development of the rows of the leaf. This is one of the morphological tests for marijuana. You take the leaf and turn it upside down and put a really strong acid on it. The acid gets at these little cystoliths and goes bubble-bubble-bubble to release its carbon—as carbon dioxide. So a microscopic test for marijuana is the presence of this system, of this structure. There are many other plants that have it. If a plant does not have it, it is not marijuana. This is the meaning of a "presumptive test."

A test is a presumptive test if it establishes a territory in which what you're looking for could lie. You run a color test for such and such a compound. You add a little bit of acidified Van Urk to the material. If it turns purple, that's consistent for indoles such as LSD. If it does not turn purple, then LSD is not there. You can use it as a way of saying, "This material is *not* LSD." If it turns black, then you cannot say LSD is or is not there. You can either accede to a presumptive test, or you can fail a presumptive test, or even have an ambiguous response to a presumptive test.

There are a lot of field test kits that have been sold, have been used, as the ways that people can go—usually policemen—into the field, stop a car, find a little bit of something that's on the floor in the backseat, put it into a little thing, add a drop of this, and see if there is sufficient reason to take the chemical to the crime lab to find out what it really is. If the presumptive test is positive, that implies it might be LSD or might be heroin or might be whatever drug the tests are set up for. It is sufficient—if you have a presumptive positive—to take the chemical to the crime lab and put the person under arrest. So it's not evidence that can be used in court to prove it is, but it's considered sufficient evidence to have a reasonable suspicion that it could be. So presumptive tests have their value because they allow this intermediate discrimination to be made.

You have little machines you blow on in Canada when you've been driving and possibly driving under the influence. It's a little hoozie-what that you blow in, and it develops color of a certain length or develops a certain electrical charge as presumption of having had too much alcohol. It would not stand up in court but is sufficient to take you to a test that you can really find out how much alcohol is in you.

So these presumptive tests have become very popular and become very useful, because it is a way to argue that a person may be a threat to society, because of a violation of law, but also to allow a policeman in good conscience to release a person if there is not that presumptive argument. So not only does it produce some awkward false arrests, but it also positively

LECTURE 23 ~ *Cannabis and LSD*

eliminates the possibility of certain inappropriate arrests. It has two sides to it, and both sides are to be considered.

How did I get onto that? I was going into presumptive tests, color tests ...

ANN: Hashish.

SASHA: No, earlier, in between, back of the marijuana leaf.

Another situation. How does a person who pulls you over and says, "I wonder if you have been using marijuana?" or "Is that marijuana that's hanging out of your shirt pocket?" or "I see a roach in the ashtray, I suspect it's marijuana." How does he know it's marijuana? These policemen and the crime lab people are not botanists, and even a botanist can be hard put to say, "This is marijuana." A good dyed-in-the-wool botanist would say, "I will not identify any plant until I see the blossom. From the blossom I will find the characterization of the plant." They are classified by blossom—not by leaf and by stem and by root structure. So how do you know it's marijuana? We've heard this argument. "I've seen a lot of it, and I'm very familiar with it, and that looks like marijuana to me."

This is very good. But then it leaves a person open to certain judicial manipulation in the courtroom. "You've seen a lot of marijuana. Tell me, how did you know the marijuana you've seen a lot of was marijuana?"

"Well, it had the label on it" or "It's been around the crime lab for 20 years as our reference." What is its source? This concept. I mentioned this a month or so ago—the legality, the concept of the reference origin. The *exemplar* is the legal word for it; the thing against which you compare this, and if it agrees with that, then this is the same as that and this is taken to be marijuana.

You go to a crime lab and say, "I understand you people have been using the seizure of 20 years ago as reference marijuana to use as an exemplar against which you check marijuana."

"Well sure, it's the thing you got in the Hays case back in 1947."

"How did you know that was marijuana?"

"Oh gee, years ago whoever was running the lab back then put it up there, and that's what we've been using."

"How do you know it is?"

"Well, it looks like it."

THE NATURE OF DRUGS

You get these little psychic reasonings. It's a tricky area. There are exemplars. I think I mentioned this in the blood-alcohol routine. You're measuring alcohol in a person's blood by putting alcohol in reference blood that doesn't have alcohol at certain levels. So you find out how much is there as a standard to tell how much is over here. You need a standard against which you can titrate quantitatively, but you need identity of a drug to qualitatively compare.

"Where is the reference alcohol?"

"Well, we got it from the Rosewood Distillery, which sells alcohol to organic chemical laboratories."

"How do you know that was alcohol?"

"Well, it looked like alcohol. Smelled like alcohol." It quacked like a duck and it barked like a duck or whatever it is, so it was a duck!

"How do you know it wasn't half water?"

You get these very strange hesitations. There, you've got to get the alcohol verified by spectroscopic means for purity and strength. You've got to verify it by titration against potassium dichromate. You say, "I titrated it according to the law against potassium dichromate."

"Where did you get the potassium dichromate?" So you can pursue this thing up to a certain point, and there's a point at which it comes to a stop, because by law it says, "It shall be taken that such and such is the foundation of authority against which everything will be compared."

The answer to this is very simply that the origin is the National Bureau of Standards to whom you write and say, "Well, I have one pound of potassium dichromate that is your documented standard." Like the standard of time, standard of length, and everything else. There is a reference thing there against which you compare it. They'll send it to you, probably at no charge. So a laboratory, if it has an open bottle of potassium dichromate, and it has a reference bottle of alcohol that has been titrated against it, then they can call that the secondary standard and use it in titration.

So what is marijuana? Well, it's a tricky one. I've had the pleasure of seeing the original sample of marijuana of Linnaeus' that was given the name of *Cannabis sativa*. It is in the Botanical Museum in London. They have shelf after shelf after shelf of these old, dried things, and I actually saw the one that was *Cannabis sativa*—looked at the original plant that had defined the

LECTURE 23 ~ *Cannabis and LSD*

genus and species. It didn't look at all like marijuana to me. It was scraggly. It had none of this lush growth that you find, a few cultivars removed from it. And it had serrated leaves. It had no blossoms there. That's a total error. And that is a reference against which things are compared? It doesn't make much sense, because it doesn't compare very well. You get into the argument that you can define a plant chemotaxonomically.

What does it have in the way of chemicals, or what does it look like? Marijuana is now more and more defined by what it has in it as chemicals. I chose this plant material, and I made an extract of it, and I got in that extract a material that was physically and spectrophotometrically indistinguishable from THC.

Well, they say, "Aren't there other plants that might have THC?" Well, yes, but THC is itself a scheduled drug. So in essence, they closed that little gap by making the component of marijuana also a scheduled drug. How do you field test for marijuana? Well, I've gotten into some very strange territories, as both a prosecution and a defense witness in this exact area.

I was at one case down in Santa Cruz, where they said, "Here is a paper bag that contains a representative sample of the 512 plants you've got out of that field." Well, how do you get a representative sample of 512 plants? That made me immediately a little shy and suspicious of the person who collected the samples.

"Well, I took a leaf of every one that I came by."

"Well, were there eight tomato plants coming by?"

"Well, there were other plants. I didn't take any leaves from them." He has already titrated what he's seen by his judgment of what was coming by, to choose leaves from what was coming by to verify and judge what was correct. A loop. And that loop in court is immediately get-in-able and undoable.

In fact, the whole case was kind of weird. They had seized all the things. There is a term, *tare*. I don't know if I have mentioned this. T-a-r-e. It's how much a thing that contains something and what it contains, weighs. Obviously in toxicology, try to get the words "tare" and "tare of" involved, so that people keep in mind that you can take a tare of something that is empty, or a tare of something that is full.

Invariably when a seizure is made—I know of one just about a month,

month and a half ago, in which a seizure was made of 215 grams of LSD. As LSD is a substance people asked about—I'm tying this together in strange ways. Invariably when seizures are made, the drug and the container are weighed. That is the weight that goes out as the weight of the drug. So 215 grams of LSD was about three grams of LSD in a 212 gram bottle. But it went out as the weight of the entire thing. The concept of tare I almost cynically think is intentionally overlooked. Then you figure out how much is a typical dose that can sell retail for how much. So you can sell 150 micrograms for five bucks. How many 50 micrograms is there in 215 grams of LSD? The answer is $4.8 billion dollars' worth. It comes out as being the largest seizure of LSD in the Sunset District in three years.

Then the trial turns out that the actual quantity that was present was only three grams, but the damage is done and the goods are done. The damage is done in the sense that there is a negative structure against this dope dealer who they're going to be able to prosecute more intensely in court. The good is done because on the statistics of narcotics arrests, they've got $4.8 billion of narcotics off the street. It turns out that's not so at all, but it invariably works that way.

In marijuana, they have seized how many plants of marijuana? How many pounds of marijuana? How many flatbed pickup trucks full of marijuana? One situation I was in—I'll get back to Santa Cruz. This was in Ukiah. They had taken out three flatbed trucks filled with marijuana. Well, it turned out the three flatbed trucks were almost entirely filled with the mud balls on the roots of marijuana. I mean, they're weighing the entire operation, and mud is not a scheduled material. Yet what went out were stems, pieces, sticks, and plants—some of them were up to six or eight feet high, some of them were little seedlings that had just been dropped as strays. But they counted the plants. In the Santa Cruz operation, they took appropriate leaves when the leaves were there and put them in a paper bag, and the paper bag was then assayed to verify what was taken was marijuana. The weight was the entire operation. The paper bag was what brought in the evidence.

That was in a situation that shows yet one other aspect of legal casualness. I was asked if I would, looking at that, give an opinion as to whether it was marijuana. I was on the stand and I was handed this paper bag. It was

LECTURE 23 ~ Cannabis and LSD

over. There was a hole in the corner of it that was filled with some kind of leaf material, and I refused to. I said, "I have no reason to assay a paper bag that had no security, had no authority, had no origin, and had been open to whoever wanted to take something out and whoever else wanted to put something in, and how anything I see in that bag has any bearing on this case whatsoever." The judge quite agreed and asked the DA to take the bag away.

There is the idea of the chain of command and possession. When you seize something that is going into a little spot and it's sealed off with initials and a date, I turn it over to you, and you initial and date it. If you open it, you take it out, you close it, and you seal it and initial and date it. You have a chain of command. You give it to the next person, the next person gives it to the person who stores it. And when it comes in to—pardon?

STUDENT: Chain of custody.

SASHA: Chain of custody. It is so often ignored, and yet every single step of that wave must be documented. You will look at that. "Where did you get it from?"

"I got it from such and such."

"To whom did you give it?"

"To so-and-so."

"I was responsible for it during that period of time."

Then they'll get the person to whom you gave it, put them on the stand. "Where did you get it from?"

"Them."

"Who did you give it to?"

"That person."

And right down the line. One break in the chain of custody, that evidence has no bearing on the case. Here's a bag that was open, unsealed, with a couple of holes in it, and I felt it would be totally irresponsible.

The same approach applied to a situation in which I was given a blood alcohol for independent verification. I got it from the forensic lab in Oakland, Western Labs, and I was asked to run a blood alcohol to back up their analysis. Well, no. It was not to back up their analysis. It was the defense's challenge on the amount of alcohol that was found. I got the vial, and the vial was sealed. It had a little forensic lab cap over the top of it, and on it, what it had was the name of the arresting officer. I don't think that was the source of the blood. It was the source of the blood, but not really the source of the blood. It was the defendant's, but I didn't know who the defendant was. There was no name of anything other than the name of the arresting officer. I just very quietly refused to run it. It was an anonymous sample. Why would you ever want to run an anonymous sample and let someone else put a name on it? Totally irresponsible. So I refused to run it. It went to court. I refused to run it. The defense said, "Why didn't you run it?" I told them why I didn't run it.

So this is not a chain of possession, a chain of custody, with the absolute knowledge of what you're running. If you're ever in a situation where you're asked to verify something or challenge something or run something in a case in which there is a criminal aspect, or any legal aspect, even a civil aspect, consider seriously, would you want to be in that anonymous position of having part of your urine or blood run by someone in which they may have been switching ground? It may have not been by intent. It could very well be by accident, but it has no bearing on what you produce in the way of blood or urine. No.

In this situation, think how others would handle you, and think how you would handle others. One of the best defenses is "I had no conviction of the authenticity of the sample." A jury will respond to that very nicely. They'll say, "Of course. If he doesn't know whose sample it is, it can't be pinned to the person." What bearing would the finding have on the case? So, this is the necessity of being extremely careful not only to the point of view of the law body to maintain that continuum of knowledge, but from you personally, if you're involved in that chain, to realize the strengths and the weaknesses of that chain. If the chain is not intact, be very cautious that you do not add authority to areas that are not observed.

Okay, back to marijuana. How do you know if the material is marijuana? Well, there are a number of tests. There are the physical tests. The visual picture of the leaf. But there are a lot of things, from Japanese maples and who knows what, that have little serrated leaves. You have the cystoliths on the back, which are determined by microscopic viewing and with a bit of acid on it that releases the carbonate. You have color tests—there's the Beam test. There is the Ghamravy test. There is the Duquenois test.

LECTURE 23 ~ *Cannabis and LSD*

One person asked about the Duquenois. The Duquenois test is the standard color test. It is an extract of the plant that is submitted to ethanol as an aldehyde, and acid, and in the course of this it develops a purple color. The color is extractable into chloroform. So in essence, the Duquenois test is the generation of a purple color from the Duquenois reagent and extract of the leaf, and the Levine modification of it is the extraction of that color into chloroform. It is said that this is a definitive test for marijuana. No, it's not. There are many spices, there are many other things—coffee grounds, if you have the right kind of coffee grounds, it will give a positive Duquenois. But it is a very good exclusionary test, because if the material does not give that color, it is not marijuana.

I remember one instance of—I don't think I've mentioned this before—voir dire. Have I mentioned the term *voir dire*? *Voir* is "see," *dire* "to speak." It's when you get an expert on the stand. Before he's admitted as an expert, the two attorneys, the prosecuting and the defense attorney, have a chance to talk to him. Ask him where he was born, where he got his education, his background, his authority to be an expert, the field in which he will be admitted as an expert if he is admitted as one.

By the way, if you are ever called in your professional life to become an expert in a criminal case, it's very important that you make your field of expertise as narrow as necessary for the case. It's great for the ego to also bring in the fact that you're a confirmed anthropologist and you've been a minister at the Unitarian Church for 10 years if it sounds like it's a good thing, makes you look a little larger. But you may get questions about the Unitarian religion and anthropology that make you very embarrassed on the stand. Figure out what your expertise is and limit yourself to that.

This was a case where a person was being voir dired by a very wily lawyer who was trying to defend a person in a very defensible case, and the case was going to be the identification of marijuana. Part of it was the Duquenois test. During voir dire you do not ask anything about the test. You ask things about how the person in principle would respond to something, or how he would run things, or what conclusions he would draw from what he saw. And the question was the Duquenois color, amongst other things. The person said, "It is a reaction of acid aldehyde with ethanol and acid on extracts of marijuana that turns purple," which is very neat. That's why I said Duquenois. It took seven hours to unravel voir dire. Questions came up like, "What is purple?"

"Well, you know, purple is the color of that shirt over there."

"Oh, I thought that was violet! Was violet a positive test on the Duquenois?"

"Well, yeah, I guess it would be."

"You're not sure?"

"Well, I guess you could spectroscopically determine the difference between them."

"Do you run a spectroscopic analysis?"

"Oh no. No need to, because if it's violet, it's positive." And boy, just this kind of opening, the lawyer went just piece after piece off the other. What was left was virtually nothing.

"I didn't run a color test. It depends upon visual identification." Orchid got thrown in there as another color. He ran this testing and he ran that.

"Why don't you run this?"

"Because that is not as good as that. That is an inadequate test."

"It's not sufficiently adequate."

"Let's call it by name. You do not believe that test." He got into these claims he'd see as being a valid analysis, to throw out TLC as being a valid analysis. He claimed all this weird background and he got thrown out. He was a state chemist, and he had sweaty palms after for seven hours of this operation. He ended up saying what he wouldn't call a valid test, and I think he excluded everything he might want to call a backup. I don't know how the case came out, but it was a very grueling seven hours of voir dire.

But the thing is with the Levine modification, the color is extracted into chloroform. "But you ran this test for years before Levine established his modification."

"That's right."

"And the test was not as good."

"That's right."

"Therefore, the test had a greater probability of being wrong. And with a greater probability of being wrong, since the test of the Duquenois is known to be wrong on some occasions, you undoubtedly must acknowledge that you probably gave wrong answers on more than a few occasions."

LECTURE 23 ~ *Cannabis and LSD*

"Well, I guess so."

"Then you are using a test to send people to prison on the basis of tests that you now know to be inadequate. Have you done anything to remedy this?" [Laughter.] Put yourself in that for seven hours!

So when you get in there as an expert, watch what's being said. A very important point in court is listening to what the question is. Often, a very serious problem is that a question is asked and an answer comes out that goes beyond the question. It goes on and elaborates upon something else in which you may be giving fantastic ideas to the prosecuting attorney, who had never thought of that area. "Oh, I never thought of that. Let's ask some questions on that idea." Don't go beyond the thing that is asked.

For example, especially in areas where you don't have the judge to defend and you don't have a lawyer to intercede, such things as giving depositions. Depositions, has anyone given a deposition? A deposition is where you are sitting with the other attorney across from you, with a little guy who goes tippy-tap on this little machine, recording every word for posterity. Your friendly lawyer, your ally lawyer sitting alongside you, he has no way of voicing anything that would stick as an objection, because that's the structure of depositions. You have no judge present. "Your honor, I would like it very much if—" There is no judge, and questions can be asked about anything they choose. It's like a grand jury questioning. You are asked anything that the questioner would choose to ask you about, except you have no way of turning to your lawyer and saying, "Is this a proper question?" The lawyer would say, "No, it's not a proper question," and the guy would say, "Enter it anyway." That's what happens in a deposition.

So they'll come up with questions. You have to be very cautious, and a very good trick is to present your answer as your opinion pending further research. The reason for a deposition is that when you get into court, then everything you've said can be glumped on the judge's desk in one clump and you don't have to go through the whole thing again. You save time in court. There are questions such as, "Do you have an opinion in this case?" Think what the question is. The question is not "What is your opinion?" The question is "Do you have an opinion?" The answer is yes. Not the answer is "I think it was hashish that had been diluted with soy sauce." No. The answer is yes. "What is your opinion?" Then they get the specific question.

This happens in real court, too. Because if you're asked by the defense attorney, "Do you have an opinion?" and you blub out the opinion, he has no way of building up the jury's ears to get that opinion as having a good quality. "Do you have an opinion?"

"Yes."

"What is this opinion based on?"

"Zz, zz, zz".

"And your authority?"

"That, that, that, and that" and all this background comes in, so when you come out with the opinion, it has a little added ringing of authority in the ears of the jury. But the answer to "Do you have an opinion?" "Yes, it was hashish diluted with soy sauce" takes away any possibility of giving added psychological reinforcement of what the opinion is.

If you're ever in a deposition, and they ask what your opinion is, you can say the following: "My opinion was hashish diluted with soy sauce, but—" At which point the person who is taking the recording waits. You've not finished your sentence yet. "I intend to do quite a bit more research in this area, and that opinion may change." Beautiful. You've taken the fangs out of the teeth of the other lawyer. Because this is what your opinion is as of now, but it's going to be presented *then* in court, and in the intervening time, you may get all kinds of additional evidence, your opinion may change all down the line. It's the only way to get out of the deposition with your ass intact. "This is where I am now, but later it may different."

Okay, marijuana. Hash is the resin that protects the seeds, protects the female aspect of the marijuana plant. Usually the more harsh the conditions, the more resin. It's the resin that contains aromatic compounds, which protects the seed from ultraviolet, absorbed ultraviolet. So if you have intense conditions, rather aggressive conditions for growing, you're apt to get more resin. Hence the idea of raising marijuana under rather spartan conditions. Yes?

STUDENT: What are those spartan conditions?

SASHA: Usually a lot of exposure to the elements that make this plant difficult to grow. They don't give it the lush richness. But if you want the hashish, you want the resin, you want to give it just short of almost not

LECTURE 23 ~ *Cannabis and LSD*

surviving conditions, such as exposure to a lot of light, which would cause the resin to be generated to protect the seed. Shorter growing conditions will make less plant but will usually produce more resin.

STUDENT: Could you explain shorter growing conditions?

SASHA: Under conditions where it is forced to mature because of the shortness of the day cycle; under unusual conditions. The genetic contribution can be quite sizable if the origin of the seed was from certain areas.

STUDENT: [Inaudible.]

SASHA: In my experiments it did not. Yet I've talked to people in Mississippi where it has. My seed sources may not have been valid. There has been a lot of work done by the Department of Agriculture on hemp polyploidy and hemp growing conditions, back in the 1920s when hemp was a material that was used commercially for rope and not a material that was used commercially for drug sources. There are a lot of interesting studies that were made at that time and are in the government literature.

There has been a lot of art put into the writing of what the growing conditions are for sinsemilla and for different forms of marijuana that I have not kept up with. But generally, the resin, which is to be taken off as resin, is the source. It is what's called hashish, and it has most of the THC content, maybe several times that of the actual marijuana plant.

Other terms used in marijuana. The so-called red oil is a hexane extract of a plant which has been evaporated down to residual oil. It will take out the THC but a lot of other things too. It's called Red Oil of Marijuana or hash oil. Under the right conditions, that oil can be distilled to get away from nonvolatile fats, and that's often called "distilled THC" or "distilled oil." Hash oil. It's a very pale amber color as opposed to a red color.

What other forms? There are isomerizers that had been used to try to convert inactive forms to active forms. I have never seen any quantitative assay of their efficiency. The idea that you have such things as cannabidiol, which is not active but can be acid catalyzed to become THC, active. In principle it's fine. In practice, I don't know if it works or not[23]. Any thoughts?

---

[23] In 1978, the DEA published their conclusion that commercial isomerizers, specifically the Iso-2, did not work as described, suggesting this was possibly due to their employment of sulfuric acid rather than hydrochloric acid as had been described in the literature. See *Microgram* vol. 11, no. 7, page 109. See also vol. 11, no. 1, page 1.

STUDENT: What about coating it with frozen carbon dioxide?

SASHA: Carbon dioxide can inspire change of sex.

STUDENT: I'm talking about after you harvest it.

SASHA: I have no knowledge of that, no knowledge of whether it's valid or not. As with many things that are sort of aphorisms of the trade, it may work. I don't know. I've never tried it and I don't know anyone who has tried it. The one thing about the isomerizers is they are now considered to be paraphernalia. It is associated with drug trade, and there are antiparaphernalia laws in both the federal and state area. If, for example, you sell cigarette paper and it is argued as being in conjunction with marijuana smoking, that becomes a crime. If it's in conjunction with smoking Bull Durham, it's not a crime. They have such things as clips, cigarette papers, a host of things, including such things as chillums. Anyone know what a chillum is?

STUDENT: To smoke out of? A little pipe.

SASHA: Oh, that's—okay. I thought that it had a name that started with a "B."

STUDENT: Bong.

SASHA: Bong, but also, what's a carved thing? A chillum is a pipe for smoking.

STUDENT: It's a straight pipe. You hold it straight up in the air like a cone. It looks like an ice-cream cone. It's used by Jamaicans.

SASHA: Chillums were specifically listed as things that could be considered—but not limited—to narcotic paraphernalia. I had no idea what a chillum was, had the wrong idea entirely. So certainly pipes become paraphernalia if they are geared for marijuana use, and possession is a crime. Pipes that are geared for other uses are not paraphernalia and the possession is not a crime, hence the intent becomes very instrumental in determining whether a crime is committed. Intent, of course, is an argument between people who talk, and not something that's intrinsic in "you like a star on your forehead" or something. So these are difficulties in the whole paraphernalia law.

LECTURE 23 ~ *Cannabis and LSD*

Okay, a question was asked about LSD flashbacks. Are they real? It's hard to say. There are instances in which the experience of a flashback is the concept of déjà vu. I think that's a familiar one for people who have pounded around either in French or psychology. "I have a feeling this has all happened before." Who was the one who made the marvelous phrase? "It was the feeling of déjà vu all over again." [Laughter.] But the feeling that this has happened before has occurred in many instances; the idea of smelling something and realizing a whole thing had reoccurred. In the drama of LSD usage—and this is especially true with LSD when it has produced a very negative or difficult to assimilate or traumatic experience—things that occur to you, sounds you remember hearing, visions you remember seeing, a conditioned situation is an example.

There was one girl who talked to me at some length about a bad LSD experience where she had found herself walking in ivy, and the ivy was totally sinister. There were faces in the leaves of the ivy, and the faces were of people or things that were out to get her. And she could not walk in the ivy, but she was in the middle of the ivy. She couldn't get out of it. She had associated this in a very bad way. It was a very difficult experience. She told me that it was about two or three months later that she was riding with a person that she was not comfortable with, back to somewhere in the Midwest, in a pickup truck, and there were splotches on the windshield. She saw in the windshield these splotches as being like the faces in the ivy, and the whole experience came back and sort of rekindled. A flashback. That would be what's called a flashback.

Is it due to LSD? The fact that the faces were seen under LSD and were traumatic means you've associated them with LSD. I believe it's associated with the trauma of the faces, which happened to be made evident by LSD. I think you can wake up from a horrendous dream, a killer dream of some kind, and you have this chill. "What woke me up? I don't know, but I'm awake." I think people had something akin to this. In that dream there were, say, faces in the ivy. Then, at some later date, you see splotches on the windshield that were faces that reminded you of that dream—a flashback too. But I don't think it is any more a flashback to a dream than a flashback to LSD.

People have argued that there is a residual molecule, maybe the molecule is not totally out of the system. No, the molecule is out of the system for all intents and purposes within a couple or three or four days of using LSD.

The flashback can occur at any time. It is a déjà vu. Something that reminds you of a traumatic event—although it can be an unassigned emotion. It's like suddenly—I had this just the other day. I smelled for the first time in 14, 15 years, I smelled slightly scalded, simmering milk. Somehow milk got on a hot plate on the stove, and I caught that smell. It had just that right concentration. I was back when I was 11 years old, when I had scalded milk at one time on a hot stove. I can't remember the details. But I suddenly knew that smell and I was back there in a flashback.

The term "synchronicity" that Jung has brought into his arguments, things that are just too coincidental to occur. The hearing of something, and feeling, the sound of dread associated with it, or the sound of uncertainty, the waking up with the certain knowledge that your mother is dying and you can't shake that knowledge. Why? Because somewhere, whatever it was that got into the unconscious from the sound or smells or conditions or the buzzing of a fly, I don't know what, was associated at one point with a dramatic awareness of the death of someone. Then the rekindling of the clue brings, at some level, the association with the problem that previously had been kindling that clue. This is the concept of a flashback. I don't think it's a drug affair. But it's very much built into the nervous system and the brain.

Yes?

STUDENT: I remember what I had heard on the street is that—I don't know. Somehow some of the chemical is retained in your spine—

SASHA: That's the lingering molecule idea. I don't think it has any merit. You can't find it. There's no reason it should be there. It washes out. You can argue an extended effect partly because the molecule takes a while to metabolize and be gotten rid of. But the recurrence of an effect at another time you would lose, because the molecule just isn't there.

STUDENT: If it were there, would it continue to be effective from when you took it?

SASHA: No. I said after a couple or three days the molecule cannot be detected. But the LSD experience is over in a few hours. So the molecule can still be there and not incur an effect.

LECTURE 23 ~ *Cannabis and LSD*

ANN: There might be some confusion with things like PCP, which deposit in the fat, and you can have continuing PCP eruptions.

SASHA: Not particularly.

ANN: Really?

SASHA: No. PCP, for example, is something that gets into the fatty tissue, very lipophilic. "Lipo"—lipid, fat. "Philic"—loving. It loves fat. It goes into the buttocks and the fat deposits in the body, and it is oozed out over a long period of time. That is one reason why THC does the same thing. It goes into the fat tissue, and it's oozed out over a long period of time. Hence, very disruptive tests that can be made for THC a week or two later. The same thing with PCP, but this does not mean the material is active for that length of time. You may have some residues from PCP or from marijuana at the 24-hour point. You sometimes have a period of some refractoriness, not too much for either of those drugs. You have a period in which you are paying a compensation for where you were; you don't feel quite together or quite with it. You may be not totally functional. This is a consequence of the impact on the body of the experience, and not a consequence of the chemicals going into fat storage. You don't have neurons in your fat. It is stored in there, but it doesn't do anything there. But it does do its mischief in long-term testing, because you will find on the sensitive instruments you have both PCP and marijuana days and days and days after exposure.

I had a phone call from a lady I worked with in a clinical lab about 15 years ago. It took a while to put her name together with who she was and who the face was. It was good to talk to her again. She said her family was all away and her youngest daughter was now a nurse, a registered nurse somewhere, and that she was going to get involved in a nursing project that involves drug rehabilitation. They have to give a urine test, and she wonders if she had used marijuana three or four days earlier if it might show up. This is the review of registered nurses in Kansas. I love it. I said, "Yes, it might." I said, "It's very simple. Do the following: Tell her, if she has the aggressiveness to carry it off, she was at a party where all the funny stuff was going on. She did not use marijuana, never used marijuana, but she had been at a party where they were maybe using marijuana, and that was three or four days ago." And let the chips fall where they may.

They say, "Well, do you have it in you?"

"Maybe. I don't use marijuana." That's the way you have to address that sort of thing, because it will linger in there. This is called "passive smoking." Passive smoke is where you are in the environment where the material is being used. You take a certain amount into you. The amount is maybe 1/20th of what it would be if you actually shared the joint and toked some. But the amount of 1/20th can be picked up, because the test is 20 times more sensitive than it need be.

So, there is no good way—although what can be done in reality may not follow the procedure—in which a presence of a particular drug in your body is inescapable evidence that you have used the drug. You might have been exposed to it but have not used it. She said, "My daughter is exactly the gal who could carry that off. I have no doubt she can carry it off." The tests that were being used, the extractive tests, were much less sensitive. Radioimmune assay is very sensitive.

Yes?

STUDENT: I remember when I was growing up, I had always heard if you dropped acid seven times you were legally insane. [Laughter.] It's just one of those things you've heard.

SASHA: Forget it. I'm not insane. [Laughter.] I've dropped acid more than seven times, and I'm a bit insane but not legally insane. [Laughter.]

STUDENT: I remember what I heard is that it's safer to take acid once a week than to take it twice in one week.

SASHA: Okay. Safety, I can't speak of. I don't think anybody should take acid. You may do things that you regret doing, and you may act out of a strange part of yourself, and your behavior patterns may not be laudable; to that extent there may be a safety factor.

The main negative aspect of using acid frequently is that you don't get the balance for your buck. You are refractory to it. I know a person by the name of Thad Ashby. He is a poet who goes around the western hemisphere. I haven't seen him in 15 years; he may no longer be with us. He goes around carrying a little satchel. Anytime he visits someone, the satchel goes with

LECTURE 23 ~ *Cannabis and LSD*

him and goes in the fridge. When he leaves, he takes the satchel from the fridge. That is his lifetime supply of LSD, and he likes to keep it cold. He's got it all marked off in there. His routine is as follows: "today I take 100 mics, tomorrow I take 200 mics, the next day I take 400 mics, and the next day I rest. Then the next day I take 100 mics again." Because he has to up it each time, double it each day to get the same impact. At the 400 point, he's going to run out of LSD. [Laughter.] So he knocks off a day and hopes to get the refractory period lost. Well, I don't think this is so. I think refractoriness of LSD lasts three or four days, and any use of it more frequently than that will not give you the effect you want. Hence, you will use more than you normally would.

The side effects of LSD—LSD, after all, is an ergot material—leads to ergotism: coldness in the extremities, a certain amount of vascular closedown. You can get vascular spasm; you can get restricted circulation. There are definite negative aspects of LSD that will come in more and more as you get to larger and larger doses. So I think you are getting into a bit more of a physical risk because you need to take more material. I'd say twice a week, if that is your desire, would be probably the most frequent you can do it and get reasonable responses. Once a week could very well be a good precaution, just for economy of material. As well as economy of your own spirit and time investment.

Yes?

STUDENT: How about something like cocaine? What would be harder on the body: doing a gram in one night or doing a gram throughout the week? What would be more toxic or harder on the body?

SASHA: I think when you use a lot in a short period of time, you're falling into a couple of traps that are probably working against you. One, with continuous usage of cocaine, you don't maintain the exhilarating power of the high that you got your first one or two times.

STUDENT: Regardless of the high, what are the aftereffects of the body?

SASHA: I can't see a difference with them.

ANN: I thought that the dangers, when they do exist, are from the impact of a large amount in a small amount of time.

SASHA: All at once. Normally you get cardiovascular problems, and you get real problems if you take the large amount all in one dose. But if you pace it out, your cardiovascular system becomes tolerant to it, and you also change your response to it. I think all drugs would be probably better if occasional and spaced out rather than used in an orgy of burning out at one time. This is true with almost all aspects of eating, of living and doing. If you can be a bit of a moderate, you tend to maintain a response, which can be from food or from sex, whatever you have, if you retain a certain amount of balance of things. I think one of the disruptive things of any of these compulsive contributors—be they drugs or be they food or be they sex or be they work or whatever it is—is that you get into an addictive pattern of handling them and you lose the facility of choosing. You get into doing things by habit, by the addictive habit, then you are less of a good judge as to the consequence. That has to be maintained as some sort of cognitive sense.

Any other questions about marijuana, hash, LSD? Yes?

STUDENT: I do have a question. How can you measure THC content?

SASHA: To measure THC content, probably the best way would be spectrophotometrically.

STUDENT: So, you have to have $150,000 to measure it?

SASHA: No, you can do it with a $6,000 one. [Laughter.]

STUDENT: Is there any litmus paper test?

SASHA: Yes, you can get a pretty good guess by running a thin-layer chromatographic thing, which is a 50-cent plate, a TLC plate. How many people have run thin-layer chromatographs? One, two. Okay, somewhere along the line I'll bring in a thin-layer chromatograph and run one here for show-and-tell just so you can see what it is and what you can do with it. Then spray it with something that will give it a color intensity that is a function of how much is there. It won't give you an accurate number, because you don't have a reference thing to match the number against, but you can say, "This is 10 times what that is . . . about" and get an estimate of quantity.

LECTURE 23 ~ *Cannabis and LSD*

STUDENT: So you can do that?

SASHA: Very casually. Well, probably with about 20 to 30 bucks for a tank, some other things, and a spray to kind of go over it with, you could probably run a pretty good assay. It would not be quantitative. You would separate things like delta-1 from delta-6 from cannabidiol, and you can tell if there are appreciable amounts of the other materials present. It would give you a good qualitative picture.

STUDENT: For the flashbacks, a friend of mine was telling me something. Instead of having the same images from the LSD trip, having different images. So it's not a flashback of the same images, but it's a similar thing to what he experienced on LSD. He was sitting in class one day and out of the corner of his eyes, the walls started moving and expanded. He hadn't had any in quite a while. But it wasn't a flashback of things he had experienced before. But it was once again—

SASHA: Something reminded him of it.

STUDENT: Okay.

SASHA: Okay, I don't know how to handle that. In fact, I will say one thing, that one person I know, a French horn player who is now down in the southern part of the state, has a son. The son got into LSD about 8 or 10 years ago, and he flipped. He really kind of went into a psychotic place. The connection with the LSD was in this man's eyes. You have what's called a pre-psychotic person, a person who is not psychotic but not that far from it. I think you could almost visualize yourself under certain circumstances as having a diffuseness of mind attention, of mind direction, that you could almost understand as being maybe a little bit troublesome.

This son, at the time he was experimenting with drugs, went through what I would call a psychotic break and stayed, more or less, in this new territory. When I met him, he had monstrous shoulders and arms, the hands are folded like this, with a perennial smile on his face. You would have the feeling that you couldn't get through to him, and you couldn't. It was a change. The emotion was not there. The affect was not there. There was this continuous smile and inability to talk about anything complex.

I just remember very vividly his folded arms, beefy arms, big body, and his continual unresponsive smile. The father told me LSD did this to him. And I thought a little bit and said, "Did it really occur at the end of the experience?"

"Nah, it came from the time he was farting around with all those drugs."

I said, "Did he use much marijuana?"

"Well, he was using a fair amount of marijuana."

So, you have an area where you are dealing with a person who may have only a marginal connection with a real world that they see as logical and usable and coherent, and something gave them a view of another world that is easier, and he went there. Now with this person, I would always keep an open mind to the possibility that what he is seeing is a bit of this easier world, and he might just have accepted that from the LSD experience. LSD is among the areas of psychedelics that have been called psychotomimetics, because they do in some ways imitate some aspects of psychosis. Mentally, most psychotic structures deal very heavily with the auditory. "I heard voices. The voices said 'kill.' I killed." LSD and most psychedelics deal with the visual: "I saw things." But I'm always suspicious of a person who has this kind of thing, if there are sounds and voices attached. That is a clue in the other direction. [To student.] What's his state now? Is he still with aspects of worry about the LSD experience?

STUDENT: No, and it wasn't a negative thing either. He takes more now. I've talked to other people, myself having a similar thing, but that was just the most extreme one of having a flashback of not a memory, but a recurrence of the effects.

SASHA: I don't know how to answer.

ANN: There are lots of different conversations that are very intense, and sometimes you get a conversation about ghosts and things that go bump in the night or something, and you feel it chill up and down your spine, and things can get a strange energy, or the room is spooked.

There are different ways in which you can suddenly get the visual effects of things moving. I've had it happen. No connection with drugs at all. I was in an altered state for a few minutes. That is different than the usual because

LECTURE 23 ~ *Cannabis and LSD*

your attention and your energy has changed, because of a lot of intense stuff. It could be exactly the same.

STUDENT: Right. It's just similar effects from different things—

STUDENT: Right.

STUDENT: I have a slightly unrelated question. Peyote. Do the Indians ever—is there any use of it for nonreligious purposes? Or is that pretty strictly religious?

SASHA: It's strictly religious, but the religious use can be called for rather casual reasons. So it's a little bit dicey. They can celebrate someone's having become a father or someone having returned from a long trip by having a religious celebration. So yes. It is only religious in that sense, but it's quite loose.

STUDENT: What type of LSD was that poet taking? Was it Sandoz or—

SASHA: Oh, I don't know what his source was. But a lot of the source material that I had seen in the early distribution in this state had been blue, goopy oil, which I'm sure is loaded with other ergot residues. Pure LSD is a white crystalline solid, either as an acid salt or as the base. But a lot of the material that was made available for experimental use was material that had been very crudely made and titrated only by the person who sells it. I'll take a drop, you take four drops, turns out it's sold as a half teaspoonful for a dose and what might be in there was not known.

STUDENT: It was liquid form?

SASHA: A liquid form.

STUDENT: Is there any difference between the different kinds, like White Lightning, Blue Cheer, Orange Sunshine? Purple Devil? Or was it just—

SASHA: Purity, identity, origin, salability, product recognition. The liquid form of LSD in solution is quite unstable. Hence, it has to be kept cold, kept out of the light, and kept out of the air. In solid form, it can be more stable. A lot of the solid forms are a liquid form put on a matrix. So it's still a liquid form, still unstable.

As a salt it is still subject to moisture and air oxidation and ultraviolet light, but it is more stable because it's in a salt. Something that is often overlooked is taking LSD, which is extremely responsive to halogenation, adding chlorine—when people will mix it up in tap water that contains chlorine—and when they get around to assaying it for whatever reason somewhat later, there's no effect. Because chlorine in tap water can destroy it. So there's a lot of little things in solutions that are very treacherous. As far as the different forms with their classic names, usually it is different dosages and they carry user recognition, imagery.

ANN: Are most of those made in Japan now? Because they used to—

SASHA: Japan, quite a large amount. Australia too.

ANN: Australia? Oh! I've never seen one of those.

SASHA: And the interesting thing about LSD usage, it certainly isn't in the newspapers much anymore, but apparently the usage in America is unchanged over the last few years. The amount that gets used, the size of the circle of people using it, is pretty much unchanged. There is just a lot less trauma from it, and I think the reason is that in the use of LSD, it has been learned how to be used.

STUDENT: Are the dosages a lot lower than they were?

SASHA: It's possible. The typical dosage of the clandestinely bought material can run 50, 75 micrograms, but the dosages of some of the Owsley tablets and things from 20 years ago could be 150, 200 micrograms. But on the other hand, a lot of people use three or four tablets of the lighter stuff. So I don't know to what extent the dosage is a valid argument.

STUDENT: Does it all come in the blotter form now?

SASHA: All forms. Tablets and blotter forms currently. The blotter form is currently quite in vogue.

STUDENT: I think I saw it once made out of gel or something.

SASHA: Oh yes. The little squares of gelatin. In that case, it had been put in the gelatin, but in the matrix of the gelatin, so that the surface would

LECTURE 23 ~ *Cannabis and LSD*

decompose, but the inside would still be intact.

Student: Can it be absorbed through the skin?

Sasha: Yes, it can be, but it's a hard material to get through, and so you have to use an appropriate vehicle. What has been used, there are such things as ethylene glycol, polyethylene glycol, a solvent that absorbs through the skin. There's a lot of talk about using DMSO, which is an arthritis material that is sort of frowned on. It's being used now, used to get into the skin. Are people familiar with dimethyl sulfoxide, DMSO? Most people I think have heard of it. It's sold at hardware stores for hardware purposes and not for medical use, but it's used for medical use. It's used largely for arthritis. It's rubbed on the areas around where the arthritis is. Characteristically, once you rub it on your skin, you'll taste it in about 30 or 40 seconds. This gives you a good argument for rapidity. It goes into the skin, into the circulation, and finally, excreted in saliva. So you actually get the taste of it.

I never figured out why it's used there. It's sold at hardware stores. I don't know any hardware use for it.

Student: It's used for cleaning. An industrial cleaner.

Sasha: It's used for cleaning? Okay. Here's a good point, thank you for bringing that up. If you're ever using the DMSO, and you're using it through the skin for any reason, arthritis or whatever, don't have dirty skin, and don't have grease on your skin, and don't have base on your skin, because the DMSO will take the grease and the base in through the skin as well. So do it from a clean point of view, because whatever is on there that's wrong will be carried into the body. One of the worst burns you can get, caustic burns, is to get sodium hydroxide dissolved in DMSO and get a drop on you. DMSO goes in and takes the sodium hydroxide in, and it burns a hole as it goes. So when using industrial cleaners, use rubber gloves. Be very careful with any use of DMSO around anything you do not want inside of your body. Yes?

Student: How about LSD in the eye, the liquid LSD? Is there any damage to scraping the lens or anything like that?

Sasha: By putting it in the eye for absorption? I don't even know if the lens would have any response to it.

STUDENT: Is that better than the mouth?

STUDENT: It's quicker.

SASHA: Quicker. Doesn't take less dosage. Actually, it's been explored in one of the reports a couple of years ago for LSD: orally, intramuscularly, intravenously, and intraspinally. There's no way you can get it in faster to the brain than going to the brain. In all cases, the dosage requirements were the same. The speed was different. It's a case where the dosage did not reflect the route. So if you're not a terrible rush, I wouldn't pursue that. [Laughter.] Yes?

STUDENT: What about the use of LSD with vitamin C or orange juice to enhance it or—

SASHA: Sounds like one of these many stories I've heard. I don't see why there should be a connection with vitamin C. Vitamin C is an antioxidant, and vitamin C could prevent oxidative decomposition of LSD in storage, but in the person, I wouldn't see how that would make any difference. Yes?

STUDENT: And are there any actual methods of something you could take to bring you down? A friend of mine said he takes Valium and it works.

ANN: Niacin?

STUDENT: Yeah, there are lots of things that are supposed to work. Do you know?

SASHA: I've heard both niacin and Valium. I think Valium would probably, or you can definitely get a person out of LSD with a good antipsychotic, a tricyclic like phenothiazine or chlorpromazine will abort the experiment, but it doesn't bring the person out of the experience. It merely makes the person not respond that much. It's a way of bringing an end to the experience from the point of view of the observer. The person is not acting in a bizarre way anymore, but the psychological turmoil is still going on. I am a great believer in not aborting an experience, even though they're getting

LECTURE 23 ~ *Cannabis and LSD*

out of hand and they're difficult to handle, they're distressful. I'm a firm believer in joining the person and going through the experience with them.

ANN: You mean without taking the drug—

SASHA: Without taking the drug. But getting with the person one-on-one, and support, be honest, contact the person, let him feel that you know where he really is, and be okay with it. Then walk and talk and whatever it is, and get the experience to work its way through. Talking a person out of it, walking a person out of it. Don't even say they are "out of it." They are in an area. The distressful response to it is very often a consequence of the environment they've been in, but put them in a reasonable environment and that distressful consequence will very often disappear. I think that a lot of the use of Haldol, for example, in the emergency room, to bring an end to a psychedelic experience, leaves psychological storms still going and is not a constructive way to resolve an experience. It's a convenient way for the physician, but is not a constructive way for the patient. Yes?

STUDENT: What do you do with somebody who is just way off on free association?

SASHA: I would go out there and free associate with them. It takes a one-on-one, and you're going to invest six hours of time, and the person is going to be surprised, they will use you as a sounding board, and you're going to get in there. If you've had some experience with psychedelics, you can find yourself going very high without any drugs whatsoever, and pretty soon you're locked in with the person. Be honest, be authentic. Don't say, "Oh, this will pass in a few minutes." You don't know if it's going to pass in a few minutes. "You'll be okay." You don't know if they are going to be okay. At some unconscious level, they'll know that you're trying to assure them and that will be a breach in that transfer of honesty. Say, "I kind of envy you. I know where you are. I've been there myself. Tell me about it." And pretty soon, you get into a deep conversation. They transfer everything over to you. You have become a trusted ally, and it's done in four, six, eight hours.

ANN: If you can get them to either write or sketch and draw on a piece of paper, it's a tremendous help, so they get it out there, whatever it is they're going through.

SASHA: But that is to me the way of handling an overdose or handling a situation that is not really going well. Be an authentic ally to them.

STUDENT: It doesn't seem like it takes all that long either, usually.

SASHA: Usually the person knows you and knows that you're familiar with the territory and kind of opens up some of the adversity. You're in there in five minutes. It can be very rewarding. But that doesn't end the experience. You're in there for the rest of the afternoon.

STUDENT: I had one experience where I was on the phone with my cousin, and I was kind of panicked, and I was obsessed with time, and he said, "Don't think about time. Think about space." [Laughter.]

SASHA: Fair enough. He let you out of your concern.

STUDENT: There seemed to be more space in the room. [Laughter.]

SASHA: Okay, other questions on—

ANN: There are some interesting things we've heard over the years about the best handling of small children who accidentally ingest LSD.

SASHA: Oh, yeah. That is something that comes up periodically. Parents keep a few sugar cubes in the fridge, a nice, safe place, and the kid knows sugar cubes are good to eat and down goes the LSD. I know one important person in San Francisco who had a three-year-old daughter who got in their residual stash of LSD on sugar cubes. One of the most traumatic things to do to a child, if the child is getting into some strange little spot, is to bundle them in a panic into a car, rush them to the emergency ward, have them strapped down, and pump their stomach. Can you imagine what that means in the way of memory to a three-year-old? It's a pretty grim association with a lot of things. This person realized their kid had taken a modestly large amount of LSD. And the kid was absolutely enjoying playing with the usual imaginary friends they had never been able to see anyway. [Laughter.] They just kind of giggled a little bit more and came through the entire thing

LECTURE 23 ~ *Cannabis and LSD*

without a single traumatic moment.

The trauma is usually in the parents and getting into an emergency situation down at the local hospital where the young physician who thinks this whole area is life threatening, does everything lifesaving that's possible. So to a large measure, this is not helpful, unless there are overt signs of neurological overdose such as convulsions or circulatory disturbance, where you know the physical body is undergoing risk. Then go ahead. If you must go into an emergency ward for a situation, treat it symptomatically. Treat it supportively. If there are circulatory problems, relieve the circulatory problems. If there are convulsive problems, quiet the convulsive problems. But don't go and try to remove the drug. By this time the drug is no longer in the stomach. Stomach pumping is pure trauma. So as a rule, children are quite resilient to LSD and come through with no scars.

STUDENT: About a month ago, the Daly City Police Department sent out a notice or whatever all over San Francisco and Daly City that it's real common right now, something that's going on is that dealers have "blue stars," I think that's what they call it. But it's LSD, and they're giving them to little kids and it's absorbed through the skin.

ANN: Oh yes, tattoo colored things.

STUDENT: Yeah, they're little tattoos that they give to the kids. And that's their new way of getting new clients or whatever for their drug dealing business, going to junior high aged kids, and that's a way of pulling them into it.

SASHA: All I can say is I wonder how it's absorbed. Is there a vehicle there?

STUDENT: It's a tattoo and so it's absorbed through the skin. They put it on as a tattoo and they have a trip and so then they come back.

SASHA: I had not heard about that. I appreciate hearing about it. I have no idea[24].

ANN: It's one of those little transfer papers, apparently.

---

[24] The myth of the Blue Star Tattoo and similar transfers has been shown to be antidrug fiction intended to frighten parents. It has no basis in reality, and there is no evidence this was done for or given to children. See https://erowid.org/chemicals/lsd/lsd_myth3.shtml.

STUDENT: After I tried LSD, one of the first things I thought was, "Gee, I wish I had known about this when I was six." [Laughter.] Yeah, with LSD, I thought, "I wish I'd known about this when I was six" because all the effects were what I always wished I could have been doing when I was a little kid. It would have been nice to do this or that.

SASHA: Yeah.

STUDENT: And that was one of the first things I—

SASHA: Oh, once the active imagination and the pure world of fantasy and hallucination are open, you can recall back to that period, and it's quite a remarkable time.

Well, I never quite got onto the environment. Anyway, we cleared up a lot of questions around the drug things.

AFTERWORD

Sasha Shulgin and the Alchemy of Education

Chemistry, the word and the discipline, derives from an older word and discipline, *alchemy*. The historical transition from alchemy to chemistry occurred several hundred years ago, occasioned by the European scientific revolution that reconceptualized what had been an animated and sentient nature as now composed fundamentally of dead matter (elements, atoms, molecules) interacting in mechanical ways according to discoverable physical laws.

The practice of alchemy in its exoteric form looked very much like an earlier version of the chemistry of today, except that to the alchemist, nature was alive. Alchemy also had an esoteric component, of interest to some practitioners, concerned with transforming the personal and collective psyche. In this sense, alchemy relates to shamanism, to yoga in its full manifestation, to modern day depth psychotherapy, and to spiritual traditions of many kinds that deepen connection with mind/spirit/psyche.

Sasha Shulgin's expertise and creative productivity as a chemist are legendary. His chemical creations were transformations of physical matter—adding and subtracting and rearranging atoms to generate new molecular forms. In creating these new forms, he was guided by questions of how the molecules might impact the many manifestations of the mind: thinking, feeling, perceiving, consciousness, connectivity, and beyond. Sasha's work in the laboratory was an alchemical practice wherein matter and mind were intertwined.

Sasha's demeanor in the classroom as a teacher was another type of alchemical practice, manifesting outwardly as transmission of factual knowledge and at the same time catalyzing an inner process that could transform one's view of the world. An alchemy of education.

Since my student days when I studied chemistry, physics, and mathematics, my deepest interest has been the nature of mind. That interest began, in part, with wondering how it is that we humans can invent or discover mathematical frameworks (like general relativity and quantum mechanics) that appear to describe the entire mysterious universe. Later, as I learned about biology and psychology, I appreciated that the chemicals we call "psychotropic drugs" are powerful probes of connections between mind—consciousness, experience, psyche, spirit—and the physiology of body and brain. Surely by following this trail, new ways to scientifically investigate the mind might emerge. And psychedelic drugs are undoubtedly the most powerful chemical probes of that mind-body connection.

In 1988, I came to San Francisco to undertake a yearlong clinical internship at the Veterans Affairs Medical Center. Not long after my arrival, I had the great fortune of meeting Sasha and Ann Shulgin, having been introduced by a mutual friend. I was a newbie to the world of high-level scientific discourse concerning psychotropic drugs, psychedelic drugs in particular, but I quickly appreciated that grace had landed me in its sanctum sanctorum.

Being a lover of chemistry, I was mesmerized by Sasha's musical descriptions of molecular artistry. I was equally impressed with his command of the history and sociology of human relationships with psychotropic drugs and his willingness to share openly and honestly about the destructive policies related to drugs, policies that had become increasingly draconian during the 1980s. It was inspirational to witness this kind of well-informed discourse, especially in that particularly dark era of this-is-your-brain-on-drugs misinformation. And I was taken as well by Sasha and Ann's humanity; they were warm, open-hearted souls who skillfully nurtured connection and community.

Sasha and Ann were at the time deeply engaged in creating what would, several years later, emerge as their classic work of alchemical exploration, *PiHKAL*. I recall their describing a narrative vision weaving together details of chemical syntheses, empirical psychopharmacology, human relationships, and love, looping back and forth in poetic ways. Wow!

Following my year at the VA Medical Center, I returned to the University of Oregon and was offered the opportunity to teach a couple of classes, one of which was on the topic of drugs and behavior. Preparing to teach would be quite a bit of work for me, as I had not previously taught a university class on any subject. It was then that I recalled Sasha telling me he enjoyed teaching classes on "the nature of drugs" at various colleges in the San Francisco Bay Area. I couldn't help but think that whatever he was doing in a class on drugs was something I would like to emulate, so I mustered the courage to write a letter asking if he had any advice he might be willing to share to help me get started.

Sasha replied to my letter with a telephone call, and a week after that I received in the mail a small package tied in twine that contained a two-inch stack of typed lecture notes—the notes for many of the lectures that have now become the chapters in this very book. I was humbled and deeply moved by Sasha's generosity in sharing this distillation of his decades of knowledge, intuition, and pedagogy on the complex topic of the nature of drugs to help a novice along their own educational trajectory within this expansive and multifaceted terrain.

The class I taught at the University of Oregon during 1989–90 drew heavily upon the materials Sasha shared with me. And it was a great hit with the students, who clearly appreciated not only the chemistry and history, but also the discourse around drug policy that was based on science and social data, rather than war-on-drugs obfuscation. I found I could teach things in my classes that many believed were off-limits in a university setting, simply by sticking with the empirical data, the science, the history, and the sociocultural and legal contexts. I never needed to be opinionated or to advocate. I only needed to be honest and accurate. The facts speak for themselves.

I have continued to teach on the subject ever since, and Sasha's philosophy of communicating the "music" of a territory that ranges over chemistry, botany, physiology, psychology, history, anthropology, philosophy, constitutional law, and public policy is deeply embedded in all of my academic work.

In the past several decades, as our society has moved away from oppressive darkness and toward the light on issues related to psychotropic drugs, access to accurate and balanced information is more essential than ever. The possibility of psychedelics moving back into the realm of legitimate mainstream discourse has finally been realized. Indeed, things are moving at a very rapid pace. My ardent hope is that we are able to proceed with care and caution, respecting the great power of psychedelics to manifest mind, thus bringing forth all the complexities associated with psyche.

Understanding our relationships with what we call "drugs," and their origins in ancient relationships with plants and fungi, may help reconnect us to an appreciation of aliveness in nature—an agency, animacy, intentionality, meaning, sentience—a consciousness of sorts. The loss of this aliveness has played a role in highly problematic relationships with nature over the last several centuries and contributed to bringing us to the brink of planetary catastrophe. In reanimating nature, reconnecting with this deep truth of alchemy, lies a powerful source of hope for us humans and all our relations.

Thank you, Sasha and Ann, for your many contributions to this great task: your chemistry, therapeutic practice, pedagogy, community, vision, and your courage. May the work be carried forth with wisdom and love.

— David E. Presti, PhD
University of California, Berkeley
Spring 2022

INDEX

abortion, 138, 410
accidents, 36, 131, 136, 141, 235, 330, 452
acetylcholine (ACh), 5, 17, 223, 225, 251–252
acute, 131, 201, 302
Adam. *See* MDMA
addiction, 111, 243, 371, 411–412
adrenaline, 2, 191, 223, 352
adrenergic, 2–4, 223, 332, 335
aerosol, 16, 150, 252, 442
Afghanistan, 61, 99, 101, 106, 415
agonist, 224–225
agriculture, 432, 456
AIDS, 132–135
alchemy, 107
alcohol, 15, 36, 49–51, 74, 80, 88, 92, 101, 106–107, 122, 127, 130–131, 136, 150–154, 156–157, 160–161, 163–176, 178–180, 187, 202–203, 219, 245, 249, 261, 301, 310, 331, 405, 414, 446–448, 451
alcoholism, 106, 151, 414
algae, 283, 436
alkaloid, 9, 16, 76–77, 91–92, 97, 112–113, 123, 129, 200, 217, 271, 277–278, 285, 340, 343, 345, 347, 349, 353, 368, 373–374, 386, 388, 439
AMA. *See* American Medical Association
Amazon, 299, 337, 341–342, 361

American Medical Association, 178, 330
amino acid, 3, 15, 17–18, 25, 223, 303, 374, 385, 401–403
amnesia, 167, 340
amphetamine, 5, 18–19, 22, 30–31, 33, 70–71, 86, 205, 225, 284–289, 292, 300, 303, 312–314, 318, 322, 329, 332, 334, 336, 397
amyl nitrite, 203–204
analgesia, 122, 124
Andretti, Mario, 308
anesthetic, 31, 50, 78, 119, 176, 187, 197, 201–203, 208–210, 215, 234–236, 238, 340
aneurism, 330
animal pharmacology, 294, 323
antagonist, 148, 224–225, 304, 407
anthropologist, 362–363, 373, 453
antidepressant, 4–5, 311
antihypertensives, 4
aphrodisiac, 21, 203
arecoline, 61, 217–218
Asia, 11, 54, 61, 89, 99, 101, 143, 181, 216, 415, 432
ataxia, 305, 433, 437
atropine, 92, 200, 225, 227–229
attorney general, 25–26, 28, 327
Australia, 38, 133, 467
autonomic nervous system, 223
ayahuasca, 341, 345, 360–361, 363, 370, 373

B12, 69–70
Banisteriopsis, 345, 347
barbiturate(s), 130, 160, 166, 177, 179–180, 182
belladonna, 200, 227, 230
Benzedrine, 30
betel leaf, 61–62, 217
betel nut, 61–63, 215–217
biotransformation, 38–39
blood, 38–39, 41–42, 56, 59, 81, 95–96, 132, 148, 151, 153–154, 156–157, 161, 163, 165–166, 168–169, 171–175, 184–186, 193, 196–198, 203, 233, 249, 331, 337, 343, 422, 447, 451–452
blood pressure, 4–6, 21, 88, 188, 203, 219, 315, 330, 334, 354–355, 370, 373, 392, 406, 408
blood-brain barrier, 225
body language, 89, 91
bolus, 38, 55, 62, 185
botany, 216, 265, 272, 297, 432, 435
brain, 2, 4, 17–19, 42, 56, 77, 115–116, 118, 120, 139, 141, 194, 198, 210, 225, 227, 340, 405–406, 410, 442, 459, 469
Bureau of Narcotics (BN), 68, 82, 145
Bureau of Narcotics and Dangerous Drugs, 416, 425

cacao, 10
cactus, 253–254, 256, 271–273, 275, 277, 334, 357
caffeine, 6, 9–16, 19–22, 30, 131, 253, 405
cancer, 42, 47, 51–53, 57, 61–62, 120, 302, 320–321, 330, 404, 406
cannabidiol, 442, 457, 464
cannabis, 411–443, 445–473
carbon dioxide, 13, 29, 446, 457
carbon monoxide, 41–43, 198
carcinogenic, 51, 53
catch-22, 326
catechol, 5
2C-B, 337–339
central nervous system (CNS), 123, 149, 303, 311, 340, 349, 442
cerebrovascular accident (CVA), 131
chemotherapy, 320–322
Chernobyl, 326, 445
China, 10–11, 23, 54, 61, 102, 104–106, 122, 150, 211, 219, 430, 432
chloramphenicol, 401
chlorine, 192–193, 205, 207, 379, 467
chloroform, 13, 77, 172, 201–203, 205, 207–209, 215, 221, 452, 454
chlorophyll, 199, 283
chlorpromazine, 470
chocolate, 10, 20, 29–30, 88, 125–126, 244–245, 284
choice, 58, 88, 135, 138, 174, 201, 241–247, 252, 372, 408, 424
choline, 5
cholinergic, 5, 223
cholinergic synapse, 5, 223
chromatography, 171, 173

INDEX

chronic, 30, 46, 59, 63, 96–97, 102, 131, 373, 409

cigarette, 16, 36, 40–41, 43–48, 50–52, 55, 57–58, 60, 62, 106, 167, 204, 212, 215–216, 242–243, 246, 259, 412, 457

circulation, 21, 203, 205, 337, 365, 368, 462, 468

clearance, 150, 439–440

clinical trials, 311

clove, 48, 50–51, 212, 214–215

coca, 10, 67, 71, 73, 91, 93–94

cocaine, 5, 10, 30–31, 45, 60, 63, 67, 69–82, 84–89, 91–97, 107, 144, 237, 250, 319, 371, 412, 421–422, 463

codeine, 16, 75, 123, 129

coffee, 4, 6–14, 20, 147, 369, 412, 452

Cohen, Dr. Sidney, 370

consciousness, 86–87, 160, 166, 188, 234, 260, 284, 302

constitution, 58, 169, 270

convulsant, 197–198

convulsions, 17, 88, 117, 198, 365, 472

crack, 48, 50, 67–68, 75–76, 79–80

DA, 170, 450

Darvon, 126

datura, 200, 212–213, 229, 234, 255, 367

Davy, Humphry, 201

DDT, 60

DEA. *See* Drug Enforcement Administration

delusion, 213, 340

Demerol, 101, 114, 126–127, 136, 372, 425

Department of Agriculture, 432, 456

dependence, 110, 151, 202

deposition, 454–455

depressant, 5, 17, 82, 96, 99, 160, 223, 310

derivative(s), 33, 119, 123–124, 129, 147, 192, 283, 285–287

Designer Drug Bill, 82, 122, 144

Designer Drug Law, 339

detoxification, 38–39

diastolic, 315, 330

dichlorodiphenyltrichloroethane (DDT), 60

diethyl ether, 117, 205

dihydrooxyphenylalanine, 3

dimoxamine (Ariadne), 311, 313

DMT, 307, 343, 345, 347, 352, 356

DNA, 385, 402–403

DOB, 313, 336–339

Doors of Perception, The, 279

DOPA (dihydroxyphenylalanine), 3

dopamine, 3, 116, 139, 141, 223–224, 340, 352

double-blind, 180, 372

dreams, 9, 190, 459

drug abuse, 8, 137, 246, 394

drug education, 391

Drug Enforcement Administration (DEA), 82, 416

drugs, 1, 5–6, 8, 10, 13, 28, 39, 41,

50, 54, 58, 67–69, 75–76, 79–82, 86–88, 105–107, 112, 122–123, 126, 130, 133, 135, 138, 143–144, 146–147, 149, 154, 157, 163, 170, 174, 178, 182–183, 185, 189, 191, 195, 204, 210, 213, 216, 223, 234, 238–239, 241–242, 244–245, 247, 252–253, 259–262, 264, 283–285, 289, 295–296, 303, 317–318, 320–321, 325–326, 331, 335, 339–340, 342–343, 351, 356–357, 369, 371–373, 383, 385, 390–394, 396, 398–401, 403, 405, 410–414, 416, 419, 421, 424–426, 428, 430, 442, 445, 460, 463, 465–466, 471
drunk, 99, 151, 153, 155, 161, 164–165, 180, 203
dyes, 330

ecgonine, 91, 93–94
Ecstasy. *See* MDMA
embryology, 404
Emergency Scheduling Act of 1984, 326–327
Ephedra, 22–23
ephedrine, 18, 22–25, 27, 147, 204–205
epinephrine, 17, 223–224
equilibrium, 46
erection, 204
ergot, 255, 364–365, 367–368, 373–374, 381, 386, 388, 462, 466
ether, 77–78, 117, 123, 172, 201–202, 205, 208–211, 215, 221, 250
etheromania, 202

euphoria, 144, 372
Europe, 9, 13, 37, 54, 89, 101, 199, 208, 211–212, 236, 260, 315, 325, 432, 440
Excitantia, 1, 189

fainting, 191
fantasy, 102–103, 236, 473
fat solubility, 184
FBI, 82, 248, 398, 430
FDA. *See* Food and Drug Administration
fentanyl, 127–128, 130–131, 141, 144–147, 149, 150, 325
field sobriety test, 156
final, 14, 81, 84, 186, 383, 432, 436, 438
flashback, 445, 458–459, 464, 466
Food and Drug Administration (FDA), 48–49, 61, 70, 129, 143, 192, 203–204, 217, 251–252, 312, 320, 327–328
freebase, 67, 75–78, 97–98, 119, 250
Freedman, Daniel X. "Danny", 390–391
freedom of choice, 58, 138
fungus, 283, 365, 374, 386, 388, 436

g% (gram percent), 155
gamma-aminobutyric acid (GABA), 17
Germany, 112, 271, 319, 322–323, 336, 356
ginseng, 218, 221
god, 7, 10, 20, 87, 138, 165, 193,

INDEX

207, 211, 358–360, 377, 414, 432
Grateful Dead, 87
green medicine, 299

Haight-Ashbury, 30, 86, 235, 279, 306, 309, 368, 373, 377, 379
Haldol, 307, 377, 470
half-life, 120
hallucination, 103, 236, 274, 473
hallucinogens, 82, 214, 220, 225, 245, 278, 288, 292–293, 295, 304, 311, 317, 323–324, 335, 340–341, 343, 353–356, 371
hangover, 51
Harrison Act, 93
hashish, 61, 99, 106, 445, 455–456
heart, 4, 31, 33, 48, 52, 79, 88, 94–95, 131–132, 162–163, 173, 181, 187, 203–204, 261, 268, 284, 303, 315, 330–331, 333, 391, 403, 440
Heffter, Arthur, 271, 273, 275
helium, 197
hemoglobin, 41–43, 197–198
hemp, 432–434, 456
hepatic portal system, 404
hepatitis, 31, 97, 131–135
heroin, 8–9, 36, 45, 85–86, 99, 106–110, 112, 119–120, 122, 124, 127, 131, 137, 144, 180, 202, 247, 250, 262, 325, 371–372, 446
High Times (magazine), 18–19, 137, 204, 319, 325
5-HT, 341, 352
Huichol, 230, 255–257, 259, 261, 268, 357
Huxley, Aldous, 279–280
5-hydroxy-tryptamine, 352
hypertensive agents, 4
hypnosis/hypnotism, 117
hypodermic, 252

illness, 23, 257, 259–260, 270, 278–279, 299, 320–321
illusions, 367
imagery, 180, 236, 467
indole, 220, 334, 339–341, 343, 351–352, 379, 446
inhalation, 30, 50, 55
injection, 30–31, 40, 77, 112, 119, 138, 173, 252, 354
insomnia, 176, 333
insufflation, 356
insulin, 406
intoxicant, 15, 61, 166, 172, 177, 187, 189–221, 271, 273–274, 310, 433, 437, 441–442
intramuscular, 116, 469
intraspinal, 469
intravenous, 30, 40, 67, 71, 95–96, 112, 147, 353, 469
Investigational New Drug (IND), 311–312, 328
irrational mixtures, 178
IRS, 83, 91, 249, 416–417, 425, 427, 430
Isbell, Harris "Harry," 372–373

JAMA. *See* Journal of the American Medical Association

Japan, 54, 61, 467
jimsonweed, 229, 255, 367
Johns Hopkins, 107, 309
Journal of the American Medical Association, 330

ketamine, 86, 213, 234–243, 245, 247–248, 251–252, 340
kinetics, 95

laudanum, 101, 107
laughing gas, 70, 191, 202
laxative, 125
lethal, 42, 80, 147, 161, 194, 196, 201–202, 251–252, 289, 305, 321, 367, 390
Lewin, Louis, 1, 189, 271, 275
Lilly, John, 240, 242
Little, Arthur D., 439–442
LSD, 68, 235, 241, 251, 255, 308–309, 324, 336, 340, 351, 368–371, 373–375, 377, 379, 381, 385–386, 388–391, 399, 403, 428, 443, 445–446, 449–450, 458–460, 462–467, 469–473
lungs, 39, 42, 44, 47–48, 50–52, 62, 107, 168, 194, 196, 201, 207, 209
lye, 76–77
lyse, 366
lysergic acid diethylamide, 308, 374, 386, 388, 403. *See also* LSD

Maehly, Andreas, 397
marijuana, 15, 74, 86, 88, 97, 106, 122, 131, 283, 319–322, 334, 385, 396–397, 413–416, 418, 422–423, 425, 430, 432–434, 437, 439, 442, 445–450, 452–453, 455–458, 460–461, 463, 465
Martinelli's Apple Cider, 308
maté, 6, 11–12, 20
MDA, 214, 235, 302–304, 314, 318, 323–324, 329, 331, 336, 339, 399
MDMA, 5, 215, 301, 314, 317–320, 322–324, 327–333, 339
memory, 209, 216, 347, 367, 466, 472
mental illness, 278–279
meperidine, 75, 101, 107, 114, 126–127, 130–131, 136, 138, 144, 372, 425. *See also* Demerol
Merck Index, 75, 340
mescaline, 9, 235, 253, 271–275, 277–281, 285–290, 292, 301, 313, 323–324, 334–336, 338, 340, 370, 373
Mesmer, Anton, 119
metabolism, 3, 39, 46, 92, 121, 184, 186, 213, 286–287, 411
metabolize, 39, 92, 141, 143–144, 286–287, 290, 304, 460
methadone, 107, 110–112, 202, 425–426
methamphetamine, 18–19, 22, 25–26, 28–31, 33, 70–71, 86, 96, 134, 225, 289, 309, 312–314, 318, 329, 332
Mexico, 10, 86, 88–89, 99, 104, 220, 230, 239, 253–255, 257, 263, 270–271, 278, 349, 353, 357–358, 365, 388

INDEX

mnemonic, 139, 177, 303, 412
molecule, 24, 53, 92, 112, 116–117, 119–120, 122–124, 126, 198, 214, 220, 225, 234, 279–281, 287, 289, 291, 298–299, 301–303, 310–311, 318, 334–336, 340–341, 379, 397, 403, 434–437, 442, 459–460
monoamine oxidase (MAO), 144, 335
monoamine oxidase inhibitor, 144, 335, 345
Moore, Marcia, 240, 242
morning glory, 255, 379, 381, 388
morphine, 9, 16, 75, 101, 106–107, 110, 112–117, 119–123, 126, 129, 131, 144, 149, 202, 385
MPTP (Methylphenyltetrahydropyridine), 127, 130, 139, 142, 144
mushroom, 199, 220, 255, 278, 330, 343, 347, 349, 351, 358–360, 365, 381
mutagenic, 61
mutations, 445
mydriasis, 289, 332

Naloxone, 148
narcolepsy, 312–313
narcotic, 69, 85, 87, 93–94, 101, 106, 110, 123–124, 129–131, 133, 136–138, 141, 144, 148–150, 157, 197, 204, 243–245, 353, 371–372, 393, 395, 398, 416, 450, 458
National Organization for the Reform of Marijuana Laws (NORML), 321–322

Native American Church, 256, 261, 270–271, 370
needle orientation, 137
nerve, 1–4, 17, 118, 140, 163, 166, 223, 296, 318, 332, 367, 413, 440
nerve conduction, 1
neuron, 2, 4, 6, 17, 333, 410, 460
neurotransmitter, 2–6, 17–18, 142, 144, 223, 224, 335, 340, 352
newborns, 36, 363, 404
nicotine, 16, 36–41, 43–44, 46–51, 55–56, 59–61, 63, 204, 215
nitric oxide, 194–195, 201
nitrogen, 9, 23, 120, 126, 193–197, 223, 225, 251, 275, 280, 285, 297, 310–311, 313, 340, 353, 439–441
nitrous oxide, 70, 78, 113, 137, 191–195, 200–202, 209, 215
noradrenaline, 2, 223, 352
norepinephrine, 2–6, 17, 223, 352
note taking, 35, 123, 298, 318, 334
nutmeg, 61, 212–215, 298–299

Oaxaca, 220, 255, 278, 349, 358, 370, 381
opiate, 16, 106, 109, 124, 129, 133, 176
opium, 8–9, 16, 86, 99, 101–107, 109, 112–113, 122–124, 129, 131, 149, 250, 432
Opium Wars, 105
optical isomers, 24, 93–94
organic chemistry, 9, 113, 125
oxidation, 39, 142, 194, 379, 467
oxygen, 15, 41–43, 124, 127, 184,

193–198, 206–207, 210, 221, 280, 302, 304, 311, 336, 347

pain, 17, 50–51, 99, 112, 119–120, 122–123, 126, 131, 188, 201, 203, 214, 234, 236, 238, 260, 355, 367
pancreas, 406
parasympathetic, 223
parasympatholytic, 212, 227
parenteral, 95, 343
Parke-Davis, 86, 231, 234, 239, 271, 398
parkinsonism, 127–128, 130, 138–141, 143
patents, 50, 182, 322–323, 442
PCP, 213, 231, 233–238, 241, 247–252, 262, 306, 340, 460–461
PDR, 340, 400
penicillin, 86, 312
peptide, 402–403, 406
peyote, 9, 116, 230, 253–281, 357–358, 360, 370, 373, 466
Phantastica, 1, 271
phenethylamine, 146, 223–224, 283–287, 298, 317–318, 334–335, 340, 352, 439
phenobarb, 166, 177, 179, 184–185
phenobarbital, 160, 177, 181–182
phenylacetic acid, 25, 28–29, 290
phenylalanine, 25, 39, 303, 318, 332
pills, 22, 24, 33, 47, 178, 203–204, 306, 393, 400, 409, 426
pineal gland, 406
piperidine, 49, 126–127, 248–251
placebo, 119, 217

Playboy, 317, 411
plumbing of the body, 317
poison, 9, 51, 172, 199, 206–207, 265, 287, 366–367
poisonous, 198–200, 206–207, 227, 265, 274, 366, 369, 433
polypeptide, 406
poppy, 99, 101, 104, 112, 123, 129
portal, 404
portal systems, 56, 404, 406
postsynaptic, 2
power, 230, 249, 259, 304, 327–328, 428–429, 463
pregnancy, 299, 368–369, 405, 407–409
presynaptic, 2–3
prophylactic, 90, 132, 304, 369, 371
prophylaxis, 374
protein, 18, 39, 402–403, 422
psilocin, 343, 347, 349, 352, 356, 371
psilocybe, 220, 255, 278, 330, 347, 349, 351, 370, 381
psilocybin, 220, 278, 343, 347, 349, 351–352, 356, 371, 373, 375, 381
psychedelics, 214, 280, 284, 287–289, 292, 298, 303, 307, 309–310, 313–314, 329, 335, 337, 357, 369, 388, 390, 401, 465, 470
psychological dependence, 411
psychopharmacology, 345, 356
public law, 416
Pure Food and Drug Act, 393
purine, 9, 412

INDEX

Quaalude, 179, 181–183, 394

radioactivity, 290, 318, 391, 413, 445
rationalization, 423
receptor, 2–4, 116, 197, 304, 403
receptor site(s), 304
recreational use of drugs, 221
Reefer Madness, 237
reflexive mydriasis, 332
research, 40–41, 50, 68, 121, 123, 150, 220, 234, 239–240, 251, 279, 281, 309, 317, 326, 328, 335, 370, 372–373, 386, 396–397, 408, 414–416, 419, 421, 424–425, 429, 433–434, 438, 441–442, 455
reserpine, 5–6
Road Chief, The, 262–264, 268–270
Russia, 45, 54, 106, 150, 199, 251, 260

San Francisco, 30, 59, 61, 86, 144, 217, 306, 395, 426, 472
Sandoz, 351, 358, 386, 388, 398, 466
Sansert, 369, 371
Schedule I, 68–69, 82, 110, 138, 149, 182, 271, 315, 319–321, 325–329, 339, 390, 415–416, 424
Schedule II, 82, 86, 182, 240, 312, 321–322, 325, 395, 415–416
Schedule III, 86, 177, 182, 240, 250, 312, 325, 327–328
Schedule IV, 325, 416
Schedule V, 68–69, 325, 395, 416
scheduling, 145, 325, 327–328

schizophrenia, 278
Schoenfeld, Gene, 137
scopolamine, 200, 225, 227
sedative, 130, 179, 186, 329
sedative-hypnotic, 166, 179, 405
seizure, 55, 80, 306, 319, 325, 369, 398, 447, 449–450
self-medication, 109
serotonergic synapse, 2
serotonin, 2, 223, 333, 340–342, 352
set and setting, 356, 358, 360–361, 369, 371
sex, 136, 242, 246, 432, 457, 463
shaman, 230, 269, 358–360, 362–363
shock, 295
sickness, 361
sleep, 2, 4, 16–17, 31, 70, 103, 113, 115, 117, 166, 176, 179, 186, 201, 237–238, 284, 312
smoke, 15–16, 36, 40–46, 50–51, 54–55, 57–59, 95, 103, 205, 215–216, 221, 244–245, 259, 264–265, 268, 347, 360, 369, 442, 457, 461
smoking, 30, 36, 38, 40–41, 43–60, 62, 96, 98, 101–104, 212, 245, 259, 265, 353, 414, 430, 457, 461
snorting, 79, 356
snuff, 47, 52, 55, 61–63, 341–343, 345
sodium ascorbate (vitamin C), 213, 219, 469
sodium hydroxide (lye), 469
St. Anthony's Fire, 364–365, 367
stars, 1, 234, 472

— 489 —

stat, 153
stimulant, 3–6, 10, 16–19, 21, 23–24, 30–31, 51, 67, 70, 82, 94–96, 99, 179, 204, 218, 286–289, 292, 303, 311, 315, 333, 340–341, 393, 442
stomach, 10, 56, 76, 80, 115–116, 156, 185, 235, 243, 297, 321, 472
strychnine, 17, 271–272
stupefactant, 437
sugar, 8, 20, 78, 244, 333, 442, 471–472
sulfur, 184, 336
sulfuric acid, 97, 457
sympathetic, 223
sympathomimetic, 274
synapse, 2, 4–5, 223, 227, 284, 335
synaptic cleft, 4
synergistic, 46, 434
systolic, 315, 330

tea, 6, 10–12, 14–16, 20, 23–24, 54, 70, 101, 105, 218, 265
teratogenic, 405
Thailand, 61, 99, 217, 250
thalidomide, 404
THC (tetrahydrocannabinol), 92, 283, 320–321, 422–423, 434, 436–442, 449, 456–457, 460, 463
theobromine, 7, 20–22, 30
therapeutic index (TI), 160–161, 172
therapist, 314–315
thioamobarbital, 184
tobacco, 16, 35–36, 38–40, 47–56, 58, 60–63, 74, 97, 101, 103, 106–107, 131, 215–216, 245, 253, 255, 258–259, 262, 264–265, 357, 430
tolerance, 40
tongue, 139, 203
toxic, 9, 13, 19, 95, 142, 161, 192, 195, 229, 274, 278, 305, 320, 324, 410, 463
toxicity, 43, 95, 142–143, 161, 172, 177, 277, 294, 313, 323–324
tranquilizers, 99, 187
transformation, 214
Trinidad, 342, 439–441
tryptamine, 335, 341, 347, 352
Tuinal, 178
Turkey, 9, 54, 89, 99, 101
tyrosine, 3, 25, 303, 318

Ups, Downs, and Stars, 1
trine, 10, 39, 59, 92, 144, 148, 156–157, 165, 168–169, 174–176, 186, 219, 233, 331, 337, 392, 406, 451–452
trine testing, 87, 186, 260, 413, 461

Valium, 166, 186, 307, 469–470
vasodilator, 21
veins, 31, 184
ventricle, 405
vesicles, 4
vitamin C, 213, 219, 469. *See also* sodium ascorbate
voluntary, 245
vomiting, 9, 35, 38, 80, 116–117, 203, 235, 267, 274

INDEX

Washington, DC, 84, 326
Weil, Andrew (Andy), 8
withdrawal, 10, 35, 40, 110, 176, 243
World War I, 177, 192, 206–207, 322
World War II, 86, 119, 270, 433, 437

xylocaine, 78

yagé, 343, 345